Cloud Policy

Distribution Matters

Edited by Joshua Braun and Ramon Lobato

The Distribution Matters series publishes original scholarship on the social impact of media distribution networks. Drawing widely from the fields of communication history, cultural studies, and media industry studies, books in this series ask questions about how and why distribution matters to civic life and popular culture.

Emily West, *Buy Now: How Amazon Branded Convenience and Normalized Monopoly*

Robin Steedman, *Creative Hustling: Women Making and Distributing Films from Nairobi*

Anne Kaun and Fredrik Stiernstedt, *Prison Media: Incarceration and the Infrastructures of Work and Technology*

Lee McGuigan, *Selling the American People: Advertising, Optimization, and the Origins of Adtech*

Jennifer Holt, *Cloud Policy: A History of Regulating Pipelines, Platforms, and Data*

Cloud Policy

A History of Regulating Pipelines, Platforms, and Data

Jennifer Holt

The MIT Press
Cambridge, Massachusetts
London, England

© 2024 Massachusetts Institute of Technology

This work is subject to a Creative Commons CC-BY-NC-ND license.

This license applies only to the work in full and not to any components included with permission. Subject to such license, all rights are reserved. No part of this book may be used to train artificial intelligence systems without permission in writing from the MIT Press.

The MIT Press would like to thank the anonymous peer reviewers who provided comments on drafts of this book. The generous work of academic experts is essential for establishing the authority and quality of our publications. We acknowledge with gratitude the contributions of these otherwise uncredited readers.

This book was set in Stone Serif and Stone Sans by Westchester Publishing Services. Printed and bound in the United States of America.

Library of Congress Cataloging-in-Publication Data

Names: Holt, Jennifer, 1968– author.
Title: Cloud policy : a history of regulating pipelines, platforms, and data / Jennifer Holt.
Description: Cambridge, Massachusetts : The MIT Press, [2024] | Series: Distribution matters | Includes bibliographical references and index.
Identifiers: LCCN 2023045519 (print) | LCCN 2023045520 (ebook) | ISBN 9780262548069 | ISBN 9780262378697 (epub) | ISBN 9780262378680 (pdf)
Subjects: LCSH: Cloud computing—Law and legislation | Telecommunication lines—Law and legislation. | Data sovereignty.
Classification: LCC K564.C6 H65 2024 (print) | LCC K564.C6 (ebook) | DDC 343.09/99—dc23/eng/20240129
LC record available at https://lccn.loc.gov/2023045519
LC ebook record available at https://lccn.loc.gov/2023045520

Contents

Acknowledgments   vii

Introduction   1

Cloud Policy: Genealogy of a Regulatory Crisis   1
Designing the Cloud   8
Visualizing the Cloud   12
The Path Dependencies of Cloud Policy   18
The Stakes of Cloud Policy   22

1   Pipelines   35

Pipeline Principles   37
AT&T and the Blueprint for Cloud Policy   48
Cloud Policy for the Convergent Era   54
The Information Superhighway and Beyond: Digital Pipelines   67
Pipeline Principles Revisited   84

2   Platforms   91

Cloud Royalty   91
Platform Governance   96
The Survival of News   131
Alternative Visions   136

3   Data   141

(Im)materiality and (In)visibility   143
Locating Control   149
"The Abyss from Which There Is No Return"   156
Private Control / Public Data   181
Data Sovereignty, Data Localization, and National Clouds   187

Epilogue: Preserving the Cloud's Future 195
   Activism and the Way Forward 199
   Creative Solutions 202
   We Could Have Been a Contender . . . 205

Appendix: Notable Investigations and Actions against Big Tech, 2017–June 2023 207
Notes 213
Bibliography 275
Index 299

Acknowledgments

This book has been in the works for a long time, and I am grateful to so many people who have helped me along the way. At UCSB, the entire Film and Media Studies Department has always supported my work wholeheartedly. I am so lucky to have them all as friends and colleagues. The campus awarded me research support in the form of the Emmons Award, an Academic Senate Award, and an Interdisciplinary Humanities Center Faculty Fellow Award. Thank you to Susan Derwin and the IHC for recognizing media policy as part of a humanities research agenda!

I have presented this work in progress at many institutions, all of which brought out brilliant and engaged people who helped me to improve it. Those wonderful experiences were some of the highlights of writing this book. They included a stay at the Swinburne University of Technology with hosts and mates Ramon Lobato, Julian Thomas, Jock Given, Angela Daly, and the late, great Scott Ewen; a trip to Queensland University of Technology with friends and colleagues Stuart Cunningham and Terry Flew; and time in residence at NYU with Anna McCarthy and the fantastic students in Cinema Studies and MCC. Especially generative were events at Utrecht University hosted by Judith Keilbach; at University of Helsinki and University of Tampere in Finland led by Minna Ruckenstein and Kaarina Nikunen; and at the Università Cattolica in Milan organized by Massimo Scaglioni, as were talks at MIT's Comparative Media Studies program, and at Concordia University with colleagues and students from Communications and Film Studies. An FCC workshop, the Future of Broadband, organized by Amit Schechter was particularly important during the early stages of this research.

Other colleagues and mentors—all of whom I am honored to call friends—who have been deeply supportive along the way include Charles Acland,

Miranda Banks, John Caldwell, Gail De Kosnik, Des Freedman, Dave Hesmondalgh, Vicky Johnson, Denise Mann, Paul McDonald, Ross Melnick, Chris Newfield, Constance Penley, Allison Perlman, Alisa Perren, Violaine Rousseau, Petr Sczepanik, Greg Siegel, Ellen Seiter, Greg Steirer, William Uricchio, Hilde Van den Bulck, Patrick Vonderau, Janet Walker, and Haidee Wasson.

Former FCC Chairmen Tom Wheeler and Newton Minow and former FCC Commissioner Nicholas Johnson all granted me interviews, and I am grateful to all of them for their time and their professional contributions to this history. Their collective commitment to fighting for the public interest is a true inspiration. Special thanks are owed to Nick for his friendship and support that has meant so much to me over the years. I am in awe of the amazing people at the Center for Democracy & Technology, including my fellow Fellows, and grateful for their work and wisdom that has informed this project. And extra thanks to CEO Alexandra Reeve Givens for her interview and leadership. I was also fortunate enough to have many lawyers, computer scientists, public interest activists, and industry executives spend time with me to discuss the landscape of cloud policy. Their generosity has left a great impact on this book.

Sincere thanks to my fantastic graduate student researchers Steven Secular, Rich Farrell, and Miguel Penabella for their important contributions and support. And to Pete Johnson for the incredible job helping me with the final stages of manuscript preparation.

I am indebted to the MIT Press dream team, including Justin Kehoe and Distribution Matters series editors Ramon Lobato and Josh Braun, who have been extraordinary partners throughout. I am also very thankful to my brilliant copyeditor and all of the anonymous readers and reviewers for their hard work and insightful comments, which made this a better book.

Thanks to Kelly Goldberg, Lisa Hajjar, and Rebecca Epstein for the kind of friendship that gets one over the finish line. And to Lisa Parks and Cristina Venegas for the workshops, the rituals, and the eternal witchy sisterhood.

My students, who are so much fun to teach when they come to class, are the ultimate reason I wrote this book. They deserve so much better from the stewards of our media and infrastructure policy, and they will be the ones to carry this fight forward. To them I say: I believe in all of you, and remember to always read the footnotes.

This book is dedicated to the memory of my dad, whose confidence in me and maniacal DNA have made all things possible. His free and determined spirit lives in these pages, too.

Introduction

> Here we have the bare bones of our future. Let us now fill in the skeleton with living flesh and see what sort of creature it is that we have created.
> —Douglas Parkhill, *The Challenge of the Computer Utility*, 1966

> We can no longer permit technological innovation to "just happen," and then attempt to "regulate away" the adverse effects.
> —US Office of Telecommunications Policy, 1974

Cloud Policy: Genealogy of a Regulatory Crisis

Every time we use the cloud, we are engaging with more than one hundred years of policy history. We have reached a point in this history where we can no longer afford to ignore its lessons.

Policy is a gateway through which all media and its infrastructure must pass, and that includes the sociotechnical systems that are together known as "the cloud." This infrastructure that allows us to work remotely, send emails, share on social media, watch or listen to streaming content, bank, shop, or attend class online is also the arena where many of our contemporary rights are being adjudicated. The cloud is now a primary locus of governance for individual and collective privacy, speech rights, information access, and data security. It has opened new terrain for industry competition. These material technological systems and immaterial "spaces" are also where sovereignty is being asserted and redefined. It is where our attention must turn in order to preserve the digital civil liberties that have endured decades of politicized attacks. To post a tweet, navigate using Google maps, or store documents on Dropbox is to be imbricated in this expansive policy landscape and its myriad forces of control.

The cloud is a vast ecosystem of privately owned infrastructure that stores and distributes data for remote access. It is reliant on a global network of servers, the pipelines that link those servers to Internet users, and the platforms hosting digital engagement. This infrastructure represents power struggles over issues ranging from the territories covered by its physical networks to the legality of a government surveilling its own citizens. All these conflicts, some of which have their roots in the nineteenth century, are deeply inscribed in the regulation of cloud infrastructure today. Indeed, the long arc of cloud policy began generations before the arrival of the Internet or social media. This history is bound to those of our railroads, highways, and even the laws of the sea. It has been informed and influenced by regulations designed for nineteenth- and twentieth-century media, including the telegraph, the telephone, radio, and broadcast television. A historical orientation reveals how the cloud as we now experience it has been shaped by more than a century's worth of cultural and political negotiations over the practices of media and communication technologies, some of which no longer exist. It further exposes the true scope of how the public has been failed in the process.

What follows is a policy genealogy focused on some of the most legible and critical elements of cloud infrastructure: Internet distribution pipelines, digital platforms (particularly those controlled by the dominant Big Tech companies), and of course, data, which serves as the raw material of the cloud. This also necessarily includes the "data centers" all over the globe that function as the cloud's storage facilities. These infrastructural elements of the cloud are rarely considered holistically in the process of policymaking. Yet, they coexist in an intricate web of relationships that has evolved incrementally over time, along with the technologies themselves and their user cultures. Examining these dimensions of cloud infrastructure as a collective site of policymaking is imperative for a properly scaled view of the massive regulatory challenge that we are now facing in the twenty-first century. It is also essential for understanding the cloud as a multilayered infrastructure of democracy, albeit one currently in crisis.

I employ a genealogical approach in order to reveal the many forgotten patterns, conflicts, conditions, and decisions that have helped create contemporary cloud policy. As David Garland has written, such a process is motivated less by the desire to understand the past and more "by a critical concern to understand the present. It aims to trace the forces that gave birth to our present-day practices and to identify the historical conditions upon which they still depend. Its point is not to think historically about the past but rather to use historical materials to rethink the present."[1] Accordingly, *Cloud Policy* concentrates on the relational values of contemporary infrastructure regulation and how these paradigms have changed over time, animating the forces that have led us to this point, along with what was previously imagined, what might have been, and what has been left behind in the process. In so doing, I hope the necessary and restorative interventions become clear, together with the alternative policy values and decisions that were once possible and can be again.

Throughout this history, there have been numerous dramatic power shifts affecting the regulation and provision of what would eventually become cloud infrastructure. For example, many of the original promises of openness and decentralization related to the Internet have fallen victim to the forces of privatization and monopoly capital. The role of regulators has been diminished by consolidated corporate power; much of cloud policy is now dictated by Big Tech cloud providers, either directly through terms of service and trans-industrial agreements, or indirectly through their networks of influence and lobbying muscle. Private sector values have increasingly determined public policy across all industries, which tracks with the global ascent of neoliberalism over the past four decades, and the attendant rise of antidemocratic political movements worldwide.[2] As Shoshana Zuboff has written, the road to our current predicament is littered with terms such as "the open internet" and "connectivity" being "quietly harnessed to a market process in which individuals are definitively cast as the means to others' market ends."[3] And yet, historically, the stewards of media policy were once devoted to safeguarding the public interest as it relates to industry competition, pricing, and service; defending personal privacy; and forestalling the profiteering of private corporations on the backs of public utilities. This is no longer in the realm of expectations for US citizens. In fact, the protracted decline of such public values is now a defining feature of cloud policy's evolution.

While cloud policy is inherently global, this book is focused primarily on the history of US infrastructure policy; however, crucial connections to and conflicts with other nations, regions, and policy regimes across the globe are addressed throughout. The varying approaches to privacy and data security among the US and the European Union, China, and Russia are some of the many examples. These growing points of disconnection have created geopolitical, financial, industrial, and moral/ethical crises of their own, on top of the specific policy predicaments they represent. The issues raised by such a history are vast—they point to the evolving role of the state in this regulatory arena, the increasing power of corporate gatekeepers and private sector governance, the fragility of civil liberties in the digital age, and the weight of policy's path dependencies on contemporary cloud infrastructure.

Scholarship across a wide stretch of disciplinary divides and areas of study have influenced this project. At its core, *Cloud Policy* is aligned with the ideals of fostering equality, democracy, and an informed citizenry. However, most formal policy debates and decisions are dominated by social scientists utilizing quantified logics and methods.[4] It is rare to ever find a humanist in the room. These conventions often preclude qualitative and conceptual arguments from being foregrounded in crucial deliberations and diminish the importance of historical perspectives regarding the inherent civic and social functions of infrastructure. However, the views and voices that emanate from the humanities are also fundamental to policy formation and reform. They are central to identifying the nagging creep of tyranny in this ecosystem, and to evaluating the sacrifices we make for mobility, connectivity, and convenience. Humanistic traditions help to reveal the vulnerabilities baked in to our desire to share, as well as the personal, collective, and cultural costs of corporate and state surveillance across time. Their inclusion is vital to the regulatory calculus. *Cloud Policy* is inspired by and written with that ethos.

Work by whistleblowers, journalists, policymakers, media historians, legal scholars, privacy experts, and computer scientists have all informed this research. My interviews with public interest advocates, regulators, industry executives, technologists, and lawyers have been essential to understanding this policy landscape. In the process of writing this book, I also analyzed many decades' worth of case law, government hearings, investigations, and reports, as well as formal policy deliberations and documents. I utilized additional archival sources including personal papers and court records, as well as terms of service and licensing agreements, annual reports ranging from AT&T's in 1907 to Google's in 2021, and other legal filings from infrastructure providers, media companies, and Big Tech corporations.

The many interrelated laws, processes, regulations, and ideologies animating this history collectively represent a domain I am calling "cloud policy." Cloud policy encompasses a combination of formal laws and regulations as well as the larger sphere of informal practices and corporate dictates that have shaped the policy terrain for pipeline, platform, and data infrastructure. This includes a complex set of political relationships and industrial protocols, as policy is not just what we read on a page. It also exists in layered, material and immaterial operations that result in the built environment, structures of power, and cultural values. Accordingly, cloud policy also extends to

1) a set of sociotechnical relations across pipelines, platforms, and data;
2) geopolitical agreements and tension/collaboration at multiple levels of governance that shape the interaction of these infrastructures in a global context;
3) social norms and user cultures affecting how the cloud is utilized; and
4) the ever-expanding partnership between the public and private sector.

Thus, I approach cloud policy as a space of historical investigation and analysis, with the central concerns of tracing how public interest and democratic principles have been progressively eroded. The cloud's inherent conveniences and affordances are by-products of policy, as are the personal and collective costs it extracts. Our current conditions were never inevitable, and this book chronicles many of the key decisions behind their construction and the formation of policy now governing cloud infrastructure.

Cloud Policy further elaborates the principle of "regulatory hangover," or the conceptual and practical inability of policy to keep pace with technological advances and cultural change. This is the inevitable and long-standing consequence of media and communications technologies evolving far beyond their policy foundations, particularly those foundations devoted to the protection of the public interest. Over time, as policy regimes historically designed to preserve values such as diversity, universal access, public welfare, and competition have changed, these ideals have been progressively forfeited, casualties of a protracted regulatory hangover that we are still living with today.

The term "regulatory hangover" was first described by economist Harvey J. Levin in 1973 as "the euphemism which refers to the failure of the regulatory framework to respond to technological change."[5] In his analysis of the 1971 Sloan Commission report on cable television and its nascent growth, Levin warned that the danger for regulatory hangover with respect to this burgeoning medium was "considerable."[6] The director of the White House Office of Telecommunications Policy (OTP), Clay Whitehead, sounded similar alarms just a year later. Also focusing on cable, Whitehead and the Cabinet Committee on Cable Communications urged President Nixon in 1974 to take note of the emergent regulatory hangover happening with regard to new networks of media distribution. The committee addressed the need to adapt policy as media technologies transformed, but their instruction received little notice at the time. "The changes in cable technology and in the economic and social importance of cable should have been accompanied by changes in the public policy that govern its regulation," the committee wrote, "yet, the regulators' perception of the cable medium has lagged far behind its evolving reality."[7]

This dynamic is also referred to as "regulatory lag," and it has persisted in media industries since the advent of radio. However, because of the speed of technological innovation and change in the twenty-first century, it has become a bona fide crisis in the digital, cloud-based era. As former FCC Chairman Tom Wheeler has noted, "When looking at the long arc, one has to recognize that the kind of assumptions that policymakers have historically made don't work anymore." Wheeler connects this point to an observation Madeleine Albright has made regarding diplomacy, and applies it to technology: "Every time a 21st century issue comes up, we define it in 20th century terms and propose 19th century solutions."[8]

Cloud policy affects everything from access to the Internet and data security to personal privacy and freedom of expression. Safeguarding these rights and principles in a policy space that is simultaneously local, national, regional, and global has created challenges that often defy existing regulations and traditional geographies of control. As a result, this massive assemblage of laws and policies governing cloud infrastructure is rife with conflicts that are steadily proliferating. The scope of such conflicts has been addressed by scholars including Sandra Braman, in the context of "information policy,"[9] and Julie Cohen, in the framework of "informational capitalism."[10] Others have tackled more sector-specific dimensions of cloud policy, presenting a regulatory landscape in flux that grows more unruly every day.

Ithiel de Sola Pool's work is also resonant here. His seminal book *Technologies of Freedom* (1983) forecast many of the regulatory crises precipitated by new technologies and media convergence. He argued that our policy failures are often those of regulatory imagination and criticized the manner in which policymakers applied "familiar analogies from the past to their lay image of the new technology, creat[ing] a partly old, partly new structure of rights and obligations. The telegraph was analogized to railroads, the telephone to the telegraph, and cable television to broadcasting."[11] Herein lies a foundational problem of cloud policy, and a failure of imagination that has proven devastating to the public interest. In addition to the familiar analogies, the historical rules, definitions, classifications, and doctrines applied to cloud infrastructure are inadequate at best, destructive at worst, when it comes to creating policy for the digital era. The endurance of outmoded legal constructs and the abandonment of vital political commitments along the way have led us to a moment in which nearly all power over the public sphere has been given to a very small number of private entities. *Cloud Policy* thus offers a reconsideration of de Sola Pool's concerns and warnings as they have manifested in the regulatory environment for twenty-first-century pipelines, platforms, and data.

Speaking to many of the conceptual threads explored throughout this book, Tung-Hui Hu's *Prehistory of the Cloud* (2015) addresses the ways in which power is also embedded in the core metaphors and imaginations of the cloud as fantasy. In so doing, Hu investigates the administration and design of railway networks and fiber-optic routes, sewer systems, and military bunkers, and interrogates artistic visualizations of the cloud for the ways they "allow us to think through historical problems of power and visibility."[12] Both the inherent invisibility and manufactured visibility of the cloud have had a dramatic impact on how we understand and value its infrastructure. Eventually, those evaluations translate to policy—and, in turn, lived experience. Examining the law and policy that emerges alongside government, corporate, and individual attempts to render this infrastructure intelligible is another way to access the dramatic stakes of cloud policy. It helps to reveal what Hu calls "the long term consequences of the cloud" from the fog of the "seductive 'now.'"[13]

In this effort, *Cloud Policy* brings together many disciplinary perspectives that have thus far remained largely siloed in their respective fields of law, policy, economics, and media studies, among others. The myriad histories that constitute cloud policy are also considered in relation to one another, as this domain is shared with that of the phone company, the broadcast and cable industries, Big Tech platforms, and Hollywood studios, along with the many human agents within, including their users and audiences. These institutions and industries have been shaped by obscene phone calls, organized crime wiretaps, wars, climate change, and civic protest. Their power is mediated by lawyers, judges, content producers, computer scientists, and the advocacy community, as well as by policymakers and regulators from the local to the global. The indelible traces of these many interconnected legacies loom large.

This book shines a light on their formative imprints and long-forgotten lessons, all of which are still alive and well in the terrain of contemporary cloud policy.

Designing the Cloud

Although the first use of the term "cloud computing" is often attributed to former Google CEO Eric Schmidt in a 2006 speech to an industry audience,[14] or to an internal Compaq document addressing networked computer services in November 1996,[15] the underlying concept was first explored many decades earlier. The idea and potential for networked computing as a distributed, regulated infrastructural public resource was already being considered in the 1960s. In 1961, MIT and Stanford computer scientist John McCarthy imagined a future in which "computing may someday be organized as a public utility just as the telephone system is a public utility."[16] The concept of "computer utilities"—arguably the original term for the cloud—described a networked system that "allowed many users to access the same mainframe computer through remote communications links."[17] Five years later, Douglas Parkhill's prophetic book *The Challenge of the Computer Utility* (1966) took up the concept and described what cloud computing might actually look like. Parkhill wrote about the features of "a general-purpose public computer utility for which no upper limit would exist for either the numbers and types of tasks to be performed or the numbers of customers to be serviced. In fact, as such a utility grew it might eventually embrace the entire nation and service not only industrial, government and business customers, but also private homes, until the personal computer console became as commonplace as the telephone."[18] He also envisioned that such a utility would, as is customary, be subject to government regulation and other customs reserved for industries "bound by the law of public service undertakings."[19]

Computer utility services were also referred to as "time sharing" technology, and, according to Paul Edwards, by 1967 "some twenty firms, including IBM and General Electric, had established computing service bureaus in dozens of cities."[20] At the same time, the US Defense Department's Advanced Research Projects Agency (ARPA) and its ARPANET project was funded to design and build a network, linking together multiple computers for data transmission across a wide geographical stretch through telephone lines.

This ultimately spawned the Internet, and some of the pipeline infrastructure for the cloud. Early innovators and designers of the ARPANET actually argued that the network should be treated as a "computer utility," and be held responsible for the level of reliability provided by the power company and the phone company.[21] Plantin, et al. have noted that this ideology endured throughout the Internet's developmental period, one marked by "heavy government investment, sponsored first by the US Defense Department's Advanced Research Projects Agency and then by the US National Science Foundation (NSF) in the public interest. In the 1980s, the NSF forced the broad provision of Internet connections in order to permit scientists at less well-resourced institutions to share time on the costly supercomputers it purchased for a few major research centers—exactly the 'computer utility' model."[22]

Electronic Freedom Foundation (EFF) cofounder Mitch Kapor held as a motto, "Architecture is politics."[23] This imagined connection between design, power, and control is evident in the blueprint for the ARPANET. It was created as an open network, without any filters, censors, or intermediaries between sender and receiver. The decentralized, nonhierarchical design of the ARPANET—especially remarkable given the formative role of the military in its design—echoes the paradigm of democracy's decentralized power and the values that system of government was intended to represent. Early US cloud infrastructure contained safeguards inherent in its end-to-end design that theoretically insulated it from various forms of corruption and control, much like a system of checks and balances. Such structures were originally envisioned as bulwarks against monopoly power or centralized authority. However, as Benjamin Peters has argued in his formative study of early Soviet efforts to construct a national computer network, "there is no such inherent connection between the designs of technological and political systems."[24] Instead of these imagined analogs of "utopian instinct," Peters explains that such direct correlations between political and technical systems, between computer network and formal state and social structure, are misleading because they "neglect actual political practices and their significant costs and consequences."[25] This caveat has indeed borne out across the history of cloud policy.

The network was initially switched on in October 1969, when the first nodes became operational and UCLA was connected to the Stanford Research Institute. In the next two months, two other nodes were added: UC Santa

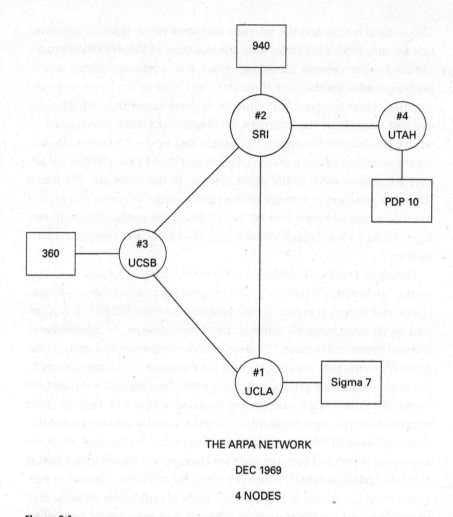

Figure 0.1
ARPA Network four node map.
Credit: Reprinted with the permission of the Computer History Museum.

Barbara and the University of Utah. Figure 0.1 shows one of the initial sketches of this originary, four-node primitive "cloud" or, as one of the project's earliest network engineers J. C. R. Licklider called it, the "Intergalactic Computer Network."[26]

With this vision in place, wires controlled by AT&T began networking computers in the US. Surprisingly, while AT&T was the state-sanctioned monopoly provider of telephone lines into the home, the company missed its chance to be the pipeline provider for the Internet in the 1970s. Their lack of foresight and inability to imagine the ARPANET computing network as a commercial communications utility was rather stunning, particularly as all users who accessed the Internet in its earliest days used fixed-line connections. However, when members of the ARPANET team took the project to AT&T in 1972 to see if the telecommunications giant wanted to become the Internet's new operator, the company declined, failing to see the network's potential at that time.[27] Furthermore, AT&T was not about to compete with itself in the arena of long-distance service.[28] Paul Baran was the RAND Corporation think-tank engineer who largely designed the distributed network based on packet switching that became the Internet's foundation. Baran had lobbied AT&T to adopt such a system in its capacity as the communications contractor for the Department of Defense back in 1965. AT&T also refused then, insisting its own network was superior and the innovations of packet switching were incompatible with the Bell System.[29]

In the 1980s, Kōji Kobayashi, former president and then chairman of Japan's Nippon Electric Company[30] wrote a prophetic book about how computing and communication—what he termed "C&C"—would merge into a single infrastructure for information processing. *Computers and Communications* was originally published in Japan in 1985 and then by the MIT Press in 1986. His vision also included a wish that "the concept of C&C will eventually bring happiness and prosperity to all humankind by transcending artificial boundaries on earth" and creating a global society that went beyond national borders.[31] Shortly thereafter, personal computers became affordable enough and storage dropped in price so that one could rely on individual networks. Eventually, servers and traffic scaled up sharply in the late 1990s, and by the mid-2000s, the "computer utility" model, or "C&C," would reemerge as the cloud (minus the happiness and prosperity for all).

Usage of the cloud skyrocketed in the 2010s with the industry around it expanding as rapidly as the digital economy. Worldwide spending on public cloud services was estimated to be close to $600 billion in 2023, which is a hundred times what it was just fifteen years earlier.[32] The global cloud storage market alone is projected to reach over $300 billion by 2028.[33] However, without corporate funding and development, the cloud would not function. The three largest providers of cloud services in the world—Amazon, Google, and Microsoft—have spent enormous sums on data centers and related public-facing infrastructure. These investments have paid off handsomely, returning billions of dollars every year and, in Amazon's case, nearly three-quarters of the company's operating profit.[34] While the Defense Department funded much of the earliest work on experimental computer networks in the 1960s that eventually became the Internet, most of the infrastructural advancement that has followed for pipelines, platforms, and data storage has been accomplished with private capital. Such public-private partnerships have existed throughout media and telecommunications history, but the private sector has been in the driver's seat since the advent of the digital era. The cloud that they built now serves practically all industries, including banking and financial services, transportation, telecommunications, agriculture, manufacturing, retail, health care, and of course media and entertainment.

The media sector has undeniably been transformed by the cloud and the "Netflix effect"—the growth of direct-to-consumer streaming services and the attendant demand for instant access to any and all forms of media. This cloud-fueled dynamic was credited for motivating Disney's $71 billion+ purchase of 21st Century Fox in 2019, and AT&T's short-lived $85 billion takeover of Time Warner in 2018 (quickly resold to Discovery in 2022), in the all-out arms race to keep up with digital content companies such as Netflix, Amazon, and Apple. Eliminating the middleman from the distribution chain and utilizing proprietary platforms to reach viewers directly has also threatened traditional cable companies, as the pace of "cord cutting" has picked up. In the US, the number of households that leave cable subscriptions behind increases each year; currently it hovers around 24 percent and is projected to approach 50 percent by 2024.[35] Moreover, the born-digital media platforms are worth, in some cases, ten to twenty times what their Hollywood studio counterparts are.[36] These data-based companies are also outpacing their studio-bound rivals in terms of spending and production volume in the digital distribution wars. Their dominance has played a central role in the formation of cloud policy.

Visualizing the Cloud

The exponential growth in cloud usage took off while most people still did not understand what cloud computing was, or recognize how much they were actually utilizing the cloud themselves.[37] That suits the tech industry just fine, because, as Tung-Hui Hu has written, "the more one learns about [the cloud], the more one realizes just how fragile it is."[38] This fragility results in incessant global data breaches—an inevitable fact of life in a cloud-based society—affecting hundreds of millions of users every year, including those resulting from high-profile hacks of major studios, social media platforms, and governments. Moreover, cyber warfare and malicious attacks by individuals and state actors is expanding at an astronomical pace. The regulatory inconsistencies that exist in the global expanse of cloud policy only serve to further undermine the security of its pipelines, platforms, and data.

Widespread awareness of this infrastructural fragility would quickly unravel the peaceful metaphor that cloud imagery creates and depends on for the necessary user buy-in to occur. Without marketing wizardry, it would be quite challenging to convince users to allow their personal data and important documents and communication histories to be housed in "somebody else's computer" at a corporate-controlled warehouse in an undisclosed location. Visions of emails, work materials, and TV episodes residing somewhere celestial, floating in space and readily pulled back down to earth for viewing on demand is a more comfortable and calming fantasy. However, the reality of remote data storage and distribution is much less sublime. In fact, it is very earthbound, material, and precarious. It is also fraught with policy landmines to navigate on both domestic and global scales. This infrastructure could have just as easily been referred to collectively as the closet, the storage locker, the trunk, the attic, the barn, the vault, the warehouse, the cellar, the stash, the depository, the garage, the safe, the silo, or the shed. Or, for that matter, it could have been called the sky, the air, the vapor, or the beyond. But the thought of putting your digital life in "the vapor" or "the shed" does not deliver the same sense of peace and tranquility inspired by leaving it "in the cloud."

Therefore, obscuring the very infrastructure on which it depends became integral to the cloud industry's success. Echoing the arguments about infrastructural literacy and practices of concealment made by Lisa Parks[39], the less the public understands about the pipelines, platforms, servers, facilities, and legal paradigms that their data travels through, the easier it is for people to hand over their personal information the cloud's unidentifiable and mysterious stewards. For the user, such data is better off in "God's habitat" as John Durham Peters has called the "marvelous clouds" that signify the glory of the divine presence[40] than in a chilly underground bunker full of computers run by massive global tech conglomerates.

Figure 0.2
Ethereal cloud promises.
Credit: Photo by Louise M. Kolff

Figure 0.3
Servers in Douglas County, Georgia, data center.
Credit: Google

This transformation of server farms from hulking storage facilities into attendants of the sublime requires imbuing the material (servers, wires, cables, buildings) with immaterial qualities, and the immaterial (digital data, the concept of remote storage) with easily recognizable and nonthreatening features. Hu refers to this as an example of the computing science term "virtualization—a technique for turning real things into logical objects," such as a warehouse of data storage servers rearticulated as a "cloud drive."[41] It allows for the projection of the cultural fantasy, as Hu elaborates, of the cloud as something "inexhaustible, limitless, invisible."[42] As cloud infrastructure is reimagined in the form of consumer-friendly constructs and metaphors, that imbued sense of unlimited abundance also works to blur the larger ecological and economic implications of the industry at large. Asta Vonderau, for one, has written about this development, explaining that "dematerialized images of the cloud obscure its infrastructural and industrial materialities as well as its problematic social and environmental consequences, including the enormous electricity and water needs of data centres, the increasing pollution through waste heat, or the low number of job opportunities the cloud industry offers to local communities."[43]

The practice of virtualizing and visualizing the immaterial components of media goes back to wireless telegraphy and the advent of radio. In his analysis of the radio spectrum and "the meeting point between materiality and the imaginary," Ghislain Thibault has pointed to the imagery of the lightning bolt and wave as instrumental in shaping the social meanings of wireless technology and representing values such as progress or sound transmission.[44] The US National Telecommunications and Information Administration map of the radio spectrum represents the visual transformation of the spectrum into specific parcels for regulatory purposes (see figures 0.4 and 0.5). Thomas Streeter has written about how Guglielmo Marconi's newly discovered "ether" that disseminated radio signals was treated as a space that "might be usefully 'bounded,' and thus given some of the characteristics of property."[45] The spectrum—which could also be understood as the original platform—thus became a space of economic exchange folded into the logics of corporate liberalism,[46] an ideology that has continued to guide the approach to regulating media infrastructure into the cloud age.

Susan Douglas has also written about the initial struggles with the "intellectual leap" required to regulate rights in the spectrum, "something that was invisible, all pervasive, seamless, and still quite mysterious as a property" in

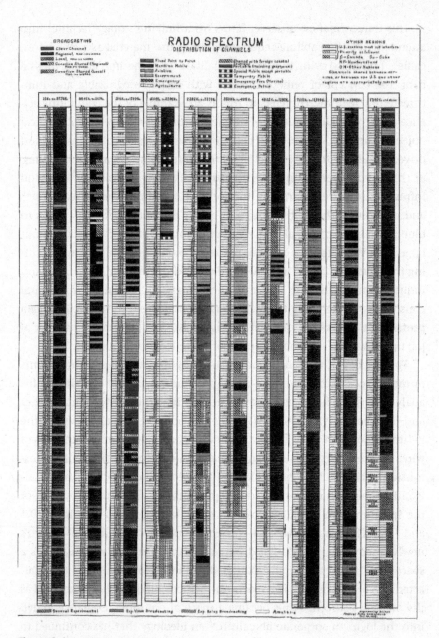

Figure 0.4
Radio spectrum map, 1928.

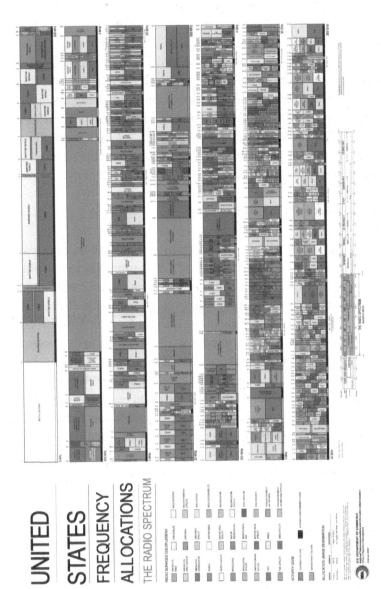

Figure 0.5
US radio spectrum frequency allocation map, 2016.

the early twentieth century.[47] After all, Douglas explains, "the air was an element Americans had traditionally associated with freedom, even transcendence."[48] This spirit of transcendence is a connection to the electromagnetic spectrum that has endured through the present era of "virtualizations" for data seemingly being transported through the air that we can't see. Silicon Valley designers and their partners on Madison Avenue have long used these associations when educating the public about how to understand the cloud. Together they have manufactured the cloud's projected cultural fantasy of an endless resource, benign invisibility, and harmless magic that makes life easier for all. This rendering of such complicated, legally fraught physical infrastructure into a palatable, cartoon-like abstraction for the masses is one of the great public relations triumphs of the twenty-first century.

The Path Dependencies of Cloud Policy

As network engineers well know, "It doesn't matter how high-tech the industry is. . . . you still have to dig in the dirt."[49] This truth was served with a powerful reminder in 2011, when an elderly Georgian woman searching for copper wires to sell for scrap accidentally cut an underground fiber-optic cable with her shovel and knocked out the Internet for all of Armenia, large parts of neighboring Georgia, and some areas of Azerbaijan. In the process of searching for a dying form of wired connectivity, she (temporarily) destroyed its replacement, highlighting the very physical and fragile nature of infrastructural path dependencies we rely on in the cloud era. The coexistence of old and new technologies and networks are embodied in this accident (even the cable itself was owned by the Georgian railway network), revealing the specter of historical infrastructures that haunt the cloud wherever it touches the ground. These traces are still alive in the many material forms, discursive frameworks, and policy histories left behind.

Much of the cloud can indeed be found in the dirt, among the layers of trenches and wires and cables that were the lifelines for nineteenth- and twentieth-century media distribution networks. It has long been true that new technologies and infrastructure are built on the skeletal remains of those that preceded them, and the cloud has similarly been constructed on the many transportation and communication networks that came before: telegraph lines were strung along railroad rights-of-way; early telephony utilized telegraph wires; DSL service was transmitted through existing copper phone lines; and coaxial and fiber-optic cables largely followed the routes of the telephone. Even some of the company names have been created out of their common past: SPRINT was an acronym for Southern Pacific Railroad Internal Networking and Telephony, and AT&T originally stood for American Telegraph & Telephone Company, representing the media technologies of the corporation's lineage. The cloud now encompasses these infrastructural layers of the copper phone wires, railway trenches, and fiber-optic cable of the past in its twenty-first-century formation.

This formation has been well documented by communications and media scholars including Wolfgang Schivelbusch, who described the strata of communications and railway lines as "one great machine covering the land,"[50] and Brian Larkin, in his argument that "urban space is made up of the historical layering of networks connected by infrastructures."[51] Shannon Mattern has further explored infrastructural layering through the concept of path dependency. She writes that tracing "infrastructural 'paths' back into deep history . . . compels the recognition that those spaces built to accommodate historic forms of communication also *inform* and function as part of today's media infrastructures."[52] Others, including Carolyn Marvin and Lisa Gitelman, have discussed the complex layering (material and otherwise) of media and communication technologies throughout history, and David A. Banks has articulated the many legal, sociocultural, and geographical mutually constitutive forces that have shaped both the railways and the instantaneous communication of the Internet.[53] In all of these examples, we are taught to "see" the embedded histories of media and communications infrastructure that are everpresent but often invisible.

To that end, Mattern has argued that our cities have been "wired" for transmission and have "hosted architectures for the production and distribution of various forms of intelligence" for millennia, and our infrastructural histories are inscribed all around us in the physical landscape.[54] In her archaeological analysis of "urban media" and the history of the mediated city, Mattern maintains that the media networks and technologies littered throughout our metropolitan environments do not merely "supplant the old" but instead can be observed as "a layering or resounding, a productive 'confusion' of media epochs" that allow new methodological approaches to writing history.[55] The former R.R. Donnelley building in Chicago is a case in point. A historic architectural landmark, it was once the country's largest printing facility and, until 1993, publisher of the Yellow Pages telephone directory. It is now one of the world's largest data centers and "carrier hotels,"[56] housing numerous telecommunications and Internet companies in over one million square feet of space. The facility is located along the Illinois Central Railroad tracks and sits just blocks away from the North American fiber-optic long-haul route running across the US.[57] It is one of many late nineteenth- and early twentieth-century manufacturing buildings that have been transformed into twenty-first-century tech hubs, thanks to a fortuitous combination of geography, economics, and industrial history. It also serves as a stark reminder of the myriad path dependencies that persist across cloud infrastructure. This particular case—of a building's conversion from processing and delivering twentieth-century data (a city's catalog of telephone numbers, aka the "phone book") to storing and distributing data in its twenty-first-century form (the 0s and 1s flowing through the Internet)—is one of the more pointed examples of how digital infrastructures can inherit the material *and* immaterial frameworks of their analog predecessors.

Throughout this book, I extend Mattern's argument about media/technology to the layering and path dependency that has been brought forth in the arena of policy. This approach opens up expansive historiographical vistas made available to us when our methods refuse the bifurcation and "reductive distinctions between 'old' and 'new.'"[58] Exploring such layering and dependencies as they exist physically and conceptually as well as legally allows us to also view policy as a "resonance chamber," wherein we hear "echoes of the past," as Mattern has described.[59] However, these echoes are now experienced as inscriptions of regulations written for an analog era, obsolete technologies, and markets that no longer exist.

Also inspired by the way Lisa Gitelman charts the "genealogies of inscription" in media technologies, forms, and cultures that "evolve in mutual inextricability,"[60] I am similarly focused on the imprints of previous infrastructure policies and their traces on our present regimes. Christian Sandvig has written about what he calls the "relational" framing of infrastructure and the manner in which such a framework represents "an infinite regress of relationships."[61] This dynamic can also be found buried in infrastructural policy and can be tracked back to regulation designed for the telegraph, the telephone, and even early video recording devices. Such comparisons are useful in illuminating the embedded legacies and the contingent, path-dependent nature of cloud policy. They are also key to understanding the threats to the public interest and democracy itself posed by the weight of outdated policy dragged into the future on the backs of evolving technologies.

While scholars such as Hogan and Shepherd have emphasized that "the materialities of data infrastructures . . . shape the politics and laws informing ownership, access, transparency, privacy, and freedom,"[62] the "infrastructural imaginaries" that Lisa Parks has described also determine the contours of regulatory discourse and action. These "ways of thinking about what infrastructures are, where they are located, who controls them, and what they do"[63] also carry a spectral trace in policy. For example, the concepts of scarcity or abundance with respect to infrastructure have a very long history of enacting limitations on provision, expansion, and competition that have little relation to actual technological capacities. Take the imagination of the electromagnetic spectrum, which has maintained the lore of scarcity since the advent of radio. Mara Einstein has explained that spectrum scarcity "was a myth almost from the time of its inception. As early as the mid-1920s, technology existed that would overcome the perceived shortage in spectrum. . . . The belief in spectrum scarcity was a government choice."[64] Nevertheless, this imagined scarcity of spectrum infrastructure has determined policy decisions for over a century, benefiting a select handful of corporations while disadvantaging any would-be competitors and shortchanging the public deserving of a more robust market.

These calcified fictions have undeniably affected competition and innovation. Thomas Winslow Hazlett argues that "we are swimming in underutilized frequency spaces" but restrictions on market rivalry "are said to be embedded in nature. Artificial policy choices are transformed into necessities."[65] He further adds that "it is a crime against science" for institutions guided by principles from the radio era to dictate the regulation and policies for today's wireless markets.[66] This mindset of scarcity has endured into the era of digital distribution, haunting the policies for cloud infrastructure despite the public relations imagery of infinite expanse. Indeed, the spectrum is still supporting cloud infrastructure, including mobile and satellite-based Internet services. The frameworks of scarcity have continued to be embraced by regulators, and as one author of the US National Broadband Plan (2010) has argued, "innovation will continue to be stifled . . . unless and until we adopt a psychology of abundance in the regulatory space."[67] Former FCC Chairman Newton Minow has also urged that "we have to find a way to reinstate a commitment to the public interest in our world of nonscarcity."[68] Yet, its specter looms large and continues to dominate the policymaking mentality, as it has for the last century.

Even the dynamic of "regulatory capture" is a deeply felt inscription of the policy process. This fundamentally corrupt dimension of regulation has been elegantly defined by Robert Horwitz as one wherein an agency "*systematically* favors the private interests of regulated parties and *systematically* ignores the public interest."[69] It is as old as regulatory agencies themselves. The Interstate Commerce Commission (ICC) created in 1887 was the first, and one of the most notoriously corrupt.[70] It was a model example of how regulators can become co-opted by special interests and those they are supposedly policing at the expense of the public they are charged with serving. Such dysfunction is inherent in a political system that allows for private money to exert undue influence on government officials through lobbying and campaign financing. When the corporations being regulated are the benefactors of public servants, the necessary separation between those creating and enforcing policy and those subjected to it no longer exists. This problem has only become more pronounced as lobbying has intensified, steadily increasing the power and sway of private firms to the point that former FCC Chairman Reed Hundt once joked that FCC stands for "Firmly Captured by Corporations."[71]

Cloud policy is a sweeping topic. Jean-François Blanchette has written that the cloud has become an object of focus in policy matters dealing with "market regulation, fairness, universal access, reliability, criticality, national security, sharing of limited resources, congestion, inter-network competition, national economic welfare, capacity planning, monopoly and antitrust, among others."[72] To be sure, this policy terrain is distinguished by its remarkable breadth. In this book, however, I have chosen to concentrate on the issues of privacy and surveillance, free speech, access, and competition/antitrust. They are among the most important matters being regulated vis-à-vis cloud pipelines, platforms, and digital data, and together their histories best illuminate why cloud policy matters.

The Stakes of Cloud Policy

Privacy and the Surveillance Society
The right to privacy has become one of cloud policy's many casualties. In addition to the formal laws and informal corporate governance practices that undermine individual privacy (addressed in chapters 2 and 3), the relationship

between the state and the private sector has become a main agent of privacy's destruction, to the point that cloud infrastructure now hosts a "public-private surveillance partnership that spans the world."[73] The collaboration of commercial infrastructure providers and the US government in the formation of the surveillance state has been a constant thread throughout cloud policy's history, and this alliance goes all the way back to the beginning of electronic communication. The partnership of US communications carriers and intelligence agencies began shortly after World War I with the Cipher Bureau in the army's Military Intelligence Division; also known as the Black Chamber, it was a precursor to the National Security Administration (NSA). Even in peacetime, this bureau worked with the Western Union Telegraph Company and the Postal Telegraph Company to intercept diplomatic and military communications. When it was first revealed that the bureau had been reading military and civilian telegrams and mail in 1929, it was defunded by Secretary of State Henry Stimson "on the grounds that it was unethical for the United States to engage in such unprincipled activities," and because "gentlemen do not read each other's mail."[74] This diplomatic gentility did not last, as, once the Cold War began, Western Union, RCA, and ITT provided the government via the NSA "with paper tape, microfilm, and later magnetic tape copies of most international telegrams."[75] Known as Project SHAMROCK, this project continued for decades.[76]

Cloud pipelines have a relationship to practices of surveillance with roots that extend back to the nineteenth century in the telephone's long, storied history as an agent of privacy invasion for the state. As Colin Agur has explained, "telephone surveillance is as old as telephony itself," and as a result of the wiretapping work by powerful, big-city police forces, "telephone surveillance emerged as a largely unregulated tool of protection and order in early twentieth-century America."[77] James Fly, FCC chairman under President Franklin Roosevelt and one of the greatest opponents of wiretapping, often reminded the president that the government was regularly violating the provision in the 1934 Communications Act that held "no person ... shall intercept any communication and divulge [it] to any person."[78] The politicized and extensive use of wiretapping by the FBI under J. Edgar Hoover in particular installed the telephone and its pipeline infrastructure as integral to policing and law enforcement activities in the US, and it has remained a state-sanctioned instrument of privacy invasion ever since.

Wiretapping was determined to be Constitutional in the Supreme Court case of *Olmstead v. United States* (1928); it was banned by the Communications Act of 1934 and then recuperated somewhat by the landmark case of *Katz v. United States* (1967), which subjected it to Fourth Amendment warrant requirements once again.[79] However, wiretapping (referred to as a "dirty business" by Supreme Court Justice Oliver Wendell Holmes) remained an extremely contentious practice, even among and between government agencies and officials. Justice Brandeis wrote a scathing and influential dissent in *Olmstead* in which he warned about the "subtler and more far-reaching means of invading privacy" that had become available to the government based on this decision. "Discovery and invention have made it possible for the Government, by means far more effective than stretching upon the rack, to obtain disclosure in court of what is whispered in the closet." Brandeis wrote that it was "immaterial that the intrusion was in aid of law enforcement. Experience should teach us to be most on our guard to protect liberty when the Government's purposes are beneficent. Men born to freedom are naturally alert to repel invasion of their liberty by evil-minded rulers. The greatest dangers to liberty lurk in insidious encroachment by men of zeal, well meaning but without understanding." He added, "Crime is contagious," arguing that if the ends justify the means, and if the government is allowed to commit crimes in order to convict criminals, it breeds contempt for law and invites anarchy.[80]

In the Cold War era, telephone lines became a significant concern for the government looking to protect state secrets from internal enemies, including their own citizens, that would undermine national security. One of the more interesting twists on the surveillance threats posed by these formative cloud pipelines can be found in a series of National Security Administration (NSA) in-house posters from the 1950s, 1960s, and 1970s. These posters exhibit tremendous suspicion of the NSA's own agents, providing a window into how "security" has been imagined and managed over time. The images repeatedly emphasized the numerous threats to national security embodied by individuals "talking too much" or "sharing secrets" with one's spouse, family, or priest. The telephone figured prominently in such warnings and was vividly portrayed as the most dangerous and threatening communication technology. According to the NSA, using such pipelines in ways that betrayed the agency's rules was hazardous to one's health, in the extreme (see figures 0.6–0.8).

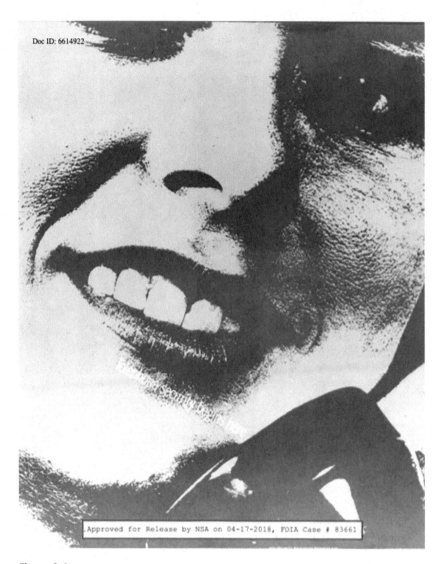

Figure 0.6
"Telephone Security Begins Here," NSA internal poster.

Figure 0.7
"Don't Talk Yourself To Death," NSA internal poster.

Figure 0.8
"Telephone Hanging in a Noose," NSA internal poster.

In the 1960s, Congress turned the tables and began to address the threat that the state and its use of computerized technologies posed to the privacy of its citizens. Much of this was in response to President Johnson's proposal for a "National Data Center" in 1965 to consolidate federal databases as part of the Great Society project. Congress became alarmed and held numerous hearings in the House and Senate between 1966 and 1967 to discuss the many potential invasions of privacy represented by government control of individual data. As explored in detail in chapter 3, the idea of a state repository of citizen data created quite an uproar, and the "National Data Center" did not come to pass. In the intervening decades between the impassioned work of the 89th Congress and Facebook CEO Mark Zuckerberg's 2010 insinuation that "privacy is no longer a social norm," the ravaging of this right became a joint effort of the government and Big Tech—conducted largely via black-boxed cloud infrastructure and abetted by corporate lobbying and the cultural appetite for and eventual dependence on all that mobile, social media has to offer. The worst abuses now stem from the dominant platforms and their business models based on the indiscriminate tracking of their users, dataveillance, and targeted advertising endemic to "surveillance capitalism." Shoshana Zuboff has explained that this economic order "operates through unprecedented asymmetries in knowledge and the power that accrues to knowledge. Surveillance capitalists know everything about us whereas their operations are designed to be unknowable to us. They accumulate vast domains of new knowledge *from* us, but not *for* us."[81] Surveillance capitalism depends on cloud infrastructure and has further cemented its Big Tech owners as our new sovereigns in the digital space.

Access and Speech Rights
The lack of affordable public access to the Internet is a distinct failure of cloud policy in the US. The absence of universal service mandates (required for telephone but not broadband networks) has had serious consequences, contributing to an intolerable digital divide in the US that has created a rolling blackout for information, education, public health, and civic engagement. Globally, it has been referred to as a "digital canyon," as two-thirds of the world's schoolchildren are without Internet connection in their homes.[82] The COVID-19 pandemic further exposed the human and societal cost of this inequality, as Internet access determined one's ability to work, attend

school, or receive health care during a global health emergency. Lower-income children in the US were four times more likely to have to do their schoolwork on a cellphone during quarantine because they did not have adequate Wi-Fi in their home, and 40 percent of these children had to resort to sitting in a McDonald's parking lot or somewhere with public Wi-Fi because there was no reliable Internet connection at home.[83] One of the lessons we must take from the pandemic is that the effects of cloud policy can extend all the way to the art of survival in the modern world. Internet access must be viewed as an essential public service, which requires policy rationales and foundations for broadband provision to match that of other utilities. The history of "pipeline principles" that have determined our access to the Internet thus far is addressed in detail in chapter 1.

In addition to the financial barriers of Internet access, citizens must also surrender their personal data and digital privacy in exchange for access to Big Tech's platforms and services, fueling their extractive culture of data mining dependent on mass surveillance. Cloud policy's assault on access and privacy is somewhat predictable, given this infrastructure is being regulated in an environment dominated by asymmetrical public-private partnerships, agency capture, data-driven business models, and diminishing state oversight. The resultant threats to our democracy and civil liberties are profound and ongoing.

The platform ecosystem is also the sector of cloud policy where the most urgent problems exist for speech rights, as discussed in detail in chapter 2. In many ways, the Internet is a daily experience with mis- and disinformation, most of which is protected by the First Amendment. It is further amplified by algorithms dependent on data extraction, inflamed by the use of AI and bots, and then left to Big Tech platform stewards to self-police, determining the boundaries of acceptable discourse and acting as the arbiters of truth. Thus far, that has not proven to be a successful model. We are left with a contemporary version of "yellow journalism," the late nineteenth-century sensationalist practice pioneered by papers such as Pulitzer's *New York World* and Hearst's *New York Journal*. Only now, it is digital, mobile, and inescapable. From this perspective, the past 130 years has been one flat circle of political hyperbole and culture wars. Addressing this problem is one of cloud policy's most critical and vexing challenges.

Competition and Antitrust

Cloud infrastructure is largely owned and controlled by a handful of Big Tech and telecommunications companies that have operated with minimal regulatory oversight. This has been enabled by more than a generation of governmental neglect, as the biggest failures of antitrust policy have been ignored since the 1980s. Existing frameworks are unsuitable for the rapidly evolving markets and industries that cloud infrastructure represents. Our current criteria for acceptable levels of market concentration and monopoly behavior are also woefully insufficient, as they primarily evaluate competition through the lens of neoliberal efficiencies. Regulators' fetishization of and reliance on the "consumer welfare standard" (interpreted as lower prices) have ignored the political, cultural, and economic consequences of mergers so long as this arbitrary and external condition is met.[84] However, the effects on labor, on local communities, and on industries and markets at large are lasting, while the benefit to the consumer is only temporary. Tim Wu has argued that "the 'consumer welfare' approach has enfeebled the law. Promising greater certainty and scientific rigor, it has delivered neither, and more importantly discarded far too much of the role that law was intended to play in a democracy, namely, constraining the accumulation of unchecked private power and preserving economic liberty."[85] This lack of appropriate antitrust measures for the digital era is a truly lamentable signature of cloud policy.

FTC Chairwoman Lina Khan, in her field-defining article "Amazon's Antitrust Paradox," has also addressed this failure of contemporary antitrust. Khan argues that the focus on "consumer welfare" instead of widespread cross-market power "is unequipped to capture the architecture of market power in the modern economy."[86] The traditional guidelines and remedies for anticompetitive behavior have been ground into dust by Amazon and its exploitation of customer data, its influence across market sectors, its vast webs of market control. As Khan points out, in addition to being a retailer, Amazon is also "a marketing platform, a delivery and logistics network, a payment service, a credit lender, an auction house, a major book publisher, a producer of television and films, a fashion designer, a hardware manufacturer, and a leading provider of cloud server space and computing power."[87] It has more than one million employees and is presently the most valuable public company in the world. Amazon also controls the cloud infrastructure (Amazon Web Services, or AWS) that its competitors rely on for their businesses to function and has more revenue than Facebook, Google, and Twitter combined.

The fact that the company has continued to sidestep regulatory action only strengthens Khan's argument.[88] This profound disconnect between antitrust standards and industrial realities has led to intensified concentrations of power that have proven devastating for market competition and, in turn, for the public interest.

Antitrust hearings brought out the CEOs of Amazon, Apple, Facebook, and Google in the summer of 2020 for one of the most extensive government examinations of Big Tech's acquisitions, policies, and competitive practices to date. Other ongoing investigations include actions by Congress, the Federal Trade Commission, the Department of Justice, the European Commission, and at least forty-eight state attorneys general. However, as of this writing, none of the US-led inquiries have led to a single antitrust remedy. Many US lawmakers are so out of touch with the various markets and technologies involved in cloud policy, they make a farce of the state's attempts to control them. For example, during the questioning of Facebook CEO Mark Zuckerberg in the 2018 Senate hearings on data privacy and Russian disinformation in the wake of the 2016 election, it became clear that some of the legislators conducting the questioning had no idea what social media's business model was, even though that was the cause of the scandal itself.[89]

SENATOR ORRIN HATCH (R-UT): Well, if [Facebook is free], how do you sustain a business model in which users don't pay for your service?

ZUCKERBERG: Senator, we run ads.

HATCH: I see. That's great.

Memes about Hatch riding his dinosaur to work at the Capitol erupted on social media, but the archaism of the Senate is no laughing matter. The inability of Congress to understand the industry they are charged with investigating, and the complex technologies it relies on, diminishes lawmakers' ability to protect the public's interests in their capacity as legislators and policymakers.[90] The institutionalization of such ignorance among the majority of senior lawmakers is the result of a center of power unwilling to renew itself or adapt to the twenty-first century, and a press that has let it go largely unnoticed. All these dynamics have directly contributed to the civic bankruptcy of contemporary cloud policy.

Economic and political structures are inseparable, and Tim Wu has argued in *The Curse of Bigness* that antitrust is the key to preventing tyranny in both realms. However, the level of consolidated corporate power that exists today in cloud infrastructure is growing beyond the ability of contemporary antitrust enforcement practices to control. Zephyr Teachout has argued that Big Tech and other monopolists represent "a twenty-first century form of centralized, authoritarian government," warning that "no democracy can survive for long once a few corporations have amassed governmental power in such massive form and scale."[91] The trusts of the late nineteenth and early twentieth centuries were largely understood to represent such a threat. Big Tech is undeniably the new millennium's iteration of this economic order that now plagues what has become the Digital Gilded Age.

The lack of relevant regulatory frameworks for the digital ecosystem, such as the failures of antitrust enforcement as identified by Wu, Khan, Teachout, and others, has proven to be a core enabler of a fundamental shift in cloud infrastructure. This shift has seen the Internet and its platforms go from being valued as a democratizing force in their development and infancy to functioning as profit machinery for corporate cloud providers presiding over a form of antidemocratic governance. As sociologist Paul Starr has argued, "The digital revolution now threatens to undermine values that it was supposed to advance—personal freedom, democracy, trustworthy knowledge, even open competition. It isn't as though the technology did this to us on its own, or that we stumbled absentmindedly into an alternative dystopian universe. Today's technological regime grew out of critical choices to ignore lessons of the past and allow private power to go unregulated."[92] The chapters that follow detail these choices regarding pipeline, platform, and data regulation that have delivered a policy landscape tragically devoid of the lessons from its past.

As our rights and institutions become further enmeshed with cloud infrastructure, the stakes for cloud policy escalate. Our present conditions call for the broadest possible constituency to assess the contemporary cultural politics of media technology and the risks posed by policy frameworks designed for conditions and practices of previous centuries. It is an interdisciplinary undertaking to contextualize the insidiousness of algorithmic culture as a discriminatory mechanism of control; articulate the cultural costs of private corporations determining the value of personal data; and argue for the importance of regulating Internet access as a public utility in order to uphold principles of equity and social justice. Such efforts require the collective perspectives of media historians, lawyers, anthropologists, cultural geographers, economists, and policymakers. Bridging their often-isolated concerns and interventions regarding technology, policy, history, and culture is the vision at the heart of this book. With this research, I hope to bring infrastructure policy alive for those who care and for those who think they don't. I also want it to expand the pathways for scholarship to inform contemporary policymaking, articulate cloud policy as an urgent cultural crisis, and identify the footprints of its historical struggles so that we may create better paradigms for the future.

The book is structured in three chapters, examining the evolution of policy for the pipeline, platform, and data infrastructures of the cloud. Their histories are neither separate nor detached from one another; and they all intersect further with those of hardware, software, code, satellites, and energy, to name just a few more dimensions of this larger domain. However, their respective trajectories do demonstrate the most critical stakes of cloud policy in ways that merit distinction. Once these interconnected, mutually constitutive legacies are put into relational context, their "big picture" impact and historical import becomes clear. Indeed, the current state of cloud policy is most legible when viewed through the lens of history. Such perspective illuminates the values dictating these infrastructures of daily life, that which has been lost along the way, and what we might want to reclaim for the future.

The trajectory of cloud policy as a whole is an ideological progression, a change in the cultural valuation of infrastructure as civic good (e.g., electricity, landline telephony, the post office) to infrastructure as a tool of corporate profit generation (e.g., cable television and broadband Internet). Cloud policy's ultimate successes or failures will accordingly depend on the regulatory philosophy and values we embrace moving forward, and whether the model will be one inspired by public services or one based on private greed. Historian Richard John has reminded us that, "for the Founders, a well-informed citizenry, not profit-making for even letter delivery, was the reason the Postal Service was so crucial to the future of the republic."[93] If we are to effectively handle the current threats facing our citizenry, cloud policy must share this spirit and be informed by the mistakes and triumphs of its past. This requires recognizing the cloud as more than merely technologies that render modern life more convenient, but as the infrastructure of tomorrow's democracy, with a long history of policy decisions that carry serious, enduring consequences. This disposition is essential to its eventual promise. With the teachings of history, it is still possible to redirect our future onto such a path, as the story of how the "living flesh" of the computer utility is regulated and, to what effect, is still being written.

1 Pipelines

> The networks that connect us are the networks that define us.
> —former FCC Chairman Tom Wheeler

The pipelines to the cloud are most commonly associated with the provision of the Internet. I employ the term "pipeline" broadly and conceptually, as cloud pipelines are conduits for data distribution that are both material *and* immaterial. They are the myriad broadband networks—both fixed and mobile—that we depend on for distributing data to our computers, devices, and smart technologies.[1] Although the cloud did not become a legible infrastructural formation until the twenty-first century, the policies for cloud pipelines have been steadily evolving for the past one hundred years. They have shared histories with nineteenth- and twentieth-century distribution networks including electricity, telegraphy, telephony, and railroads. These analog lineages are ingrained in today's pipelines, and their complex legacies are key to resuscitating cloud policy's public values in the digital age.

The history of cloud pipelines is steeped in the ideology of the "modern infrastructural ideal" as articulated by Graham and Marvin (2001). This model promotes networked infrastructure as the binding, connective tissue of urban planning and development, dependent on "efforts by governments and states to support the shift to regulated, near universal access to infrastructure networks across cities, regions and nations."[2] Throughout Western economies, this type of infrastructural rollout took place from roughly the mid-nineteenth century through the 1960s, with an implicit acknowledgment that monopolies—either privately or publicly owned—were best suited for managing such services that were viewed as the nation's "circulation systems," critical to its power and growth. These networks provided cohesion, standardization, efficiency, and "universalizing norms of access" for utilities such as gas, electricity, water, transport, and communications.[3] In addition to AT&T's telephone lines, the many New Deal projects in the US creating roads, sewers, and waterworks are examples of this ideal, which held that infrastructure was a public good, so important for all members of society to access that their provision could not be left to the vagaries of market forces. Belief in the modern infrastructural ideal underpins much of the ideological foundation for early pipeline policy. However, its promises collapsed in the late 1970s under the weight of ascending neoliberal forces and the politics of privatization. This took place in a perfect storm of great technological change requiring massive investment amid major economic crises of the state.

The history of cloud pipelines is thus one of ever-increasing privatized, monopolistic control over public infrastructure and deteriorating public protections and access. This chapter is concerned primarily with these developments vis-à-vis the "last-mile" pipelines: the infrastructure that delivers the Internet the final leg to users, as opposed to the much larger terrestrial, undersea, and satellite networks (backbone providers) spanning the globe.[4] America's serious last-mile problem has been compared to the early struggles over electrification and its history of inequitable provision, with monopolists offering overpriced, poor service or none at all to rural and other so-called "unprofitable" communities.[5] The consolidated power of early twentieth-century holding companies for electricity along with government-sanctioned monopolies in telegraphy and telephony helped establish the blueprint for contemporary Internet service in the United States.

Their histories have also shaped what I view as the cloud's formative "pipeline principles"—the foundational concepts of common carriage, public utility, universal service, and natural monopoly guiding core elements of telecommunications policy, in some cases for hundreds of years. These principles are directly linked to the values, social ideals, and civic visions that are at the heart of pipeline policy's greatest struggles and finest achievements. They are also key to redeeming this infrastructure as a public resource and resurrecting the values that have been progressively erased from its regulatory frameworks.

Pipeline Principles

Common Carriage

Common carriage was imported from English common law and became part of the US legal system in the nineteenth century. In England, certain businesses were considered to be "common callings" and, consequently, subject to various parliamentary impositions and restrictions on behalf of the public, at that time including "bakers, brewers, cab drivers, ferrymen, innkeepers, millers, smiths, surgeons, tailors and wharfingers."[6] It was established that an obligation to the public trust for industries that purport to "serve all," and a requirement to actually do so—indifferently and indiscriminately—is attached to the distinction of being a common carrier. Common carriage principles were designed to prevent discrimination on the part of service providers and ensure that all paying customers received equal treatment. This designation and the attendant policies that have been evolving over the past two hundred years have had one of the most significant roles in regulating distribution pipelines, beginning with transportation and followed by communication.

One key point is that common carriers are private businesses, not government services. However, the US Supreme Court declared in 1876 that when a business or property is "affected with a public interest" it is within the power of the government to regulate it.[7] This was first applied to railroads, with the help of the Interstate Commerce Act of 1887. The ICA gave the Interstate Commerce Commission, the first independent regulatory agency, jurisdiction over common carriers and essentially declared railroads to be common carriers themselves. Legal scholar Philip Nichols further explains that "although the Interstate Commerce Act dealt exclusively with railways, it is integral to the history of communication common carrier law because it served as the initial basis for Congressional regulation of communications and provided many of the definitions found in the Communications Act."[8] Moreover, as Susan Crawford has noted, "Both railroads and telephones had been given access to extensive public lands and had benefited from the power of the state to condemn property for their use; in exchange, they had to offer their services without discrimination to all comers, and their rates would be set by the ICC."[9]

Later, other transportation and distribution pipelines including ports, elevators, ferries, amusement park rides, airlines, buses, and telecommunications providers were also regulated as common carriers. Telegraph companies were found to be common carriers by the Supreme Court in 1901.[10] In 1910, Congress passed the Mann-Elkins Act as part of an intense period of Progressive reform legislation, which empowered the ICC to expand their authority to telephone, telegraph, and wireless services and declared such providers to be common carriers as well. Once the Communications Act passed in 1934, the Federal Communications Commission assumed jurisdiction of these industries from the ICC. For the next sixty-two years, until its revision in 1996, the 1934 act largely dictated pipeline regulation for common carriers such as telephones (as well as for broadcasters and satellite transponders, which were declared to be "private carriers" instead of common carriers).

Despite its lengthy history and usage, a widely agreed-upon legal definition for a common carrier has proven to be somewhat elusive.[11] The Communications Act of 1934 is of little help, defining the term as "any person engaged as a common carrier for hire, in interstate or foreign communication by wire or radio or in interstate or foreign radio transmission of energy."[12] Radio broadcasters were a specific exception. A 1976 court of appeals case involving the FCC identified the basic characteristic that differentiates the common and private carrier: an implicit requirement of indiscriminate and indifferent service for all.[13] Subsequently, the conditions of nondiscrimination and equality came to characterize the fundamental qualities of a common carrier pipeline in most regulatory debates. Robert Britt Horwitz has simplified it as a "*commerce-based* notion of the public interest," which guaranteed access to the means of transmission as applied to telegraphy and telephony.[14]

If today's broadband pipelines to the cloud were regulated as the common carrier communications networks of the twentieth century, cloud policy would be serving and protecting the needs of the public more so than those of corporate pipeline stewards. However, broadband is *not* currently classified as a common carrier, inspiring a fight between activists, private Internet service providers, the FCC, the courts, and Congress that has been ongoing since the passage of the Telecommunications Act of 1996. Nor is broadband service classified as a public utility, the next pipeline principle of consequence. Herein lie some of the major cracks in the regulatory foundation for pipeline infrastructure that have collectively cast off the ideologies—and benefits—of the "modern infrastructural ideal."

Public Utility

A public utility is a service so central to the operation of society (e.g., electricity, gas, water, telephone) that it must be available to all at a reasonable rate. Subsidies are often provided for those who cannot afford or access public utilities easily. Utilities are usually monopoly providers of crucial public services with the same makeup of private ownership and public control found in common carriers. While the two terms are frequently confused, utilities receive more government concern, attention, and public protections than common carriers in the regulatory space, and they represent a smaller group of services. As Jack Balkin has explained, "the standard reasons to treat an enterprise as a public utility flow from its quasi-monopoly power and the universal need for what it produces. The goals of public utility regulation are to maintain reasonable prices, to secure universal access, and to ensure the quality of continuous service for consumers."[15] Harold Feld, senior vice president at the public advocacy organization Public Knowledge, further clarifies, "While most common carriers are not public utilities, just about all public utilities are common carriers."[16] Common carriers (of passengers, of communication, of cargo) must provide equal, indiscriminate service to all because of their role in society; however, unlike the more highly controlled category of utilities, they are not subject to the additional government regulation of their rates and other obligations to the public.

Many media policy experts, lawmakers, and public interest advocacy groups including the ACLU, Public Knowledge, and Free Press have long argued that broadband should be treated as a public utility. The fact that the provision of Internet service with nondiscriminatory, reasonable rates has been such a struggle to achieve is a vivid testament to the power of pipeline lobbyists in the halls of the US government. Crawford has called the lobbyists for Comcast and AT&T "our era's railroad lawyers," noting that carriers "will litigate unceasingly in support of their claim that any form of regulation will destroy their incentive to invest in infrastructure and innovation."[17] Their resistance to regulation in practically any form becomes baked into law and policy after billions of lobbying dollars change hands—in the past ten years alone, telecommunications services have spent roughly $100 million *every year* to influence the federal government; the industry consistently ranks in the top twenty for total lobbying expenditures, ahead of the financial industry and the airlines.[18] Meanwhile, the public calls for more affordable, accessible broadband are drowned out by these corporate funds in an endless cycle of antidemocratic practices rooted in Washington, DC. This corrosive dynamic of regulatory capture has been affecting the course of pipeline policy since early telephony.

Historically, the power of public utilities has not been effectively controlled by market forces because their providers are monopolies, which themselves follow a timeless, predictable pattern of exploitation, corruption, and lack of regard for public welfare.[19] The scandals among monopoly utilities are ever-present and ongoing. It was determined in the late nineteenth and early twentieth centuries that the government must step in to compel their stewardship of the public good. Economist Horace Gray famously wrote in 1940 that the "public utility concept" has not lived up to its promise, as relevant laws have "all followed the delusion that private privilege can be reconciled with public interest by the alchemy of public regulation. Consequently, none of them disturbed in the slightest degree the underlying structure of special privilege; they merely reared upon it a superstructure of restraint. Monopoly capitalism, secure in its privileges, shook off the petty irritations of regulation and continued its aggressions against the public welfare."[20] Gray criticized the way such services were inevitably declared to be "natural monopolies," and argued that the designation of public utility had become corrupted from its original intention of being "a system of social restraint designed primarily . . . to protect consumers from the aggression of monopolists." Instead, he claimed, "it has ended as a device to protect the property . . . of these monopolists from the just demands of society."[21] While an absence of government intervention has borne this out in the histories of telephony, railroads, and water in numerous municipalities (most publicly in Flint, Michigan), the pendulum swing to actively deregulating utilities has proven similarly disastrous. The deregulated electricity market and its stunning failures in Texas vividly underscore the importance of regulated utilities, and echo Gray's call to reform the concept of natural monopoly that figures so prominently in their provision.[22] As Feld has argued, "critically when we designate a service as a utility, that means it has become too important to leave to the benevolence of corporations, the kindness of kings, or the cold indifference of the market. We must guarantee fair access for all under a rule of law."[23] That is possible only through strong public-interest-focused legislation and modernized, updated policies for utility regulation.

The inherent contradiction between the profit motivation of private companies and the compelled public service requirements of a utility has been the source of debate since the age of the telegraph and has continued unabated into the cloud era. Richard John wrote about Samuel Morse and his fears about both the state and corporations monopolizing the network. According to John, Morse envisioned a "hybrid public-private network that combined the oversight of federal control with the energy of private enterprise could guarantee the 'checks and preventives of abuse' necessary to prevent the misuse of this otherwise dangerous power."[24] Viewed by some as a "prophet of regulation," Morse still wanted a measure of "proper governmental control" over the telegraph, as he fully foresaw its transformation into "a great public utility."[25]

Dan Schiller reminds us that at the turn of the twentieth century, reform groups advocated for nationalization as an alternative to the corporate control of the telephone and telegraph, with the post office and its efficient, nondiscriminatory, universal service providing "an attractive template."[26] It was the most serious consideration ever given by the US government to the idea of public ownership for such utilities. Woodrow Wilson's postmaster general even submitted a 150-page report in January 1914 recommending that Congress immediately "declare a Government monopoly over all telegraph, telephone, and radio communication" and nationalize them "to be operated as an adjunct to the Postal Service." Among the primary reasons given were that "the monopolistic nature of the telegraph [and telephone] business makes it of vital importance to the people that it be conducted by unselfish interests, and this can be accomplished only through Government ownership."[27] That mandate was not taken up in the US as it was in Great Britain, France, and most other industrial nations (other than the brief exception of the telephone system being nationalized at the end of World War I for a year[28]). Five-time Socialist Party of America presidential candidate (running once from prison) and labor activist Eugene V. Debs believed that monopolies were inevitable and argued for a model of state supervision and collective control; at the turn of the century, he stated prophetically, "What is true of the telegraph is true of the telephone. It is true of railroads. The people should own them, or they will own the government."[29]

President Franklin Delano Roosevelt staked much of his early political career on containing the unmitigated power of the consolidated holding companies also known as "private utilities" in electricity. In stark contrast to the rhetoric of political campaigns today, FDR regularly spoke about the intricacies of common carriage requirements, the social obligations of public utilities, and the corruption in the ranks of monopoly providers in his stump speeches. He talked about the root of public utilities and their obligations to society, which stemmed from the early seventeenth century and legal proceedings by King James and his court (most notably his adviser Judge Lord Hale) regarding the role of ferry boats. The ferries were monopoly operators who could charge customers whatever they wanted, regardless of the service level, which was inevitably poor. Lord Hale determined that every ferry should be "under a public regulation" to "take but reasonable toll" and "give good service for a fair return on his labor and his property."[30] Roosevelt viewed this standard as key to government authority over public utilities in the twentieth century as well. He even advocated for the government ownership of power utilities to stem the tide of corruption, an idea that took hold in the wake of the Great Depression:

> Judge me by the enemies I have made. Judge me by the selfish purposes of these utility leaders who have talked of radicalism while they were selling watered stock to the people and using our schools to deceive the coming generation. My friends, my policy is as radical as the Constitution of the United States. I promise you this: Never shall the Federal Government part with its sovereignty or with its control of its power resources while I'm President of the United States.[31]

This was the last time that the US had presidential leadership willing to aggressively take on the privatized ranks of infrastructure providers on behalf of the public. Roosevelt's campaigns and the New Deal ultimately led to the breakup of holding companies and the establishment of the Tennessee Valley Authority (the first publicly owned power company in the US). He created the Federal Power Act and the Rural Electrification Administration, thus allowing for federal regulation of the market for electricity. Roosevelt was also the first to recommend that communications services be treated as utilities. In his 1934 letter to Congress urging the creation of the Federal Communications Commission, he wrote, "I have long felt that for the sake of clarity and effectiveness the relationship of the Federal Government to certain services known as utilities should be divided into three fields: Transportation, power, and communications."[32] Despite clashing state and federal forces,[33] this distinction was ultimately realized for the wireline telecommunications services of the twentieth century, but the same safeguards have not yet been afforded to the cloud pipelines of the twenty-first.

Currently, forty-two million Americans are without access to high-speed Internet.[34] As the digital divide expands, regulators and Internet service providers continue to dance around the crucial concerns represented by the designations of common carrier and public utility status: the equal, indiscriminate, and affordable access to essential pipeline infrastructure for all. This precedent of neglect has fed into recurrent patterns, including government protectionism of monopolists; the perennial kicking of the can down the regulatory road, leaving fundamental policy problems unresolved; historical cycles of performative regulatory theater leading nowhere; heavy and costly government lobbying by infrastructure providers subverting the democratic process; and expensive private pipeline build-outs and upgrades subsidized by taxpayers who are themselves left behind by the stewards and values of cloud policy.

Universal Service and Natural Monopoly

The concepts of universal service and "natural monopoly" have long been inextricable from one another throughout the history of pipeline policy. According to the FCC, "universal service is the principle that all Americans should have access to communications services" and was a cornerstone of the Communications Act of 1934.[35] While this definition is undeniably vague, universal service policies made telephone service more readily available for rural and high-cost areas, and for low-income households at reasonable rates.[36] Universal service as a policy evolved from an early twentieth-century emphasis on interconnection and efficiency (Mueller 1997) to a focus on ubiquity and affordability and, ultimately, became part of the remit of the FCC to provide for the public. However, it was AT&T's corporate strategy, implemented in 1908, that conceptually wed the pipeline principles of universal service and "natural monopoly" for regulators and the public alike.

The idea of "natural monopoly" dates back to the nineteenth century (see John 2010). Since the 1840s, Congress had debated whether telegraph service and the "transmission of intelligence" (the mandate of the post office) should be a government-owned public service or the purview of private industry.[37] The prevailing sentiment was that if such an industry or service was a monopoly, much like the post office, then it should be government run, and the decision "must rest upon which is better for the public welfare."[38] After seventy years of debates among shifting political winds, the concept of "government regulated monopolies" was forged, and AT&T was the chief beneficiary—due in no small part to the work of the company's first president, Theodore Vail.

American Telephone and Telegraph was established in 1885 in order to connect the thousands of independent, local "exchanges" licensed by Bell Telephone into a single national long-distance service. Theodore Vail had worked for the post office—and for the industrial revolution's other biggest monopoly common carriers, the railroad and the telegraph—before his time in telecommunications. The late nineteenth century was a (brief) time of great competition in the telephone industry, but by 1907, AT&T was under the control of financier J. P. Morgan and on the path to becoming a true monopoly provider. Vail was brought on by Morgan to build an empire of wires, which he quickly went about doing, buying up the local competition and purchasing Western Union telegraph in 1909. This extended the company's network of shared long lines across the country and overseas and made AT&T the dominant force in long-distance service.

Vail saw competition as "duplication, inconvenience, inefficiency, and barriers to interconnection." He argued—and convinced regulators—that only a single, "universal" system could provide the type of comprehensive service the country has always deserved.[39] Monopoly control and universal service were packaged by Vail as interdependent, and necessary for coordinated lines across the country. When making his case, Vail drew on common references to the venerated post office when he wrote to his shareholders, in 1910,

> It is believed that some sort of a connection with the telephone system should be within reach of all. It is believed further, that this idea of universality can be broadened and applied to *a universal wire system for the electrical transmission of intelligence (written or personal communication)*, from every one in every place to every one in every other place, a system as universal and as extensive as the highway system of the country which extends from every man's door to every other man's door. It is not believed that this can be accomplished by separately controlled or distinct systems nor that there can be competition in the accepted sense of competition.[40]

Vail also "launched a public relations campaign that was by far the largest effort of its kind to have ever been mounted by a U.S. corporation."[41] In his expansive study of corporate PR, Roland Marchand argued that the campaign's "primary purpose was political—to protect a corporation with an odious public reputation against threats of public ownership or hostile regulation."[42] It was wildly successful. In ad after ad, AT&T equated itself and its telephone service with public empowerment and democracy: promoting the company as a "telephone democracy" (1911); a "business democracy"

Figure 1.1
AT&T ad, *Western Electric News*, 1912.

AMERICAN TELEPHONE AND TELEGRAPH COMPANY
AND ASSOCIATED COMPANIES
One Policy *One System* *Universal Service*

Figure 1.2
AT&T company logo, 1908.

(1915); an "industrial democracy" (1919), and "a new democracy of public service ownership" (1920) in which AT&T was "owned directly by the people—controlled not by one but controlled by all."[43] Milton Mueller has further characterized the concept of universal service as "an expression of liberal egalitarianism. More than just a telephone in every home, the phrase implies that a ubiquitous communications infrastructure can contribute to national unity and equality of opportunity."[44] AT&T's iconic motto of "One Policy, One System, Universal Service," adopted in 1908, drew on this connotation and effectively rebranded monopoly as a democratic gift to America. At this point, AT&T became "the People's Telephone," and its wires were thus positioned as democracy's infrastructure.

Adam Thierer writes about this stunning triumph of early corporate marketing and branding, explaining that, "once AT&T's motto was adopted as the nation's de facto regulatory policy, no other firm was in a position to adequately extend service in accordance with the new federal and state mandated social policy. The Bell monopoly was here to stay."[45] As a result of this unqualified PR success, the company was initially allowed to continue expanding without government interference. Rather than facing antitrust action, AT&T was instead treated by Congress as a "natural monopoly" utility provider in order to preserve the public goals of universal service, thus eliminating its competition in exchange for a measure of price and service regulation. The embrace of "natural monopolies" would ultimately prove to be one of pipeline policy's original sins. The flawed approaches to antitrust and anticompetitive behavior that followed only extended the damage to the public interest. The monopoly on telephone service as sanctioned by the state led to a measure of uncontrolled corporate power that has continued into the cloud era. But at the turn of the century, Vail accepted the very long leash of proposed oversight attached to the privilege of being a natural monopoly, writing to shareholders in 1907, "It is contended that if there is to be no competition, there should be public control."[46]

AT&T's *1910 Annual Report* is a historical monument to Vail's corporate philosophy. Along with anticipating the trope of the "information superhighway" and other metaphors for broadband that were used in the 1990s, it also reads like a manifesto for industrial monopolists. The report promised utopian levels of connection and service while extolling the dangers of competition among utility providers: "On the assumption that a perfect telephone system must afford this direct highway of communication between any two desiring to converse, this system must reach everyone; must be universal, comprehensive. . . . To the extent that any system does not reach everyone it is not perfect."[47] It went on to note that such a vision could be realized only by allowing AT&T to maintain its monopoly, explaining how a provider "must have absolute control of the wires" over the entire distance between the points of communication in order for satisfactory communication to take place. "To do this efficiently and economically means the combination of every kind of electrical transmission of intelligence into one system in order that new and additional uses may be developed and that the wire plan and other facilities may be utilized to their fullest extent. Cheap service comes from full loads."[48] This quote also reveals the extent to which the commitment to universal service was further tied to the mandate of capitalistic maximization and profit.

Telecommunications Reports editor Fred Henck and Bernard Strassburg, longtime member and former chief (1963–1973) of the FCC's Common Carrier Bureau, wrote of AT&T's early history that "government regulators at all levels embraced the natural monopoly concept as the essential framework of their policies. . . . It was truly a symbiotic relationship. The regulated monopoly operated in what was considered to be the public interest and, in turn, was shielded against incursions by rivals and competitors, including the possibility of government ownership."[49] Its widespread acceptance by policymakers notwithstanding, the mythological concept of "natural monopoly" is of course dependent on sustained government intervention and actions at the federal and state level. As Robert Crandall has underscored, "despite the popular belief that the telephone network is a natural monopoly, the AT&T monopoly survived until the 1980s not because of its naturalness but because of overt government policy."[50] Economist Adam Thierer adds that "at no time during the development of the Bell monopoly did government *not* play a role in fostering a monopolistic system."[51] By winning the crucial support of the government and the privileged treatment of a "natural monopoly," AT&T pulled off a virtuoso public relations victory as it "convinced policymakers and the public that a national telephone network could not have been constructed in any other way."⁵² Moreover, the company's efforts ensured that the concepts of universal service and natural monopoly were woven into policy rationales that served AT&T's own financial interests. Broadband providers wishing to emulate the power of the Bell System returned to this marketing magic of equating monopoly provision with the benefits of universal service, equality, and democracy when it came time for their own bid to rewire the nation for the digital age.

AT&T and the Blueprint for Cloud Policy

While the concepts of common carriage, public utility, universal service, and natural monopoly have not been simply mapped directly from telegraph and telephone infrastructure to broadband Internet service, their application in the case of AT&T have provided the foundation for most debates over the regulatory frameworks for cloud pipelines. The company's formative interactions with regulators were also central to the history of antitrust policy. After each successive attempt to tame or break up the company, AT&T would instead reemerge in a more dominant form. Its indelible legacy of influencing regulatory practice and policy to benefit the interests of corporations over those of the public has served as a long-standing strategic plan for future cloud pipeline providers. The company also set the gold standard for regulatory capture that existed long before the era of Big Tech. AT&T's behavior inspired Lily Tomlin's long-running comedy sketch character Ernestine, the self-satisfied and wholly incompetent telephone operator infuriating her helpless callers and laughing all the while. Summing up the popular view of AT&T on *Saturday Night Live* in the 1970s, she cackled, "Here at the phone company, we handle 84 billion calls a year, serving everyone from Presidents and Kings, to the scum of the earth. We don't care—we don't have to! We're the phone company."⁵³

Before the company's service had sunk to the level of iconic parody, there were substantial struggles with federal regulators over its anticompetitive and abusive practices. Two early clashes in particular cemented AT&T's status as a government-sanctioned monopoly, creating one of the most lasting and successful corporations in US history. These conflicts also had a considerable impact on the types of communication the company was allowed to traffic in until the 1980s: voice, the twentieth century's payload; or data, the communication currency of the twenty-first.

The first arose out of a trust-busting investigation of the company that had roots in both the Taft and Wilson administrations. It took place just as the railways, Standard Oil, and American Tobacco were being broken up. At this point in US history, large industrial monopolies were treated by the state as a threat to democracy. AT&T's pursuit of its own monopoly flew somewhat under the radar but did not fully escape notice during the Progressive Era. By 1911, once the company had started slowly gaining control of the telegraph industry in addition to the telephone, it was being referred to in the press as a "quiet octopus."[54] In addition to expanding its grip on telegraphy, AT&T would not allow independent telephone companies to attach to its long-distance service. This exposed the emptiness behind the very public "commitment" to universal service as a public good so loudly trumpeted in the company's famous PR campaigns. AT&T had maintained control of Western Union since 1909, a level of concentration in telecommunications that set off antitrust alarms. However, the ICC and the state public utility commissions were extremely weak at the time.[55] Nevertheless, the independent exchanges were demanding action, and the Department of Justice finally began preparing a lawsuit in 1913.

Under the threat of being dismantled or nationalized, the company made a "gentleman's agreement" that reverberated for the next century. In advance of federal action, AT&T's vice president Nathan Kingsbury wrote a two-page letter in December 1913 that would become known as the Kingsbury Commitment. This letter effectively took the air out of the government's case, as AT&T agreed to divest Western Union and allow independent companies to use its long-distance lines in exchange for becoming a state-sanctioned monopoly. It also promised not to further expand its empire without prior government permission. This reestablished the separation between the telegraph and telephone that had existed since 1879 and normalized relations among the DOJ, AT&T, and its local competitors.[56] John Brooks has argued that the Kingsbury Commitment was the moment that "AT&T formally abandoned its dreams of monopolizing all forms of telecommunications in the United States."[57] However, that would prove to be merely a temporary condition. As Milton Mueller has explained, it was just a "hiatus in the march toward monopoly rather than a victory for the competitive principle."[58]

The Kingsbury Commitment is nevertheless critical for two reasons: first, it led to the sanctioning of AT&T's "natural monopoly," fixing that pipeline principle for the company's foreseeable and extremely profitable future.

Second, it erected barriers between the provision of voice (telephone) and text (telegraph) that were a continual spoiler for the company on its way to becoming a cloud provider. For AT&T, it was the first of two prohibitions against carrying text in addition to voice over its lines. However, the Bell System would not be contained for long. As Tim Wu pointed out, "the spirit of the Kingsbury Commitment was, you get special advantages, you get to monopolize this industry—but you're supposed to behave yourself."[59] To nobody's surprise, AT&T was unwilling to behave itself. In fact, it continued to expand its holdings, even dabbling in radio and film sound technology, with federal and state regulatory agencies remaining on the sidelines. Schiller has stated of this juncture in the 1920s that, "In the face of the AT&T juggernaut, regulation seemed a palpable failure."[60]

That changed when the company's monopolistic character attracted the unwanted attention of the newly established FCC in 1934. The commission's first action was authorized in March 1935 by President Roosevelt and a joint resolution of Congress: a comprehensive investigation of the telephone industry in the New Deal spirit of reform, known as the "Walker Report" after FCC Commissioner Paul Walker, the report's primary investigator and author. The main focus was on AT&T, which controlled more than 85 percent of the industry at that time.[61] The investigation—which encompassed everything from AT&T's corporate and financial history, capital structure, patents, service contracts, and accounting methods to its intercompany relationships, investments, rates, revenues, and expenses—cost the government $1.5 million and took almost four years.[62] The resulting report is a stunning seven-hundred-plus-page document chronicling the Bell System dating back to 1875. It concluded, among many other things, that "there is no competition . . . with the unified Bell System."[63]

In addition to acknowledging the telephone provider's role as a steward of the public interest and social well-being, and its status as a public utility and a "natural monopoly," the report was clear that "the importance of the telephone industry and the magnitude of telephone operations demand actual and not nominal regulation. A coherent and constructive program of regulation must be developed and placed in operation in order to protect the interests of the public."[64] It went on to argue for the importance of an expert trained regulatory staff that was granted broad powers, and for amending the Communications Act to clarify those powers in relation to interstate regulation. The call for change was clear and unqualified. In response, AT&T

wrote a 280-page denial and protest when the proposed report was released in December 1938.[65] The final Walker Report, formally released in 1939, was markedly toned down in the wake of AT&T's criticism, its recommendations notably blunted. As Schiller described it, "the Walker Report's call for legislation to engender both massive structural alterations to AT&T and much heightened regulatory powers for the FCC had been left by the wayside."[66] Its initial impact was negligible, a regrettable outcome that can be attributed to a myriad of factors, including New Deal agency politics, the effects of corporate lobbying, the size of a bureaucracy like AT&T—and, of course, World War II broke out a few months later in Europe, and the US was drawn in by the following year. Nevertheless, the report stands as a historical accounting of AT&T's long-standing monopolistic control of the industry, and the schism that has existed since the Kingsbury Commitment between the company's special privileges and the "good behavior" that never materialized.

Ten years after the Walker Report, AT&T experienced its first formal antitrust lawsuit, a case prosecuted in 1949 by the Truman administration. At this time, the Cold War was heating up, and AT&T, coming off a close partnership with the army and navy during World War II, was deeply engaged in national defense work through its Bell Labs and Western Electric subsidiaries. President Truman personally asked AT&T to manage the Sandia nuclear weapons laboratory in New Mexico (known as "Los Alamos' less glamorous sister,"[67]) for the Atomic Energy Commission. He positioned it as "an opportunity to render an exceptional service in the national interest."[68] Because the company was concerned about the pending antitrust litigation, AT&T accepted on a "no-profit, no-fee basis."[69] Moreover, AT&T's manufacturing arm, Western Electric, was the prime contractor for some of the US Army's main anti-aircraft missiles in 1950 during the Korean War and, later, more advanced weapons with nuclear warheads.[70] In fact, AT&T, Western Electric, and the government were so entwined "that for virtually the entire Cold War, it was the phone company that managed the country's nuclear weapons laboratories."[71] Longtime head of Bell Labs Mervin Kelly devoted half of his working hours to military and government affairs and had the same level of security clearance as the head of the CIA.[72] In his exhaustive history of Bell Labs, John Gertner has described AT&T as one of the main pillars of the post–World War II military-industrial complex. He argued that, for the company, government contracts offered not only a source of revenue but also a true strategic advantage, as "they gave the company strong allies within the government that the company would need as the twentieth century reached its midpoint."[73] AT&T's indispensable status undeniably compromised the government's willingness and ability to regulate it objectively.

Consequently, when the DOJ put together its antitrust case against the company in 1949, it was the Department of Defense that repeatedly lobbied on behalf of its ally—the defendant, AT&T—for the case to be dismissed in the name of national security. The Department of Justice nevertheless sought to have Western Electric and AT&T broken apart, claiming Western Electric was "a chosen instrument of monopoly."[74] The government was at odds with itself, and the multiple conflicts of interest undermined the integrity of the DOJ's antitrust case throughout. President Eisenhower's defense secretary, Charles Wilson, actually wrote to the attorney general's office in July 1953 expressing his department's "serious concern regarding the further prosecution of the antitrust case now pending against Western Electric Co. and the AT&T Co. [seeking] that Western Electric be completely separated from the Bell System."[75] Secretary Wilson advised that "the Armed Services and the Atomic Energy Commission have entrusted the Bell System with highly important responsibilities in the development and production of new weapons and system which are essential to this country's plans for national defense"[76] and that the pending antitrust case "seriously threatens" that critical work.[77] Incredibly, it was publicly revealed in a later congressional report that this letter was in fact "ghostwritten" by AT&T—indeed, by Kelly himself, who refused to acknowledge the conflict of interest in working as a Defense Department consultant while advocating directly for AT&T. He testified, "I was working for the Bell System and also I felt and feel very strongly that I was also working for the Nation."[78] The extent to which the case was corrupted was subsequently detailed by the House Antitrust Committee's report, as they documented multiple "instances where Defense Department officials abdicated to the Bell System their official responsibility."[79]

Despite the Defense Department's numerous attempts "[to inject] itself into the very merits of the litigation,"[80] the Justice Department held fast and pushed the case forward. Nevertheless, the breakup of AT&T and Western Electric did not happen. Instead, the case resulted in a consent decree that was settled in January 1956 with the condition that AT&T could keep Western Electric (again sanctioning its monopoly in equipment manufacturing), but the Bell System had to agree not to engage in any business other than "common carrier communications." That meant Western Electric could continue its government defense work, but had to refrain from engaging in any new markets, such as computing.[81] It was said that the lawsuit ended only when Eisenhower took over the presidency from Harry Truman, and his attorney general, Herbert Brownell Jr., was persuaded by an AT&T lobbyist to drop it.[82] The major impact of this consent decree was that AT&T was restricted to doing business as a common carrier—and therefore prohibited from engaging in unregulated industries, including carrying data. It was an effort to control the company from "cross-subsidizing" its position in other markets, such as computing, with the benefits that came from being a "natural monopoly" in telephony.

The Justice Department spent seven years and millions in taxpayer dollars on this case, only to see AT&T's monopoly reinforced and its investigation severely undermined by the government's own collusion with the defendant. The 1959 report written by the House Antitrust Subcommittee concluded that "the [1956] consent decree entered in the A.T. & T. case stands revealed as devoid of merit and ineffective as an instrument to accomplish the purpose of the antitrust laws."[83] This failed attempt to break up AT&T is a significant moment in the long history of collaboration between the government and pipeline providers for military and law enforcement purposes. It is equally important for the way it shifted AT&T's relationship to regulators and, in turn, further politicized the monopoly provision of telecommunications infrastructure. The resulting consent decree served mainly to anger the Department of Justice for a generation, as revelations of government agents having unethical meetings with AT&T executives, along with the extent of the Defense Department's interference, came to light.[84] This eventually led to another dramatic antitrust case two decades later, with a very different outcome.

Although AT&T had undeniable and unqualified successes in monopoly expansion while fending off regulatory showdowns, the company made some serious miscalculations when it came to predicting the future of its own wires in relation to the Internet. Most surprisingly, it refused to participate in the creation of the original digital packet switching network for the Department of Defense that would become the foundation for the Internet. This network was to be a distributed one, not centralized like AT&T's. But AT&T failed to understand the potential, seeing only a threat to its own existing long-distance business. Pioneering Internet architect and engineer Paul Baran recalled, "In retrospect, it looks like AT&T not only behaved badly in terms of blinding themselves to what could be significant innovation, they also behaved stupidly in setting their company on a path of aversion to the world that was going to replace them. And Bell Labs, which should've been used as a way to prevent them from making that kind of mistake, wasn't."[85]

AT&T has since been the subject of numerous sustained government investigations and one consequential legal action in the 1980s. These battles are meaningful for documenting policy history as more than simply a chronicle of specific legislation and agency decisions; their aggregate details present an evolution of infrastructural ideologies, and the public obligations of its stewards. The case of AT&T further prefigured the loosely regulated, consolidated corporate power that shapes the cloud and access to its pipelines today. It provided a vivid introduction to the considerations of antitrust and the privatization of governance that have come to dominate cloud policy. The difficult lessons from the failed 1949 case and the 1956 consent decree that followed figured prominently when regulators turned their attention to the next communications technology taking hold: computers.

Cloud Policy for the Convergent Era

In the 1960s and 1970s, electronic computing technologies were expanding their cultural presence. They were becoming part of public communications infrastructure and no longer solely the purview of the military, the academy, and corporate data scientists. The telephone industry, among many others including airlines, finance, and news organizations, began acquiring and relying on early computers to inject more efficiency into their networks. This presented the FCC with some difficult questions, and the commission was forced to contend with the growing interfaces between the unregulated world of computing and its customary domain of regulated communication. As a result, the agency began to explore the best way to classify and regulate these digital technologies that were becoming part of telecommunications distribution in a series of investigations known as the "Computer Inquiries," which one FCC senior counsel labeled "a necessary precondition for the success of the Internet."[86] The FCC identified this dynamic as "convergence" in 1966, and it is here that computing becomes a part of pipeline policy for the cloud of the future.

While the term convergence has been used widely since the early 2000s, it has a much longer history than is often acknowledged. In fact, it has been part of regulatory discourse since the first Computer Inquiry in the 1960s. We now live in a world defined and designed by convergence—the always-evolving integration of media, computing, and communications. Propelled by developments in technology, political economy, business models, and user culture, convergence has continually presented new conundrums and challenges for regulators. At the outset of the Computer Inquiries, the FCC noted that convergence had already "given rise to a number of regulatory and policy questions within the purview of the Communications Act" as they began to tackle some of these fundamental issues of classification and regulatory design.[87] The subsequent struggles over pipeline governance were magnified by the lack of new policy vision created for the realities of this changing landscape. That is to say, it is never enough to simply remap policy developed for existing technologies, infrastructures, social practices, and markets directly onto the new, and expect the same rights to be protected in a completely different environment.[88] This miscalculation is one of the enduring, core struggles for cloud policy that has yet to be resolved.

Computer I (1966–1971)

As communications began its long-term relationship with computing, AT&T was still barred from engaging in new markets beyond common carriage, including data processing, thanks to its 1956 consent decree with the Justice Department. Nevertheless, the FCC quickly saw what was on the horizon as telephone lines began to combine with computerized technologies over the next ten years, and the regulatory challenges would be significant. Consequently, the commission began a series of explorations in November 1966 to analyze the "the convergence and growing interdependence of the computer and communications."[89] These three distinct proceedings, collectively known as the Computer Inquiries, took place over the following two decades. Some of the discussion anticipated the net neutrality debates during the Bush, Obama, and Trump administrations yet to come. In a statement foreshadowing the arrival of the cloud many decades later, the FCC began with the following:

> The modern-day electronic computer is capable of being programmed to furnish a wide variety of services, including the processing of all kinds of data and the gathering, storage, forwarding, and retrieval of information—technical, statistical, medical, cultural, among numerous other classes. With its huge capacity and versatility, the computer is capable of providing its services to a multiplicity of users

at locations remote from the computer. Effective use of the computer is, therefore, becoming increasingly dependent upon communication common carrier facilities and services by which the computers and the user are given instantaneous access to each other.[90]

As former FCC bureau chief Bernard Strassburg characterized it, the commission began their inquiries in the role of "marriage counselor," helping to negotiate what became an enduring partnership between communications and digital computing in its earlies stages.[91] There were no provisions for regulating data services as such when the Computer Inquiries began. In 1966 at the start of Computer I, there were rules for keeping telecommunications *out* of data markets (most notably AT&T's 1956 consent decree) but nothing determining how data services would be regulated in and of themselves. Now that telecommunications and data services were both beginning to utilize and depend on one another, it was important to establish new rules. Accordingly, from this point onward, data was a prominent feature in telecommunications policy in one form or another.

There main questions animating Computer I during the late 1960s and early 1970s centered around whether common carriers should be allowed to provide data processing services, and vice versa. The costs of such services and the issue of data privacy were also under consideration, but the bulk of the FCC's attention was devoted to determining whether common carriers would be permitted to enter the *unregulated* markets of data provision, and what the potential ramifications would be for communications policy.[92] Kevin Werbach has described this as the process by which the commission "defined the terms of engagement between the regulated world of telephone companies and the innovative world of computer-based applications."[93] Strassburg, who also authored the first report, explained that Computer I was about establishing rules and guidelines and definitions "largely focused on the what the telephone companies could or couldn't do in the data processing field.... We didn't want AT&T or the other telephone companies to so intermix their operations and make a mess of both markets."[94]

The FCC's Final Decision and Order, adopted March 10, 1971, consisted of three decisions that established the regulatory framework for the relationship between telecommunications and data moving forward. Primarily, the order officially distinguished "communications services," which were common carriers (e.g., telephony), from "data processing" services, which were unregulated and would remain so (e.g., computing).[95] The essence of this decision has also been characterized by Robert Cannon, former senior counsel at the FCC, as the agency's attempt "to divide the world between 'pure communications' and 'pure data processing,'"[96] Such a framework "assumed a world of centralized computing processing distinct from the communications lines that linked to it,"[97] which grew increasingly antiquated as computing technology itself became more decentralized, networked, and distributed. This conceptual division between communication and data services was the defining mistake of the Computer Inquiries, because it was entirely impractical, almost instantly obsolete, and wholly unsustainable. The involvement of computer processing in both telephony and data servicing was only growing, and making such distinctions required the imperfect science of determining whether and at what point one type of content (voice) ends and the other (data) begins. In the rapidly developing "convergence of modes," as de Sola Pool called it, there was no longer a one-to-one relationship between a medium and its use,[98] as pipelines were combining variously regulated industries together for distribution now that they were responsible for all that would be datafied.

The FCC's order also determined that common carriers *could* enter the unregulated data services market in the future, provided they did not jeopardize the public interest, convenience, and necessity by doing so. At this point, data was treated as a resource of importance to the public interest, but this valuation would be progressively destroyed by the rise of privatized policymaking yet to come with the advent of Big Tech platforms. Computer I was also significant for the future of cloud computing, as telephone companies now had a green light to carry and process data. However, because of the potential dangers relating to their ability "to favor their own data processing activities by discriminatory services, cross-subsidization, improper pricing of common carrier services, and related anticompetitive practices and activities,"[99] the FCC enacted a "maximum separation" rule "to insure that the public was offered efficient and economical communication services."[100] This mandated that any phone company planning to engage in data services as such had to establish a distinct corporate subsidiary using separate equipment, officers, accounting, facilities, operations, and so on for that business.

While the maximum separation requirement applied only to carriers with revenues over $1 million, it nevertheless had little effect on AT&T and the Bell System at the time because of the 1956 consent decree, which specifically barred them from the data market entirely. The FCC also reiterated multiple times throughout the order that it was "not proposing, at this time, to regulate data processing."[101] Dan Schiller has characterized this as the missed opportunity of a co-opted commission, arguing that, when faced with such a pivotal moment for convergent media policy, "the FCC perversely responded by paring back its own jurisdiction. Taking express cues from the Nixon White House and its successors, the FCC—which is supposed to be an independent agency—relaxed and withdrew regulation over the explosively dynamic new industry of computer communications."[102] Explicit policy for a convergent media landscape was still yet to be written.

At the time, the FCC was also very concerned about the problem of information privacy created by "a trend toward concentrating commercial and personal data at computer centers. This concentration, resulting in the ready availability in one place of detailed personal and business data, raises serious problems of how this information can be kept from unauthorized persons or unauthorized use."[103] The agency went on to caution in the Computer I inquiry that as "the fragmentary nature of information is becoming a relic of the past . . . both the developing industry and the Commission must be prepared to deal with the problems promptly so that they may be resolved in an effective manner before technological advances render solution more difficult. The Commission is interested not only in promoting the development of technology, but it is at the same time concerned that in the process technology does not erode fundamental values."[104] Meanwhile, a series of congressional hearings, "The Computer and the Invasion of Privacy" (discussed at length in chapter 3) kicked off in July 1966, just four months before the Computer Inquiries started. These hearings sounded the alarm about the threats posed by digital databases, particularly those controlled by the state. It was a brief period of great energy around this topic in Congress as well as the FCC, and sadly, it would remain a high point of government attention to defending the public's digital privacy.

Carterfone and Antitrust Shifts
As the Computer Inquiries were ongoing, the history of cloud policy was also being written by a stubborn Texas cowboy who was unafraid of AT&T and its lawyers and willing to bet the farm to pay for twelve years of litigation as he took them on.[105] Thomas Carter's case, which the *New York Times* described as the one that "broke the American Telephone and Telegraph Company's monopoly on telephone equipment,"[106] was not one of

the FCC's more high-profile decisions. It is, however, important for the way it addressed the issues of interoperability, network decentralization, portability, and innovation that were key to the future development of cloud infrastructure and pipeline policy.

In what became known as the Carterfone case, cattle rancher Thomas Carter developed an application called the "Carterfone" that allowed users to connect a two-way radio to their telephone to extend its reach and quality of service.[107] He created this device in order to communicate while out on his ranch when he was without the use of a phone. However, AT&T refused to allow this connection to their network, so in 1965 Carter filed an antitrust suit against AT&T and GTE (General Telephone and Electronics Corp., the largest independent telephone company during the Bell System era[108]). In 1968, the case (referred to the FCC) was decided in Carter's favor when the agency ruled that non-Bell equipment could legally be attached to the telephone system. This was quite a change of pace for the FCC, which had been extremely supportive of AT&T throughout the Cold War period. As Matthew Lasar wrote, this was a time of great political upheaval and drama, Robert F. Kennedy had been buried just two weeks earlier, and "almost no one noticed as the FCC's Commissioners quietly rebelled against the world's biggest telco, unleashing the future."[109]

It has been called one of the FCC's best rulings[110] and a David versus Goliath victory.[111] Sandra Braman has further characterized Carterfone as a "landmark decision [that] opened the way to the series of antitrust measures and deregulatory steps through which AT&T's grip on the market was loosened and more genuine competition was made possible."[112] The decision allowed for "the use of [customer-provided] interconnecting devices which do not adversely affect the telephone system" and determined that AT&T's prohibitions on such devices were unreasonable and unduly discriminatory.[113] In so doing, the Carterfone case forced the issues of interoperability and decentralized control into the forefront of telecommunications policy considerations, and ensured that the public—telephone subscribers—could use these protected, state-sanctioned monopolistic networks as they wished, instead of according to rules determined by the phone company. The corporation was in many ways funded by the public, and this case was a stark reminder of that fact. AT&T, after all, had received more than a century's worth of government aid in the form of its protected "natural monopoly" status and the guaranteed monthly service fees that followed, as well as the unfettered access and rights-of-way to public lands to install and expand its networks, and untold billions in tax dollars financing its government contracts.

The Carterfone decision also ensured that consumers were eventually able to purchase their own modems and connect them without AT&T's permission in order to invite the Internet into their homes over the telephone network. The ruling was a small victory for Thomas Carter, as only 3,500 Carterfones were sold; however, it was a giant win for the future of cloud policy and for the public. It permitted users to continue expanding the possibilities for the nation's telecommunications system, as opposed to accepting the terms dictated by its corporate owners. It also allowed for modem connections to the telephone network, which ushered in widespread Internet access twenty-five years later.[114] And it chipped away at some of AT&T's authority over telephone network regulation and put it back in the hands of the FCC—a transfer of power from private sector terms back to regulators that was relatively short-lived in the long arc of cloud policy. Carter died in 1991, before he could see the full unfolding of what his invention helped to create, but his lawsuit had a lasting impact on the convergence of computing and communications infrastructure that far exceeded any possibilities that existed when it was filed in 1965.

The year after the Carterfone decision came down, as the ARPANET—the Internet's first skeleton and technological foundation—was being switched on, the Justice Department launched one of the most high-profile antitrust suits of the era. In 1969, on the last working day of the Johnson administration, the DOJ began proceedings against IBM, the fifth-largest corporation in the US at the time. The government alleged that it had monopolized the "general-purpose digital computer" market and violated the Sherman Antitrust Act. At that point, IBM's market share was in the range of 70 percent.[115] Their chief counsel, Nicholas Katzenbach, had served as President Johnson's attorney general just two years earlier.

The Justice Department's second antitrust suit against AT&T came five years after the IBM case began, yet the two remained politically entwined for their duration. The AT&T lawsuit was filed just a few months after President Nixon resigned in disgrace, and the Antitrust Division under President Ford was determined to "shake off its Nixon administration image of laxness in treatment of big business."[116] As part of that mission, it attempted to finish what the DOJ had begun back in 1949, and end AT&T's monopoly on telecommunications services and equipment. At the time, AT&T was the country's second largest public firm. The case against the company was filed on November 20, 1974. It was the DOJ's second attempt to break up the Bell System, and the fourth major investigation of AT&T by the US government. The pipeline giant had thus far escaped serious government intervention, but found itself in federal court once again, this time alongside the country's biggest computer hardware manufacturer. They were both facing cases built on Section 2 of the 1890 Sherman Antitrust Act, just like those levied against robber barons like Rockefeller with Standard Oil and Duke's American Tobacco monopoly two generations earlier.

The government was seeking divestiture of AT&T's manufacturing arm Western Electric (once again, since the 1949 antitrust case initiated by the Truman administration had failed in their attempt) as well as the separation of the company's long-distance service from its twenty-two local operating companies. And yet, there was ongoing and intense pressure at every turn for the Justice Department to drop the case, including from Ronald Reagan's secretary of defense, Caspar Weinberger, and secretary of commerce, Malcolm Baldrige. In 1981, Weinberger, like Defense Secretary Charles Wilson before him in 1953, wrote to the attorney general, strenuously advocating for the case to be dismissed.[117] Even President Reagan himself opposed the breakup.[118] The Defense Department had also returned to their dire warnings about the threats to "national security" posed by such a breakup, which they had employed on behalf of AT&T in the 1949 case. History seemed to be repeating itself.

However, the DOJ had momentum on its side, thanks to all the years the department had already devoted to the case. The new antitrust chief, William Baxter, was committed to separating the regulated and unregulated components of AT&T, declaring early on in his tenure that he intended "to litigate it to the eyeballs."[119] Moreover, the Justice Department had grown more independent of the White House than it had been in the past. This was in part thanks to a scandal during the Nixon administration involving the president's interference in antitrust cases against ITT after the company donated funds to underwrite the 1972 Republican National Convention.[120] As economic historian Peter Temin has explained in his chronicle of the AT&T breakup, this scandal "made the point that antitrust cases and consent decrees were the province of the Justice Department, not the rest of the administration."[121] And so, despite opposition from the president and multiple cabinet members, Baxter forged ahead with the most aggressive antitrust case AT&T had ever faced, threatening to dismember the company piece by piece.

It is important to point out that the cases against IBM and AT&T were taking place as the interpretation of antitrust law was transforming. The history of "trustbusting" in the late nineteenth- and early twentieth-century Progressive Era had long receded. The influence of the relatively activist approach of the 1960s and 1970s was beginning to wane, along with its principled application of antitrust law and explicit connection to the preservation of democratic norms. Up until that time, Tim Wu argues, "a broad political, legal, and intellectual consensus saw excessive economic concentration and monopolization as both economically dubious and politically dangerous. . . . The antitrust laws and an anti-concentration mandate were broadly accepted as part of a functioning democracy."[122] Historian Richard Hofstadter had famously eulogized the antitrust movement in 1964 as "institutionalized" and "one of the faded passions of American reform."[123] The ascendance of the Chicago School of antitrust that had begun in earnest by the late 1970s had the greatest impact on contemporary cloud policy. Named for the University of Chicago, where it was largely formulated, this neoliberal approach to antitrust emphasized market-driven solutions and nonintervention. Chicago School theorists (such as Robert Bork, Richard Posner, and Frank Easterbrook) asserted that markets operate most efficiently and competitively without regulatory interference. Their philosophy advocated minimal antitrust enforcement while being extremely tolerant of large firms and industry consolidation.

The Chicago School successfully promoted the maximization and sole consideration of "consumer welfare" (defined by lower prices) at the expense of all other cultural, political, and economic values that had been central to antitrust for the previous ninety years. As Matt Stoller has characterized it, "the Chicago School was a reconstruction of the thinking of the nineteenth century, when opposition to concentrations of capital seemed as foolish as opposing the creation of clouds or the flowing of rivers."[124] Consequently, in industry after industry—from health care, telecommunications, and finance to transportation, energy, and insurance—crucial consumer protections and services were eradicated, and the scale of corporate power exploded. High-profile antitrust lawsuits were a dying breed; the IBM/AT&T action was their last gasp before the full embrace of the Chicago School and the ethos of deregulation in the Reagan era effectively killed off the majority of antitrust enforcement in the US.

Computer II (1976–1980) and Computer III (1985–1986)

Meanwhile, it was becoming very apparent that Computer I was conceptually insufficient to deal with the developing interdependencies of computing and communications. The FCC itself admitted later that "even as the Computer I rules were being implemented, technological developments rendered them nearly obsolete as it became harder to distinguish communications from data processing or computing."[125] The problems were immediately obvious, as even a single email in its earliest form posed a question of policy categorization that was impossible to answer using the Computer I framework. The separate communications and data processing categories that the FCC had established were not functional, their definitions were already outdated, and the impending problems were certain. Only five years after Computer I was signed, the FCC decided to renew its exploration into regulating the provision of computing technologies and pipeline infrastructure. As Queen Elizabeth was sending the first royal email over the ARPANET in 1976, the agency began the inquiry process of Computer II. The following year, Apple started selling the first mass market personal computer, further highlighting the gap between policy and the transformation in communication technologies already under way.

After almost four years of deliberation, the FCC concluded the Computer II inquiry in 1980. The commission decided to create a new framework for regulating convergent media that went beyond Computer I's brutalist distinctions of communications versus data processing services for pipeline infrastructure. The result was the creation of new categories of "basic" services and "enhanced" services to classify the different types of networked communication taking place. Basic services were those simply transmitting either voice (such as a telephone call) or information (e.g., a fax) using a common carrier. They would be regulated. Enhanced services also utilized a common carrier, but they were unregulated as they encompassed more interactive, layered, or convergent media that "combines basic service with computer processing applications that act on the format, content, code, protocol," and so on, altering the basic service in some way.[126] These categories are illustrative of the embedded path dependency that is so pronounced in policy history, as the new is continually built on the old—physically, politically, legally, and conceptually.

Despite the prominence of past frameworks in these new regulatory distinctions, the FCC knew that "enhanced" services were the future; in the final decision of Computer II, the commission acknowledged the "rapid technological and market developments affecting communications and data processing services, the ever-increasing reliance upon common carrier transmission facilities in the movement of all kinds of information, and the need to tailor communications-related services to individual user requirements."[127] This would soon bear out, as enhanced services like email, voicemail, and Internet access were delivered over the common carrier telephone lines into the home, combining the regulated pipelines with unregulated computer applications and data.

Common carriers were newly allowed by Computer II to provide the deregulated "enhanced" services—including AT&T, which had been previously prevented by the 1956 decree from entering those markets.[128] The largest telephone providers, AT&T and GTE, were given a new set of restrictions in Computer II that mandated "structural separation"—essentially the new version of "maximum separation" that was enforced by Computer I for carriers with annual revenues over $1 million. In Computer II, structural separation was designed to allow the two largest providers to participate while also preventing them from engaging in anticompetitive behavior. Accordingly, the FCC required AT&T and GTE to establish separate corporate entities for their basic and enhanced service provision.[129] Former FCC counsel Kevin Werbach has explained that this "was an attempt to draw a bright line between computers and the [telephone] network."[130]

It was also clear throughout this set of inquiries that AT&T figured heavily in the agency's considerations, as the FCC signaled their full support of the corporate giant throughout the proceedings. The *Computer II Final Decision* mentioned AT&T no less than 150 times. There was a clear emphasis on providing reassurances that the company would be able to participate aggressively in the enhanced services market, and that consumer prices would not be affected by structural separation restrictions. In fact, among the very first sentences in FCC Commissioner Quello's statement was the acknowledgment that Computer II was a "first step along the road to full participation of AT&T and GTE in the provision of 'enhanced' telecommunications services."[131] Commissioner Ferris viewed this removal of AT&T's blanket prohibition on entering the data services market as a decision that represented "a giant step forward for consumers and for the industry," stating that when faced with "either extending or reducing government regulation, we have chosen to reduce regulation. As a result," he wrote, "I believe the information age will arrive sooner."[132] The gospel of deregulation had arrived at the FCC.

In the *Computer II Final Decision*, the FCC also emphasized the difficulties and "the practical impossibility of drawing coherent lines in the face of such explosive technological evolution."[133] They described the long process of attempting to "delineate a distinction between communications and data processing services and failing to arrive at any satisfactory demarcation point," concluding that "further attempts to so distinguish enhanced services would be ultimately futile, inconsistent with our statutory mandate and contrary to the public interest. . . . It is apparent that, over the long run, any attempt to distinguish enhanced services will not result in regulatory certainty."[134] Consequently, enhanced services and their data—despite relying on basic, regulated services for transmission—remained indiscriminate and unregulated. Like so many policy frameworks, this one cemented over time and arrested the development of new paradigms. Once such scaffolding is in place, it is rarely transformed; instead, it merely calcifies into flawed future policy. While Computer II was largely successful during its six years of operation, the "bright lines" between telephone and data service were not permanent, and they would quickly blur and fade.

Two years after the Computer II decision, antitrust enforcement in the pipeline industry and beyond had its last hurrah. On January 8, 1982, the DOJ finally succeeded in breaking up AT&T despite the public objections coming from the White House, Congress, and much of the cabinet. After eight years of litigation and negotiation requiring hundreds of millions of dollars, AT&T agreed to a consent decree. With that, the world's largest and most successful telecommunications network was subsequently dismantled. Its twenty-two local subsidiaries were restructured into seven Regional Bell Operating Companies, or RBOCs, that became known as the "Baby Bells."[135] AT&T remained the sole long-distance provider, effective January 1, 1984. It was the largest divestiture in US corporate history, reducing AT&T's assets by two-thirds as it was required to spin off about $80 billion worth of holdings.[136] The company was allowed retain ownership of Western Electric, just as in 1956.[137] At the time, Western Electric was still manufacturing 80 percent of US telephone equipment and most of AT&T's 827 million miles of copper wire.[138] They also kept Bell Labs, and AT&T was released from the 1956 consent decree terms and thus fully allowed to reenter the unregulated data services market. While this new consent decree might have seemed promising to those who viewed AT&T's monopolistic position as anticompetitive, it once again proved to be only temporary, just as with all other past regulatory actions that limited the company's growth or power. AT&T was soon back to reporting record earnings and profits on sales of $65 billion by the early 1990s,[139] and by the early 2000s, the company had remerged with almost all of the Baby Bells it had been forced to divest in 1984.

Despite this somewhat dramatic government intervention, the approach to antitrust enforcement had already begun a foundational shift. During this time, economists within and beyond the Reagan administration were arguing that "the antitrust laws should be scrapped all together."[140] "Big is no longer bad," the *New York Times* declared in 1981.[141] As the subtitle of Robert Bork's highly influential book *The Antitrust Paradox* (1978) read, antitrust was "a policy at war with itself." Bork, a prominent Chicago School theorist, had labeled the IBM case "the antitrust division's Vietnam." Indeed, the Department of Justice finally abandoned its case against IBM after thirteen years, four administrations, and sixty-six million pages of documents because they determined it was "without merit" and "a bad mistake."[142] In many ways, the case did fail because of its protracted length; by 1981, it was apparent that the government's argument hinged on "a market situation that existed two or three technological generations ago," rendering it "an historical curiosity"[143] by policy standards. According to William Baxter, President Reagan's antitrust chief, the government ended the IBM case after all that time because of mounting costs and diminishing odds of victory, given what he characterized as the government's "flimsy" evidence.[144] The case was dropped on the very same day the AT&T case was settled. January 8, 1982, thus marked the end of an era for antitrust law, and the dawning of a new one, embodied by the paradigm-shattering breakup of AT&T and the unqualified victory for IBM.

These monopolies in pipeline provision and computing hardware had drastically different outcomes in their battles with the DOJ, but the combined future of pipelines and computing was practically assured. The IBM case was aligned with the FCC decision in Computer II, emblematic of the deregulation or, perhaps more accurately, antiregulation trend taking hold in policymaking. AT&T was dismantled, but its long history as a government-sanctioned monopoly provided nearly a century's worth of built-in advantages that allowed the company to rebound almost immediately. Thanks to the Chicago School's newly dominant influence in antitrust circles, the government grew increasingly tolerant of corporate concentration, industry consolidation, and large-scale mergers with each passing year. Moreover, the FCC had formally acknowledged the inevitable convergence of telecommunications and data processing in Computer II. Cloud policy was moving out of the analog age, and its digital contours were coming into focus.

The FCC launched their final Computer Inquiry as the dust was settling in the AT&T and IBM cases; Computer III began in August 1985 and finished in 1986. This last inquiry for the most part reaffirmed the conceptual thinking underlying Computer II and maintained the basic versus enhanced service framework for regulating voice and data services.[145] Just as the Computer Inquiries were coming to a close, the Internet was beginning to open up. Kōji Kobayashi's book *Computers and Communications* was published, and his vision for what he termed "C&C" merging into a single infrastructure for information processing was fast becoming reality. Pipelines—and their regulatory conundrums—were on the lead end of this transformation as the Internet began its next phase: the transfer to private control.

The Information Superhighway and Beyond: Digital Pipelines

Private Control Redux

The first commercial Internet service provider (ISP) providing Internet access (The World) was established in 1989, quickly followed by CompuServe and America Online (AOL). Soon thereafter, pipelines were at the center of the power struggles among (and within) the government, the private sector, and even technically advanced users and activists with their own ideas for the ARPANET, which many of them understood "not as a computing system but rather as a communications system."[146] In the history of pipeline policy, the transfer of ownership and responsibility for the network that would become the Internet from the province of government, military, and civilian research interests to the private sector was among the most significant and transformative.[147] It might also have been one of the most surprising for the network's pioneers. Katie Hafner has argued that, "for all their genius, [the Internet's creators] failed to see what the Net would become once it left the confines of the university and entered the free market."[148] Historian Janet Abbate put an even finer point on it: "They thought they were building a classroom, and it turned into a bank."[149] ARPANET managers had unsuccessfully tried to secure a private operator, specifically AT&T, as far back as 1972. Once the ARPANET was decommissioned, on February 28, 1990, the government effectively relinquished control over the network they had been building with a community of academics, computer scientists, and military contractors since the late 1960s, and the transition commenced. So began one of the most momentous decades in cloud policy, as regulators began to once again grapple with the fundamentals of regulating privately owned communication pipelines, this time for a digital era.

In 1991, as the World Wide Web began affording the general public an interface to get online and establishing protocols for global interconnection, then-Senator Al Gore introduced the High-Performance Computing Act. This law, signed by President George H. W. Bush, began to direct government funding toward computing infrastructure for education and research and upgrading the Internet backbone; it also created the National Information Infrastructure that was referred to as the "Information Superhighway" in the early days of public Internet usage. This metaphor invokes AT&T's annual report of 1910, itself replete with figurative language anticipating the information superhighway; it extolled the virtues of futuristic interconnection offered by AT&T's network that afforded a *"universal wire system for the electrical transmission of intelligence (written or personal communication)*, from every one in every place to every one in every other place, a system as universal and as extensive as the highway system of the country which extends from every man's door to every other man's door."[150] Shortly after taking office, President Clinton and Vice President Gore began to call heavily on the private sector for the creation and expansion of that infrastructure.

The Clinton administration was the first to truly contend with the development of the public Internet, and the first to produce major revisions of telecommunications law since the 1930s. During the 1992 presidential campaign, Bill Clinton devoted part of his platform to science and technology, and to creating "a 21st century infrastructure." In their plan, titled "Technology: The Engine of Economic Growth," the Clinton-Gore campaign called for a renewed prioritization of technology for civilian use over military spending, and a reliance on the private sector in partnership with the government to accomplish their infrastructural goals. Referring back to DARPA, among other military-funded tech developments, Clinton wrote that "America cannot continue to rely on trickle-down technology from the military. . . .

Civilian industry, not the military, is the driving force behind advanced technology today."[151] It was clear that the reins—and the funding—would be handed over to corporations. As Matthew Crain put it, "The revolution would be commercialized, and the private sector would be in control."[152] Gore was tasked with coordinating private and public sector activities related to technology and establishing "a forum for systematic private sector input into U.S. government deliberations about technology policy and competitiveness." The plan proposed that at least 10 percent of the $76 billion that the government spent each year on research should be redirected from the Pentagon's budget to tech efforts, including the development of computer chips, robotics, and fiber optics.[153] This was the moment that the government's tech agenda publicly pivoted from a focus on Cold War military R&D to a more industrial and civic-oriented one.

The partnership between the government and the private sector was the defining feature of the earliest official discussions about the Internet. The emphasis on economic growth was tied to the transfer of power to corporate providers; this established an official climate of prioritizing commercialism over civic values at the critical point in time that the public Internet was coming to life. In 1993, MOSAIC was introduced by researchers and programmers at University of Illinois.[154] The launch of this browser, the first to display inline graphics, "was a landmark moment in the evolution of both the Web and the Internet" because of the way it vividly demonstrated the potential for the web as a commercial and entertainment medium, and also "triggered a sudden surge in the demand for Internet connections among the general public."[155] As popular Internet usage began to grow, Netscape's invention of "cookies" and the release of their Navigator browser in 1994, along with developments in data encryption, propelled credit card usage online. The commercial, corporate Internet took off. Amazon was founded in July 1994, and one month later, a *New York Times* headline declared, "Attention Shoppers: The Internet Is Open."[156]

That same year, Congress and the White House established their digital presence as well. They also passed the Communications Assistance for Law Enforcement Act (CALEA), which helped law enforcement conduct legal wiretaps in a new technological age, now that telecommunications networks were going digital.[157] Ultimately, pipeline infrastructure received its first dedicated act of legislation for the digital era in the form of the 1996 Telecommunications Act. With it, President Clinton handed the keys to private providers with a full tank of gas, after heavily consulting with the industry to create his Internet policy agenda.[158] However, in this new regulatory landscape, the government failed to afford the Internet's pipelines the status of public utilities and common carriers enjoyed by their analog counterparts. As a result, the safeguards for access, equity, and public service values were also left behind.

The 1996 Telecommunications Act

For decades, regulators had been arguing that the Communications Act of 1934 was hopelessly outdated. Many called for its revision in light of technological changes and expanding convergence in the media and telecommunications industries. The process of updating this foundational law took place in fits and starts, beginning in the 1960s. President Lyndon Johnson convened a task force in August 1967 to study communication policy in the context of emerging technologies such as satellites along with ongoing private sector providers. The report, issued at the very end of his administration, had a healthy skepticism about two key pipeline principles: public utility and common carrier regulation.[159] One of the committee's recommendations was to break up AT&T, but that suggestion was immediately buried when the president of AT&T heard about it and complained to Johnson, who made sure it was removed from the final report.[160] Another recommendation was to expand the role of the executive branch in telecommunications policy, allowing the president greater control and the ability to "add to the efforts of the FCC" with capabilities "of taking the long view of policy and developing data and recommendations on a host of technological and economic aspects of telecommunications problems."[161] This was subsequently taken up by President Nixon, who established the White House Office of Telecommunications Policy in 1970.[162] In 1974, OTP director Clay Whitehead testified before the Subcommittee on Antitrust and Monopoly that "the 1934 regulatory apparatus works reasonably well for the purpose for which it was designed, namely, regulating basic telephone service; but that same regulatory apparatus has become a barrier to competition and innovation required for the future direction of communications."[163] It was very apparent that an update was necessary.

Revising the 1934 act was not officially completed until 1996, but numerous efforts had been devoted to doing so decades earlier, including the Communications Act of 1978 and the Telecommunications Act of 1980. These proceedings and bills emphasized deregulation, and both included provisions for lifting AT&T's 1956 consent decree. The phone company's designs on entering the cable business also loomed large in these hearings.[164] The 1980 bill in particular was extreme, putting an end to all government regulation of commercial broadcasting within ten years, and it also included a plan to abolish the FCC. That proposal, said AT&T Vice Chairman James Olsen at the time, "moves us in a direction of a bill we can live with."[165] Along the way, AT&T began to participate in crafting legislation, going so far as to propose its own law to Congress in 1976 known as "the Bell Bill," derisively called the "Monopoly Protection Act" by critics, which was endorsed by 179 members of Congress.[166] Even though the company was under antitrust investigation by the Department of Justice at the time, it was still presenting its own terms for regulation to lawmakers. However, the momentum of the late-1970s rewrites was halted when Representative Lionel Van Deerlin, the Democratic chairman of the House communications subcommittee involved, lost his election in 1980 and Ronald Reagan won the presidency, changing the complexion of the FCC and relevant congressional committees. Revision efforts stalled for more than a decade, to the point that even President Kennedy's FCC chairman, Newton Minow, said in 1995, "Today, the Communications Act stands as a monument to the mistake of writing into law vaguely worded quid pro quos. Because the act did not define what the public interest meant, Congress, the courts, and the FCC have spent sixty frustrating years struggling to figure it out."[167]

During the Clinton administration, Congress again returned to the project of revising the 1934 Act. The GOP midterm takeover of Congress in 1994 put the Republican party in full control of the legislature for the first time since President Eisenhower was elected in 1952. This shift introduced major battles over the next bills put forward in 1994 and 1995. The Republicans once again advanced the proposal to eliminate the FCC altogether in the wake of their victory. Numerous debates about new telecommunications legislation ensued in 1993 and 1994, and the lobbyists were out in full force, as billions of dollars were at stake. In addition to the political infighting in the House and Senate, a bitter, antagonistic rivalry had been developing between cable and telephone companies for decades, as each wanted a piece of the other's pie. The monopolistic Baby Bells were far and away the most profitable industry involved, and they were not willing to cede any of their power. In 1992, the Baby Bells had revenues of $82.3 billion—far outperforming long-distance service ($67.5 billion); newspaper publishing ($44.2 billion); cinema ($26 billion); broadcast ($25.8 billion); and cable ($21.5 billion).[168] Their lobbyists were instrumental in killing the bills that would have required the Baby Bells to relinquish their monopoly status.

Telephony and cable had been prevented from entering and owning one another's business since 1970; at that time, an FCC ruling was enacted to protect the budding cable industry from the phone companies eager to begin carrying television on their existing infrastructure. The ban on cross-ownership was reinforced in the 1982 AT&T consent decree and subsequently codified in the Cable Communications Policy Act of 1984, with Congress also banning the telephone companies from delivering any video programming in their local service areas. In all, regulatory agency rules, consent decree provisions, and legislative mandates enforcing the separation of telephone and cable were enacted between 1970 and 1984. Nevertheless, the Baby Bells had been going after legal relief from the decrees since they took effect, in order to get a foothold in the video market.[169] Ultimately, the issue went all the way to the Supreme Court, in 1995, where it remained until Congress changed all the rules with the 1996 Telecommunications Act. Until that time, however, the industries remained separate, and both cable television and Internet provision were off-limits for the phone companies.

The run-up to the 1996 Telecommunications Act was quite a dramatic political spectacle. It featured the lofty imaginings of libertarian techno-utopianists alongside public interest promises coming from the executive branch, both clashing with one another and further stoking the ideological disconnect between the two main political parties. Patricia Aufderheide has noted that, at this time, "deregulation was at the center of a kind of holy war in policy."[170] This holy war that would eventually help define cloud policy had been raging since the 1970s, when antitrust philosophy began to turn away from traditions of robust enforcement in the name of enabling and promoting competition, and industries successfully organized into more visible lobbying groups in Washington, DC. This was also in part thanks to campaign finance reforms allowing for the creation of political action committees (PACs).[171] By the time the debates over the Telecomm Act were taking place, AT&T was the biggest corporate PAC in the country, and phone utilities made up about half of all PAC money being funneled to lawmakers.[172] The politics of deregulation found support in economic, judicial, and regulatory theories during the 1980s, and the neoliberal state has never looked back.

As this process played out in the legislative and cultural debates over infrastructure policy in the early 1990s, the future of public interest principles hung in the balance. Two main rhetorical and ideological positions were circulating. One was voiced by the increasingly reactionary Republican Party and groups like the Progress and Freedom Foundation (PFF), a "market-oriented think tank" with close ties to Newt Gingrich that advocated for technology policy "based on a philosophy of limited government, free markets, and individual sovereignty." In their 1994 corporatist manifesto, "Cyberspace and the American Dream: A Magna Carta for the Knowledge Age," the authors argued that, "if there is to be an 'industrial policy for the knowledge age,' it should focus on removing barriers to competition and massively deregulating the fast-growing telecommunications and computing industries."[173] Fred Turner has called the PFF's Magna Carta "arguably the decade's most potent rhetorical welding of deregulationist politics to digital technologies."[174] Its strategic and rhetorical equivalence of individual and corporate freedom was enshrined in the 1996 act as near-total deregulation, with additional giveaways to monopoly providers.

A countervailing ideological position was promoted by nonprofits, activists, and the Democrats, and used by Vice President Al Gore as he rallied around government-mandated public interest values for the "information superhighway." Mitch Kapor and Jerry Berman of the Electronic Frontier Foundation were among those calling for the guarantees of universal phone service to be extended to universal access to the Internet. "We need laws to protect consumers when competition fails," they argued. Anticipating the corporate censorship that would inspire the net neutrality fights and reform movements on the horizon, they demanded new laws that required providers to carry any and all content into the home, and to protect service for those with limited resources like schools and libraries. In their call for "a new social contract between the telecommunications industry and the American people," they also articulated the need to protect user privacy in the online environment.[175] Gore had been very publicly advocating for a version of universal service and public stewardship of Internet pipelines from the campaign trail and the office of the vice president, even arguing that the network should be constructed by the government. Otherwise, Gore said in 1993, we could expect to get a "private toll road open only to a business and scientific elite."[176] In his 1994 article "Innovation Delayed Is Innovation Denied," Gore had shifted to accepting that the network infrastructure would be privately owned and operated, but publicly rejected the notion that monopoly providers were the best or only option. "The only viable path is toward competition," he stated, urging that "we must make sure that our national information highway bypasses no one. We cannot allow this country, or any community within our country, to become a communications ghost town. To be left off the beaten track in the information age is to be cut off from the future."[177] The trope of competition was held up as the prize that could be saved only by the vision of one party, or face ruination by the other. In the end, as Clinton's "New Democrats" embraced neoliberalism and Third Way governance principles, regulatory reality did not line up with their rhetoric.

In September 1995, for example, Gore called out the House and Senate versions of the Telecommunications Act under debate, both of which were Republican-led rewrites. "The telecommunications bills pending before the Congress . . . and especially the House bill, represent a contract with 100 companies," he repeated often in the press. "The highest bidders, not the highest principles, have set the bar." He argued that the bills protect monopolies, disdain the public interest, and represent a historic abandonment of the "creative partnership" between government and the private sector that helped spawn the Internet. "America's technological future is under attack by shortsighted ideologues, who pretend to understand history, but in fact have no understanding whatsoever."[178] These sentiments, however progressive they sounded at the time, were nowhere to be found when the final bill was delivered. Representative John Conyers, one of the main sponsors of the Telecommunications Act, complained during the last stage of the hearings, "We have heard from the industries involved in this bill, oh, have we heard from the industries. We have heard from the consultants that the lobbyists have hired, oh, have we heard from the lobbyists. We have heard from the law firms, we have heard from all of them. . . . What did you hear from the consumers? Oh, them? Well, what did you hear from the citizens? . . . Have you received any visits from their lobbyists? I do not think so. So what we are doing, ladies and gentlemen, in broad daylight, and I know we are sober, we are giving corporate welfare."[179]

In the end, the forces of big capital prevailed once again. The Telecommunications Act was more like what Computer IV might have been; instead of reimagined public interest principles or a renewed strategy for regulating the corporate providers of infrastructure and access, it merely solidified their power without delivering on any of the bill's earlier promises.

President Clinton signed the Telecommunications Act of 1996 into law—electronically, with a digital pen—backed by almost unanimous congressional support, on February 8, 1996.[180] The signing ceremony even included an appearance by Lily Tomlin as Ernestine, her derisive telephone operator of many comedy routines, sending televised jabs at the vice president. The Telecommunications Act launched a new era for media and communication industries, allowing broadcasters, cable TV providers, local phone companies, and long-distance carriers to compete with one another. The major deregulatory measures in the law ultimately resulted in levels of industry concentration and consolidation that reached new heights. Newly formulated empires of media and communication soon followed.

On the day the Telecommunications Act was signed, poet, activist, and cofounder of the EFF, John Perry Barlow, published his own "cyberlibertarian" vision for how the digital realm should be regulated. "A Declaration of the Independence of Cyberspace" rejected any state authority, jurisdiction, or ownership over the collective creations and dimensions of the Internet. It was addressed to "Governments of the Industrial World" and read, in part, "Our world is different. . . . We are creating a world that all may enter without privilege or prejudice accorded by race, economic power, military force, or station of birth. . . . Your legal concepts of property, expression, identity, movement, and context do not apply to us. They are all based on matter, and there is no matter here."[181] While Barlow's "utopian narrative of Internet exceptionalism"[182] rightly articulated the myriad challenges of applying jurisdictional and other statist-oriented "meatspace" policies to "cyberspace," today it mostly stands as a widely critiqued relic, or an artifact of 1990s techno-utopianist rhetoric. However, it did reject corporate gatekeepers—of content, of accessibility, or of the rights to privacy. It imagined a network that honored the end-to-end architecture of the Internet's original design in its policies and practices, and a digital landscape inspired by "the dreams of Jefferson, Washington, Mill, Madison, DeToqueville, and Brandeis."[183] And it provided a polemic for alternative thinking about the role of the public in infrastructure governance, all of which tragically presented as more radical than corporate control of Internet distribution without mandated public interest measures.

But the Barlow Internet was not the Internet we got. Instead, as Jill Lepore characterized it, "the Internet we got, under the terms of the 1996 Telecommunications Act, [was] a Gingrich-and-Gilder travesty . . . that shielded the

Internet from government regulation and made it a commercial free-for-all."[184] George Gilder, the fallen "prophet of the New Right" was a cosignatory on the PFF's "Magna Carta." He was also a friend and consultant of Speaker Gingrich and advocated massive accumulation of wealth for men, and a life of homemaking and service to one's husband for women.[185] The neoconservative ideology that permeated the Telecomm Act's values echoed Gilder and his outsized political influence at the time. The divide between the government's commitment to the stewardship of public interest obligations for the telephone seen in the 1939 report on the Bell System, and the wholesale abandonment of those principles sixty years later in the Telecommunications Act could not have been more striking. The public was now at the mercy of pipeline companies and their drive for efficiencies, competitive advantages, and profits above all else.

The Telecommunications Act and its full embrace of deregulatory principles also marked the official end of the "modern infrastructural ideal" for cloud pipelines and their policy foundations. No longer would the public be assured of government protections in these markets, as the Clinton administration embraced the privatization of policy and promoted self-governance measures for the corporate stewards of the Internet. In their 1997 "Framework for Global Electronic Commerce," the administration declared that "the private sector should lead." They also stated that "governments should encourage industry self-regulation wherever appropriate and support the efforts of private sector organizations to develop mechanisms to facilitate the successful operation of the Internet. Even where collective agreements or standards are necessary, private entities should, where possible, take the lead in organizing them."[186] The administration took a permanent backseat to the private sector, and corporations began formally dictating the terms for cloud policy.

History has shown that a lack of government regulation does not lead to increased competition. The market for Internet provision has been a perfect case in point. By 2021, two companies (Comcast and Charter) had 81 percent of the 73.7 million broadband subscriptions in the United States—31 million and 29.2 million, respectively. The number three company, Cox, had just over 5 million.[187] As a result, prices for Internet access and cable rise at nearly double the rate of general inflation.[188] The pain is felt most acutely by low-income consumers, and there is rarely more than one ISP option in any location in the US. Moreover, customer service is so poor and unreliable thanks to the lack of competition that ISPs are "the most hated" industry in America, even beating out health care and the airlines, which is quite an achievement.[189] Contrary to the marketing imagery of openness and abundance, the cloud is not a competitive space.

Also of primary import for cloud policy is the fact that the 1996 act allowed phone companies to also operate as "information providers." That is to say, they were no longer restricted to providing only telecommunications, and they could begin to deliver cable and Internet services as well. The act also created new terminology out of the remains of Computer II and III—it kept the distinction between the "basic services" (subject to common carrier regulations) and the far-less regulated "enhanced services" and essentially mapped them on to the newly established categories of "telecommunications services" and "information services,"[190] respectively. This dragged the framework created back in 1980 for Computer II into the new millennium, tethering digital media policy to technology and concepts from an analog era. As of 2023, broadband service is being regulated as an information service under Title I of the Communications Act of 1934, as amended by the Telecommunications Act of 1996. Information services are distinguished from and regulated less stringently than telecommunications services, which fall under Title II of the Communications Act and have the important distinction of being common carriers. This is the crucial distinction on which the next major battle over cloud pipelines has been waged.

Net Neutrality
The classification of Internet service pipelines—which also confers their societal value, their obligations to the public, and the responsibilities of their providers—has been one of the more contentious fights in the history of cloud policy. It is also the one on which "net neutrality" hinges. In essence, net neutrality is about applying the core principle of common carriage—the condition of nondiscrimination—to internet service providers. It is the requirement that ISPs treat all data that they carry equally, regardless of the sender or receiver, therefore making "throttling" (slowing down service for heavy users) and "paid prioritization" (speeding up service for those who pay more) illegal. Tim Wu, the originator of the term in 2002, has framed the net neutrality debate more recently as "a restatement of a classic question: How should a network's owner treat the traffic that it carries? What rights, if any, should a network's users have versus its owners?"[191] Understanding the

history of net neutrality and all of its questions and complexities is fundamental to understanding this struggle over technology policy as a fight for the freedom of access and free speech, for consumer rights, and for public protections from monopoly providers of cloud infrastructure.

Technically speaking, it is the fight over whether to regulate Internet provision under Title I or Title II of the Communications Act of 1934. Net neutrality advocates want ISPs regulated under Title II of the Communications Act, classifying them as telecommunications services, which all have the common carriage requirements of nondiscrimination. Those who oppose net neutrality have argued for Title I, which categorizes ISPs as information services, a designation that carries less burdensome regulation and allows for discriminatory practices in service. However, privileging data based on financial considerations is a serious threat to democracy. Pickard and Berman have argued that this practice has "the potential to dramatically alter our political, social, and civil lives. . . . Allowing ISPs to divide the internet into fast and slow lanes will inevitably amplify the voices, ideas, and worldviews of those with power and resources, while marginalizing those without them."[192] Laura DeNardis wrote back in 2012 that "Internet governance is now the central front of freedom of expression."[193] As President Obama's FCC Chairman Tom Wheeler succinctly explained this most critical issue of cloud policy, Internet access is "too important to let broadband providers be the ones making the rules."[194] These statements, and the principles of net neutrality, are aligned with the goals of the original "end-to-end" decentralized architecture of the Internet, which was designed to afford maximum control to its users as opposed to the owners and gatekeepers of infrastructure.

Net neutrality commanded a great deal of space for cloud policy in the years following the passage of the Telecomm Act, with the bulk of debate devoted to the classification of broadband and the authority of the FCC over these pipelines. To begin with, ISPs were reclassified by federal agencies and the courts four times in five years between 2000 and 2005.[195] The issue finally ended up in the Supreme Court in 2005, in what became known as the *Brand X* case.[196] The court's ruling in this case classified cable broadband via cable modem as a Title I information service, thus rendering the FCC powerless to treat Internet providers as common carriers and enforce net neutrality rules, and putting the bulk of control into the hands of commercial ISPs as gatekeepers. The FCC nevertheless adopted their own Internet *Policy Statement* in August 2005, using the Telecommunications Act as a foundation and establishing what the agency viewed as a form of net neutrality principles. The *Policy Statement* also asserted the agency's authority and jurisdiction "to preserve and promote the open and interconnected nature of the public Internet."[197] However, the 2005 Supreme Court decision and the regulatory approach by the FCC drew the agency into an "existential crisis" as characterized by the media reform group Free Press, because the agency was operating without any legal basis to enforce net neutrality rules.[198] The political dysfunction and polarity was also contributing to policy inertia: in 2006 alone, Congress rejected five different bills that would have granted the FCC necessary powers to police net neutrality violations.

This was all a protracted introduction to the last two rounds of the net neutrality fight, both of which took place during the Obama administration in 2010 and 2014. The prelude to those battles included a lawsuit by Comcast against the FCC, which led to an appeals court ruling in 2010 stating that the FCC did not have the authority under Title I of the Communication Act to enforce net neutrality rules. The agency was in a bind, as President Obama's FCC Chairman Julius Genachowski was unwilling to forge the more politically fraught but legally necessary path to Title II for protecting broadband access. In August 2010, Google and Verizon offered up their own "legislative framework" for the FCC to consider in August; the remarkable arrogance of two tech giants purporting to help the FCC establish policy to regulate themselves hearkens back to AT&T's "Bell Bill" in the 1970s; much like that earlier beacon of regulatory capture, Google and Verizon's proposal was also given serious consideration by lawmakers.

Four months later, the FCC passed the Open Internet Order of 2010, containing new "net neutrality-adjacent" rules that were remarkably similar to those proffered by Google and Verizon. These rules relied on a very precarious Title I authority (already rejected by the Supreme Court) and required transparency and banned the blocking of websites or unreasonable discrimination of network traffic on the part of "fixed" or wireline broadband providers. Yet, they exempted wireless networks from the requirements of nondiscrimination and blocking at a time when Internet traffic was relying more and more on mobile broadband. As Lawrence Lessig observed at the time, "policymakers, using an economics framework set in the 1980s, convinced of its truth and too arrogant to even recognize its ignorance, will allow the owners of the 'tubes' to continue to unmake the Internet—precisely the effect of Google and Verizon's policy framework."[199] FCC Commissioner Michael J. Copps released his own statement expressing serious concerns about the precarious interim fix offered by the order:

> This plan *can* put us on the right road—*if* we travel that road swiftly, surely and with the primary goal of protecting consumers foremost in our minds.... We should welcome this step toward bringing broadband back under the Title II framework where it belongs. It was a travesty to move it in the first place, and those decisions caused consumers, small businesses and the country enormous competitive disadvantage. The devil will be in the details as we work to put the Commission back on solid legal footing.

Copps also noted the "warp speed" of technology shifts, adding, "As we address the short-term legal problems before us, I hope we will have the good sense to avoid boxing ourselves in on our ability to react to future changes in technology and the economy. The path we start down today must do more than just put this agency's authority over broadband back on life-support—it must ensure our going-forward, healthy ability to protect consumers. One near-death experience is enough."[200]

However, there were more to come. In 2014, Verizon turned around and sued the FCC over the 2010 Open Internet Order (despite it being modeled on the company's own framework). The DC appeals court sided with Verizon, throwing out the rules for broadband carriers, as they were still classified as Title I information services. This forced the FCC's hand: designate broadband as a Title II telecommunications service so common carrier requirements could be legally imposed or accept that the agency's rules would continue to be overturned in court. Former cable lobbyist Tom Wheeler was now FCC chairman, and his long-standing industry ties and initial proposal for Internet "fast lanes" led many to believe that true net neutrality would remain a pipe dream under his tenure. President Obama had come out publicly to endorse regulating broadband under Title II, and the public interest advocacy community was also very actively involved, despite knowing they were heavily outmatched in this fight. Big Tech had powerful lobbyists, resources, and the history of pipeline politics on their side. As Gene Kimmelman, CEO of Public Knowledge, has explained, "nothing was ever definitively determined even when you won (or lost) a regulatory ruling. It was an ongoing dispute because the natural political and economic forces were driving toward friction, despite what the baseline legal standard was.... The big dominant firms, even when they were losing, they were fundamentally winning. Even when they didn't get their definition in law and regulation,

Figure 1.3
Netflix splash page for Internet Slowdown Day, September 2014.

the enormous legal fight and economic fight around discrimination was always favoring the dominant transport companies."[201] Still, the public was engaged, and grassroots infrastructure activism was as vibrant as ever, with millions calling and sending letters to Congress and staging protests and rallies outside the FCC. The "Internet Slowdown Day" online campaign in September 2014 even elicited the participation of numerous tech companies, including Netflix, Mozilla, and Reddit.

Incredibly, in February 2015, the FCC delivered a stunning surprise and passed the 2015 Open Internet Order, adopting the first net neutrality rules to reclassify both wired and wireless broadband Internet access as a Title II telecommunications service. Broadband pipelines were now officially treated as common carriers, and the order banned paid prioritization (or "fast lanes" for those with the resources to pay for better service); banned providers from blocking lawful content; and banned the practice of throttling (or downgrading) access for anyone. As Russell Newman wrote, "It reversed fifteen years of policy at the FCC to subject broadband networks to the same regulatory regime reserved for telecommunications networks, reestablishing a notion of 'common carriage' (albeit not an exact replica of the old telephone regime) for data."[202] The public service rationale for net neutrality revolved around the dangers presented by corporate gatekeepers of cloud pipelines that had already shown themselves to be unfair and unreasonable in monopolized markets. This regulation was characterized as critical to preserving competition for consumers and smaller companies. Chairman Tom Wheeler announced the decision, stating, "There are three simple keys to our broadband future. Broadband networks must be fast. Broadband networks must be fair. Broadband networks must be open."[203] This declaration of principles was a tremendous boon for all Internet users and for the core public values of cloud policy. It also sanctioned the principle of openness on which the network was founded—a principle that set the table for innovation, and did not require any permission from AT&T (cf Carterfone) or other corporate providers.

Unfortunately, the victory was still challenging to translate for the general public. Even the "paper of record" failed the pipeline principle test: the *New York Times* trumpeted in its headline "F.C.C. Approves Net Neutrality Rules, Classifying Broadband Internet Service as a Utility."[204] This was not exactly true. It was an important step on the way to utility designation, which is the ultimate goal so that all users could enjoy nondiscriminatory access with reasonable, affordable rates and with built-in consumer protection measures for things like network resiliency (especially during emergencies) and service quality. However, the reclassification of broadband under Title II meant only that it would be treated as a common carrier, thus ensuring that ISPs had to deliver all content in the same manner, at the same speed. The true change necessary for the democratic and socially equitable provision of infrastructure is to also treat these pipelines as public utilities—just as the telephone has been. That is a more appropriate manner in which to regulate and protect broadband service, and a more accurate recognition of the role that these essential pipelines have in twenty-first-century society. In essence, cloud policy needs to catch up to that erroneous *New York Times* headline of 2015.

Moreover, the victory for the public was only fleeting. In December 2017, net neutrality was swiftly repealed by Trump's FCC, led by chairman and former Verizon lawyer Ajit Pai. The new Orwellian-named "Restoring Internet Freedom" Order reversed the FCC's 2015 Open Internet Order and reclassified fixed and mobile broadband services once again as information services under Title I of the Communications Act. The FCC's stale, hackneyed rationale was that regulation stifles competition and private investment (by the monopolist stewards of pipeline infrastructure), spurious arguments based on myths long exploded by critical law and policy scholars. Indeed, there is nothing that approximates competition in the broadband space. Markets are monopolized, and the industry is currently dominated by Comcast, Charter, AT&T, and Verizon in a mafia-style arrangement of noncompetition that they have created for their own benefit. ISPs are among the most lucrative, well-funded, and politically connected companies in the United States, and they spent $110 million lobbying the federal government in opposition of net neutrality in 2017 alone, before the FCC vote.[205] Along the way, Chairman Pai ignored hundreds of thousands of public comments (99.7 percent of which were in favor of keeping net neutrality rules[206]) and claimed a cyberattack took down the FCC's servers during the proceedings. In fact, millions of citizens overwhelmed the system as they asked the agency to retain net neutrality, but according to activist group Fight for the Future, the FCC under Pai's leadership "sabotaged its own public comment process" and "recklessly abdicated its responsibility to maintain a functional way for the public to be heard."[207] This was a major victory for the telecomm lobbyists and a defeat for reformers, the public, and the integrity of the agency.

The degree to which the FCC immolated its own power in the broadband policy arena is difficult to comprehend. However, until there is federal legislation, net neutrality will remain a political football that consumes more resources and generates unnecessary drama, with each successive administration trying to undo the work of the last. Much of the energy around net neutrality has now turned to the states: as of this writing, seven states have adopted their own regulations[208] (which Pai tried to block, but was prevented by the courts from doing), with nine more introducing some form of net neutrality legislation. Just weeks after the repeal of net neutrality took effect in 2018, Verizon throttled the communication of Santa Clara County's firefighters during one of the largest wildfires in California's history. The company forced the first responders to move to a data plan that was double the price in order to communicate with one another in the middle of an active emergency. It should thus come as little surprise that California has subsequently led the way in net neutrality laws for individual states. The 2018 California statute prevents ISPs from blocking or throttling lawful traffic and from prioritizing the traffic of those who pay for faster service. It also prohibits the practice of exempting certain content from data caps, known as giving a "zero rating," which gives those services (often owned by the ISPs themselves) an unfair advantage. This combination of both technological and pricing strategies has gone further than any other state or federal proposal, and has been labeled the "gold standard" in state net neutrality laws.

In addition to repealing net neutrality in 2017, the US government also dismantled some of the last remaining privacy protections still required of Internet pipeline providers. Just three months after Trump took office in 2017, Congress passed a law allowing personal information obtained by ISPs to be sold to the highest bidder. Responding solely to their corporate donors and lobbyists, Republicans repealed the Obama-era Internet privacy rules that required ISPs to get explicit consent before gathering their customers' data and prevented them from selling it. Those requirements had been modeled after provisions in the Telecommunications Act prohibiting telephone companies from collecting "customer proprietary network information" or CPNI (e.g., call histories) and selling them to a third party.[209] The current law now allows cable and wireless providers the ability to gather, share, and sell their customers' browsing history, online activity, geolocation, and essentially their entire digital footprint—all without the customer's knowledge or consent. As a result, cloud pipelines have helped to expand the markets for dataveillance by assisting in the transformation and sale of citizens into commodified data subjects. This has been accomplished through a combination of agency rulings, legislation, and corporate terms of service (TOS) (as discussed further in chapter 3). When lamenting the passage of this bill enshrining ISPs as constructive agents of surveillance empires, Gene Kimmelman, president and chief executive of Public Knowledge, commented on the fact that this was a combined effort on the part of Big Tech that practically guaranteed the public would come out on the losing end. "This collaboration between Silicon Valley and cable companies has never been done before. . . . Their united, massive economic and political power was insurmountable."[210] That insidious and blatant attack on its own citizenry, the government's casting aside of the public interest for the private gain, which is becoming painfully routine and familiar, is the true regulatory crisis from which we may never recover.

Pipeline Principles Revisited

The digital inequities resulting from a century of flawed pipeline policy have had tragic consequences in American society—for education, income equality, health care, civic engagement, and democracy itself, especially in the lives of those who cannot afford or access high-speed Internet service in the home. Almost one-third of Native American tribal lands fall short of the (extremely low) basic FCC broadband standards for speed and performance, and one in five tribal households have no access to the Internet at all.[211] Thirty percent of rural homes do not have a broadband internet connection in 2021.[212] Such discrepancies have led to the pope himself weighing in on cloud policy, directly addressing Big Tech from the Vatican in 2021, calling on them to reform their policies in a show of respect for human rights and social justice. "In the name of God, I ask the telecommunications giants to ease access to educational material and connectivity for teachers via the internet so that poor children can be educated even under the quarantine."[213] With that one statement, Pope Francis has been more vocal about the urgent need for updated pipeline principles in the US than most elected officials and candidates for public office in America.

According to its original designers, the inspiration for the Internet can be distilled down to a handful of core values baked into its infrastructure: adaptability, interoperability, minimalism, flexibility, decentralization, neutrality, and longevity.[214] These rather prophetic choices inherited from the design of the ARPANET have been critical to the cultural durability of pipeline infrastructure. Particularly important is the architecture of the Internet as an end-to-end, modular, distributed, multilayered network.[215] Katie Hafner is one who has articulated the widely held view that the Internet's architecture is the reason for its long-term health and viability, writing, "It was designed as a distributed network rather than a centralized one, with data taking any number of paths to its destination. This deceptively simple principle has, time and again, saved the network from failure."[216]

Yet comprehensive policy designed to support civil liberties, to deliver true end-to-end and universal access at reasonable rates, and to respond to contemporary public needs remains elusive. The very best of the formative pipeline principles (e.g., common carriage and nondiscrimination, the classification of public utilities, and the mandate of universal service) belong in the regulatory standards for broadband pipelines. However, it is time to reject the concept of natural monopoly that has benefited only a handful of private companies—most notably AT&T—at the public's expense. Adapting pipeline principles for the digital era is necessary to achieve and preserve social equality, mitigate the gatekeeping power of corporate owners, and revitalize public interest values for this infrastructure.

Legislating net neutrality and expanding the umbrella of public utilities to include broadband Internet is key, but we also need to think more broadly. For example, Christian Sandvig has written convincingly about the importance of being attuned to the specific language of policy. He argues against using the term "neutrality" in the context of broadband regulation, instead advocating for articulating "a normative vision of what public duties the internet is meant to serve."[217] Sandvig says we need transparency and analytical tools that are not bound by distracting and false tropes such as "neutrality," but that are instead connected to a more nuanced understanding of public values in an already imperfect space—particularly in an environment where service providers still regularly discriminate and manipulate Internet traffic through technological means. Some of these means include, for example, the use of content delivery networks (CDNs), which are intermediaries that maximize the speed and efficiency of data delivery to end users. Critically, they are not subject to net neutrality regulations and have served as an effective loophole in policies designed to prevent so-called fast lanes for internet traffic.[218] Julie Cohen has also noted that the rubric of "net neutrality" is itself "a neoliberally inflected way of answering questions about economic power and public access" because it assumes that the quality of service and access will ultimately be produced by market forces as long as providers are prevented from blocking or throttling.[219] Pickard and Berman, on the other hand, have written that net neutrality is "a necessary but insufficient policy for creating a more democratic internet. It is designed to curb the abuses of large internet service providers, but it does not fundamentally challenge their market power. Nor does it directly confront ongoing inequalities in internet access."[220] These critiques highlight the need for policy and policy frameworks specifically focused on the values of nondiscrimination, equality, public responsibility, and the rejection of strictly market-based solutions or monopoly pipeline provision.

Such analysis also calls our attention to the vital relationship between information dissemination and governance. This connection has been a part of policy debates as far back as those establishing the US Post Office and the federal government's obligation to distribute and allow access to information for all. The mandate to disseminate "knowledge and intelligence" to every single home via the postal system was thought to be "one of the 'conditions of civilization.'"[221] This spirit of public stewardship was seen again in the government's report on the 1939 investigation into AT&T, but rarely since.

These moments are key lessons for the future of cloud policy, as they allow us to identify the evolving iterations of infrastructure's public values so that we may reclaim them once again.

Russell Newman has further argued that the information about net neutrality and the dominant narratives circulating in the US debates contain numerous critical erasures. These include an absence of discussion regarding the market power of broadband giants, early activism over open access, the role of neoliberal capital, and the politicized "production of knowledge" in policy circles.[222] These erasures further distance the public from the powerful civic stakes inherent in broadband regulation and erode the democratic process on which policymaking rests. Centering them in historical accounts is crucial to their restoration in tomorrow's cloud policy.

The current consolidation of pipelines and platforms is exceedingly dangerous for journalism, an informed citizenry, and the survival of an independent, noncorporatized media culture. Looking back on the Carterfone case, former FCC Commissioner Nicholas Johnson (who wrote the majority opinion for the agency in 1968) emphasized the importance of disconnecting the ownership of content and conduit. He reflected on the case in 2008, writing, "Those who sit astride the quasi-monopoly conduits of information and opinion in this country, those who suck profit out of both ends of the wire (or wireless connection) should be content with the riches that position provides—riches beyond their wildest dreams of avarice." He continued, arguing against the growing, anticompetitive power of media empires that control both distribution and content creation. "They should not be able to close off the entrances and lock the exits—like the owners of the Triangle Shirtwaist Factory in 1911—to exclude those who wish to contribute to, or draw upon, the Internet's content and capabilities."[223] And yet, this is increasingly the case. The country's largest cable and Internet provider, Comcast, currently owns NBC and Universal Studios, along with dozens of other content creators and platforms, including Dreamworks Animation, Focus Features, Peacock, and Telemundo. Comcast was also the first cable company to buy a broadcast network and a film studio with the NBC Universal merger in 2011. All the existing legacy Hollywood studios are now part of media conglomerates that own numerous distribution networks, content providers, and digital platforms, reflecting the turn toward cloud-based strategies for servicing the streaming, mobile consumer. Big Tech companies such as Facebook, Amazon, Apple, and Google have also integrated content production, distribution, hardware, and even points of sale in their climb to the top of the list of most valuable corporations in the US.

Even the separation of content and conduit, enforced in the earliest days of radio specifically for AT&T, was undone by the digital era, proving that "the quiet octopus" is the undisputed corporate champion of playing the long game. The world's largest telecommunications company entered yet another new chapter of expansion when AT&T bought Time Warner for $85 billion in 2018, folding major media players such as Warner Bros. Studios, the Turner Broadcasting collection of networks (e.g., TBS, TNT, CNN), HBO, and the DC Comics publishing and IP empire into its holdings. With this acquisition, the breakup of AT&T that happened just thirty-four years earlier became a distant memory, as the company now controlled a massive content factory to funnel through its legendary pipelines and their value was hundreds of *billions* of dollars more than it was in 1984. AT&T even took the concept of "toll broadcasting" (selling airtime in exchange for advertising) that they originated in 1922 into a new millennium, combining subscriber data from their cable, phone, and broadband pipelines with their newly owned networks to create a targeted advertising model that digital entertainment companies are now built on. As former CEO of AT&T Randall Stephenson explained the thinking behind the merger with Time Warner, "If you don't create a pure vertically integrated capability . . . from distribution all the way through content creation and advertising models, you're going to have a hard time competing with these guys."[224] Nevertheless, the merger didn't last long. In 2021, AT&T spun off WarnerMedia in a deal with Discovery, and the pipeline giant went back to focusing on telecommunications and recovering its stock value.

The policy landscape has also contributed to the inferior state of physical infrastructure with respect to cloud pipelines. Susan Crawford has made this case in her sweeping arguments for fiber, a rollout she has compared to the community-based electricity efforts of the nineteenth century. Crawford has explained that, then and now, the lack of equal, affordable, upgraded service for all in a competitive environment can be traced to the fact that "essentially unregulated privately owned cartels are in charge of data transmission. Because they often have effective monopolies in their geographic footprint, it is not in their interest to sell inexpensive world-class services, to serve rural areas, to upgrade their lines to fiber (unless under pressure from a community system), or to open their transmission lines to distribution competitors (as the energy companies have been forced to do)."[225] Just as Robert McDougall argued that "AT&T's long dominance of the telephone industry depended less on tech imperatives than on its ability to shape the political and cultural context of telephony,"[226] so too has the power of ISPs been forged. Fiber is cheaper, faster, and ubiquitous in other parts of the world. In the US, only 15 percent of homes currently have a fiber subscription.[227] The widespread proliferation of this superior pipeline has been obstructed by the ISP industry that has managed to block municipal networks while convincing regulators that upgrading their own lines is too expensive. The FCC has thus determined that the public does not deserve better than outrageously expensive Internet service that has been labeled by experts as "among the worst in the developed world."[228]

In the end, pipeline policy discussions and debates must begin with the fact that cloud pipelines are society's gatekeepers. To return to former FCC Chairman Tom Wheeler's statement, upon announcing the passage of the 2015 Open Internet Order that briefly codified net neutrality, Internet access is a right too vital to be determined and defined by corporate providers. Pipeline provision can no longer fall victim to the red herrings of the mythological "free market" that will supposedly deliver salvation to the public, or the canard that "regulation stifles innovation," which only benefits infrastructure's monopolistic corporate stewards. The key design principles that inspired the original architects of cloud pipelines—such as interoperability, participatory end-to-end design, flexibility, decentralization, neutrality, and longevity—must be foregrounded in policy discussions so that they may be reintegrated into future versions of the cloud. Only then will it be possible to realize the pipeline policy that the public has so long deserved.

2 Platforms

In a lot of ways Facebook is more like a government than a traditional company.... More than other technology companies we're really setting policies.
—Mark Zuckerberg

Content is the king maker, it's not the king. The king is the platform.
—Jeffrey Katzenberg, former CEO of DreamWorks Animation and former Chairman of the Walt Disney Company

If we will not endure a King as a political power we should not endure a King over the production, transportation, and sale of the necessaries of life.
—Senator John Sherman, author of the 1890 Sherman Antitrust Act

Cloud Royalty

Platforms are unquestionably the kings of the twenty-first-century cloud ecosystem. They are a category of cloud infrastructure that commands inordinate power over the digital public sphere. Platforms are the engines of the digital media economy—performing critical functions and services across industries and cultural sectors. They have also become the main agents of spreading mis- and disinformation in addition to exercising unchecked control over speech rights online. They further enable and normalize routine invasions of individual privacy and the culture of surveillance capitalism built on data extraction models designed by Big Tech corporations. Platforms maintain a chokehold over vital access points to media, information, and so many other "necessaries of life." The dominant platforms are run by an unelected and oligopolistic profit-seeking cabal. The entire landscape of cloud policy is replete with highly concentrated, largely unregulated infrastructural power, but the platform sector might be the most egregious. Their global reach and influence expand far beyond the national borders that contain the physical range of pipelines and the legal boundaries of policy.

In many ways, platforms are the most recognizable "players" in the cloud policy environment, particularly the central platforms under discussion in this book, which are the dominant firms controlling social media, search, e-commerce, and digital communication and advertising. They also provide much of the data processing and storage, as well as payment, identification, and other services for modern life on which the rest of the platform ecosystem depends. While companies like Facebook/Meta, Amazon, Apple, Google, and Microsoft (and their billionaire CEOs) are household names, the policy history that helped to create them is far less visible. Nevertheless, it holds the key to their outsized economic, cultural, and political influence. In September 2020, the combined valuation of Amazon, Apple, Facebook, and Google, was more than $5 trillion.[1] As the US House of Representatives wrote in their 2020 report "Investigation of Competition in Digital Markets,"

> companies that once were scrappy, underdog startups that challenged the status quo have become the kinds of monopolies we last saw in the era of oil barons and railroad tycoons. Although these firms have delivered clear benefits to society, the dominance of Amazon, Apple, Facebook, and Google has come at a price. These firms typically run the marketplace while also competing in it—a position that enables them to write one set of rules for others, while they play by another, or to engage in a form of their own private *quasi* regulation that is unaccountable to anyone but themselves.[2]

The challenges in the arena of platform regulation are monumental and, in many ways, unprecedented. The most prominent firms are global corporations that participate in (and often dominate) numerous markets at once. They act as gatekeepers across economic sectors, and exploit their power, according to US lawmakers, "to dictate terms and extract concessions that no one would reasonably consent to in a competitive market."[3] The regulatory command of US antitrust law is increasingly ineffective in this realm, as it was not designed to address the myriad and expansive multisided operations of twenty-first-century digital platforms. Consequently, their size and economic influence has grown to the point that they occupy a noncompetitive space that is for the most part beyond accountability. Amazon, for example, controls roughly 50 percent of US e-commerce between its first- and third-party sales. The company routinely uses its monopoly position to demand unfair terms from third-party sellers and suppliers. It does so while expanding its advantages, according to the House Subcommittee on Antitrust, by "avoiding taxes, extracting state subsidies, and engaging in anticompetitive conduct."[4] There is also its scale to consider, along with Amazon's infrastructural and inscrutable algorithmic command over its competitors.

The dominant platforms have collectively outgrown most policy solutions to rein in their size and influence. Together, Google and Meta account for roughly 50 percent of all digital ad revenue in the US and even more worldwide, with Amazon a distant third.[5] The 2021 earnings of just three major platform providers (Alphabet, Amazon, and Facebook—$845.34 billion) were more than triple that of the largest global media and entertainment conglomerates combined (Comcast, Disney, Warner Media, News Corporation, Fox, and Viacom—$264.7 billion).[6] The growth of the dominant platform companies throughout the 2010s has been nothing short of astonishing—with the largest players increasing their market capitalization 373 percent (Google), 454 percent (Facebook) and 1,348 percent (Amazon) in just ten years.[7]

Their cultural force is staggering. Meta has nearly three billion Facebook users all over the globe, two billion on their WhatsApp messaging service, and another billion on Instagram. In many parts of the world, Facebook is *the* Internet, often through its Free Basics program providing free data to more than three hundred million people for a (Facebook-controlled) online experience, while expanding its advertising base in the process.[8] Most Americans get their news from social media, and a third say they get it specifically from Facebook.[9] Google Search effectively controls access to information online for 4.3 billion users. Thanks to Google, Amazon, and Facebook, Matt Stoller has argued that today "we find ourselves in America, and globally, with perhaps the most radical centralization of the power of global communications that has ever existed in history."[10] According to their former employees, the dominant platforms are designed to be as addictive as painkillers or cigarettes.[11] Their business models have created what Shoshana Zuboff has characterized as "surveillance empires powered by global architectures of behavioral monitoring, analysis, targeting, and prediction."[12] As a result, platforms have annihilated digital privacy in the twenty-first century. Additionally, our freedom of speech is now increasingly dependent on the judgment and policies of a handful of tech corporations. In short, our digital civil liberties are now mediated through privately owned platforms with a measure of control that rivals that of the elected government, but carries with it none of the public service obligations or traditions.

Moreover, they have translated their astronomical profits into political influence, to the point that platform companies are now among the world's major power brokers. Denmark appointed a career diplomat to be the world's first foreign ambassador to the tech industry in 2017, treating Silicon Valley as if it were a sovereign country. Ambassador Klynge explained part of the logic behind Denmark's innovations in "techplomacy": the dominant platforms "have moved from being companies with commercial interests to actually becoming de facto foreign policy actors."[13] They are among the top twenty industries for lobbying in the US, spending nearly $85 million annually (and rising) in order to sway Congress.[14] In a definitive case study of tech industry lobbying, Pawel Popiel (2018) explained that, in addition to the *amount* of their spending, platform companies lobby on an extraordinarily *wide range* of issues. These range from competition, privacy, renewable energy and aviation laws to trade, health-related initiatives, retirement, education technologies, and budget reform, "reveal[ing] the embeddedness of the tech sector in economic, political, and social life."[15] Their unlimited resources and relentless presence in the halls of government inevitably impacts the discursive frameworks and terms of policy debates as well, leaving the true contours of regulatory capture totally uninterrogated and unspoken. This has led to a predictable cementing and expansion of the dominant platforms' influence, a result acknowledged in former CEO of Google Eric Schmidt's candid admission, in 2010, that "the average American doesn't realize how much of the laws are written by lobbyists."[16] This recalls Cory Doctorow's elegant description: "Monopoly [is] converted to money, money to power, power to policy."[17]

The control that platform companies now have over the culture of democratic life for citizens continues to expand largely unchecked. Laura DeNardis for one has argued that "social media platforms are choke points that individuals essentially must pass through to participate in significant parts of the online public sphere" including accessing news, information, and public debate.[18] As Alan Rozenshtein reminds us, "our technology giants are the railroad companies of the twenty-first century. They create and govern our networked space and thus control our lives to an extent unmatched by any other private entity."[19] Their roles in designing an "attention economy" characterized by an inescapable digital enclosure of neoliberal surveillance capitalism, reliant on black-boxed corporate algorithms, targeted advertising, and data harvesting, have been propelling an exploding research agenda in the humanities, law, and the social sciences.[20] Scholars have further identified how these operations have created a world permeated by "data colonialism" that has two superpowers (the US and China) capturing and datafying the behaviors of everyday life and translating them into profit, leaving platforms an agent of "the capitalization of human life without limit."[21] They are further plagued with AI-reinforced racism, toxicity and abuse, and destructive disinformation campaigns.[22] The result is a platform ecosystem that is eroding our digital civil liberties with little resistance from US regulators.

Platforms are now widely understood as privately owned forms of infrastructure that provide the foundation for much of our economy, information flow, and social interactions. Rendering platforms intelligible *as* infrastructure has allowed for lines of inquiry regarding their values of provision and use, and relates platforms to the long line of sociotechnical networks we have come to understand in this regard. It has further positioned them as part of the growing domain of scholarship focused on the cultural histories of media infrastructure, which provides critical perspectives for policy analysis. Unlike cloud pipelines, however, which are privately owned public infrastructure with similar monopoly control over their respective markets, platforms have no analogous regulatory classification, and are thus not subject to the same types of scrutiny or restraints. Nevertheless, this sector of the cloud has been attracting increasing attention from regulators all over the world, which has the dominant firms spending a great deal of time and money fighting to maintain the status quo in this policy landscape.

To contextualize the status quo, this chapter of *Cloud Policy* explores the history of privacy and speech rights, market competition, and access to information in relation to digital platforms. This analysis necessarily includes discussions of content moderation, antitrust policy, and the news industry, as well as the platform business model of targeted advertising dependent on dataveillance and indiscriminate tracking of their users. Addressing these issues separately is a losing proposition. Any "fix" requires systemic and relational change across policy regimes, market sectors, and sociotechnical systems, demanding a holistic approach to policy. This chapter concludes with discussion of potential modifications to platform policy that is mindful of such interrelationships, and carves out a more sizeable place for public values in the regulatory paradigm.

The role of Big Tech companies in their own governance and the shrinking influence of formal regulatory mechanisms are critical threads throughout. The power that dominant platforms have amassed now affords them the uncanny ability to elude most regulation despite patently anticompetitive practices and scandals involving global data breaches, privacy violations, and election interference, to name just a few. Opening the congressional hearings into online platforms and their market power, Representative David Cicilline stated, in July 2020, "As gatekeepers to the digital economy, these platforms enjoy the power to pick winners and losers, shake down small businesses, and enrich themselves while choking off competitors. Their ability to dictate terms, call the shots, upend entire sectors, and inspire fear represent the powers of a private government. Our founders would not bow before a king. Nor should we bow before the emperors of the online economy."[23] Here we explore how and why we find ourselves doing just that in the realm of platform policy.

Platform Governance

Platforms are considered to be distributors or neutral "intermediaries" as opposed to "publishers." This legal distinction matters, because it removes liability for the third-party content that platforms distribute. This issue of online intermediary liability was definitively addressed in an amendment to the Telecommunications Act of 1996 (more on that below), and over time, the courts have interpreted the law as creating a model of "broad immunity" for platforms. This view was further echoed in global legislation such as the European Union's e-Commerce Directive of 2000.[24] It has also been embraced by other international bodies, including the signatories for the 2011 Joint Declaration on Freedom of Expression and the Internet. This declaration had four global rapporteurs on freedom of expression recommending that "no one should be liable for content produced by others when providing technical services, such as providing access, searching for, or transmission or caching of information."[25]

The question of whether platforms are either media or technology companies has also become a discussion of interest to cloud policy. The platform companies themselves insist they are most definitely *not* media companies. Mark Zuckerberg has repeatedly stated that, at its core, Facebook is a technology company "where the main thing that we do is have engineers and build products."[26] It is worth noting that Facebook paid content creators over $1 billion in 2022.[27] Amazon's longtime chief technology officer has argued that Amazon "has been a technology company since day 1 . . . that just happens to do retail."[28] Also of interest is the fact that Amazon owns the Prime Video, Freevee, and MGM+ streaming services, as well as Amazon Studios and MGM studios, both of which develop, finance, and create film and television content.[29] Scholars have refuted such arguments made by social media platforms claiming to be simply tech companies, including those based on their functions in the market, the interests of their founders ("we are computer scientists!"), and the primacy of algorithms in their operation.[30] After all, the imposition of algorithms and AI to maximize engagement and profits necessarily imparts values, categorizations, and hierarchies in relation to the content being carried. Still, the dominant platforms continue their efforts to be positioned discursively, politically, and, most of all, in the policy space as tech rather than media companies—a sign of the regulatory implications that these labels carry. As Dwayne Winseck has explained, "defining digital communications platforms as media firms would bring them under the authority and broader policy remit of broadcasting regulators such as the ACMA, CRTC, FCC, Ofcom, etc.,"[31] which all enforce much stricter policy regimes than these platforms currently face.

Recognizing and documenting what this all means for industry competition, the public, and corporate power was part of the mission of the 2020 House hearings on Competition in Digital Markets. To begin, the committee chairs addressed the highly problematic function of platforms as "the underlying infrastructure for the exchange of communications, information, and goods and services"[32] when introducing the investigation of Amazon, Apple, Facebook, and Google. "Each platform uses its gatekeeper position to maintain its market power. By controlling the infrastructure of the digital age, they have surveilled other businesses to identify potential rivals, and have ultimately bought out, copied, or cut off their competitive threats. And, finally, these firms have abused their role as intermediaries to further entrench and expand their dominance."[33] Among many examples that the report pointed to was Google Chrome, the world's most popular browser, that the company has used "to both protect and promote its other lines of business." It also addressed Google Maps which "captures over 80% of the market for navigation mapping service—a key input over which Google consolidated control through an anticompetitive acquisition and which it now leverages to advance its position in search and advertising."[34] Amazon was a focus for its dual role as a dominant provider of cloud infrastructure and as a dominant firm in other markets dependent on the cloud, creating a conflict of interest that Amazon alone has the incentive and ability to exploit.[35]

The policy lag in catching up to the complexities of platforms' economic function and conflicting definitions have in part allowed platforms to thus far escape serious regulatory constraints in the US. Policy for this sector of the cloud is also tied to geographically bound legal frameworks that are insufficient tools to confront Big Tech's dominance, given their global operations and footprint. This has led to a platform economy characterized by monopolistic power and the installation of Big Tech companies as sovereigns governing many facets of our digital lives. The current imbalance of power is compounded by the fact that the public has largely surrendered or grown numb to the insidiousness of platform control, which now feels inescapable in twenty-first-century existence. Such failures of cloud policy have directly contributed to the erosion of democratic norms, and they must be remedied if we are to reclaim our culture from these unchecked forces of private capital.

While all cloud infrastructure is best understood as being subject to a wide range of regulatory influences, platforms are often enmeshed in more multifaceted and informal policy regimes than the numerous state-mandated frameworks created for the common carriers and cloud pipelines of chapter 1. Therefore, the concept of governance is particularly important here.[36] Flew and Martin have distinguished between digital platform *regulation* and digital platform *governance* by noting that "the concept of regulation typically refers to actions by governments and public agencies on private actors that are enabled by binding laws and which have negative sanctions for noncompliance," as contrasted with governance, which offers a more decentralized model of control that includes both state actors and the self-policing of the private companies themselves.[37] Philip Napoli's framing of media governance as a broader sphere and a more inclusive concept than formal regulation and policy alone is similar in that it includes the a "range of stakeholders participating in the process of designing the rules and norms" such as "policy makers, industry stakeholders, NGOs, civil society organizations, and even media users."[38] Software (and hardware) code and technical standards should also be viewed as dimensions of platform governance. As Lessig argued in *Code and Other Laws of Cyberspace*, these architectures of the digital space also perform regulatory functions, by either protecting values that a culture believes to be fundamental or allowing them to disappear. In so doing, code and programming serve to moderate human interactions, define allowable freedoms and controls online, and enable specific forms of culture.[39]

Such privatized regimes are a consistent and defining feature of cloud policy. The role that platforms have assumed in their own governance has increased to the point that, in the US, their own power over policy formation has begun to transcend that of the state. The congressional hearings about the 2018 Cambridge Analytica data breach affecting eighty-seven million people presented this dynamic in spectacular form. For two days, the Republican-led Congress asked Facebook CEO Mark Zuckerberg about what type of regulation he might be amenable to, treating him as a colleague as opposed to someone representing an industry violating consumer rights and privacy that they have the responsibility to regulate on behalf of the public. Some of the more stunning examples of government submission to corporate authority during the questioning of Zuckerberg included these:[40]

- Hatch (R-Utah) Whenever a controversy like this arises, there's always the danger that Congress's response will be to step and overregulate. Now, that's been the experience that I've had, in my 42 years here. In your view, what sorts of legislative changes would help to solve the problems the Cambridge Analytica story has revealed? And what sorts of legislative changes would not help to solve this issue?
- Wicker (R-Miss) We don't want to overregulate to the point where we're stifling innovation and investment. Do you think we need consistent privacy protections for consumers across the entire Internet ecosystem that are based on the type of consumer information being collected, used or shared, regardless of the entity doing the collecting, reusing or sharing?
- Graham (R-SC) You do not think you have a monopoly?

Zuckerberg: It certainly does not feel like that to me.

Graham: OK, so it doesn't. . . . Do you, as a company, welcome regulation?

Zuckerberg: I think, if it's the right regulation, then yes.

Graham: Would you work with us in terms of what regulations you think are necessary in your industry?

Zuckerberg: Absolutely.

Graham: Okay. Would you submit to us some proposed regulations?

Zuckerberg: Yes, and I will have my team follow up with you, so that way we can have this discussion across the different categories where I think that this discussion needs to happen.

- Upton (R-MI) "What kind of policy regulation—regulatory environment would you want?"
- Sarbanes (D-MD) "Facebook is becoming sort of a self-regulated super structure for political discourse. And the question is, are we the people going to regulate our political dialogue, or are you, Mark Zuckerberg, going to end up regulating the political discourse?"
- Welch (D-VT) "Do you believe that the Federal Trade Commission or another properly resourced governmental agency with rulemaking authority should be able to determine on a regular basis what is considered personal information to provide certainty for consumers and companies what information needs to be protected most tightly? . . . Who gets the final say? Is it the private market, companies like yours, or is there a governmental function here that defines what privacy is?"
- Loebsack (D-IA) "How can we be guaranteed that, for example, when you agree to some things today, that you are going to follow through and that we are going to be able to hold you accountable, and without perhaps constructing too many rules and regulations? We would like to keep that to a minimum if we possibly can. But I do understand that you have agreed that we are going to have to have some rules and regulations so that we can protect people's privacy."
- Ruiz (D-CA) "Mr. Zuckerberg, would it be helpful if there was an entity clearly tasked with overseeing how consumer data is being collected, shared, and used and which could offer guidelines, at least guidelines, for companies like yours to ensure your business practices are not in violation of the law, something like a digital consumer protection agency?"
- Walberg (R-MI) "I would say the best thing we can do is have these light-of-day hearings, let you self-regulate as much as possible, with a light touch coming from us, but recognizing that, in the end, your Facebook subscribers are going to tell you what you need to do."

As numerous legislators admitted they were "trying to help" Zuckerberg (Sullivan, R-AK) while commending him for being "open to some regulation" (Hassan, D-NH), Zuckerberg felt empowered to repeatedly warn

Congress that "you have to be careful about what regulation you put in place" and cautioned legislators that things "need to be thought through very carefully when thinking through what rules *we* want to put in place" (emphasis added). Zuckerberg had the temerity to further add that "Facebook is absolutely committed to working with regulators, like Congress, to craft the right regulations. Facebook would be happy to review any proposed legislation and provide comments." During these hearings, lawmakers seemed to forget that the answers to their questions about responsibility, accountability, and the most desirable values for public infrastructure are not supposed to come from a corporate wish list, but instead from deliberative process among elected officials with a vision for how to best serve the public.

The ideologies foundational to the historical development of computing and information technology have been formative influences on platform policy. They have been shaped by universities, the government and the military, activist communities, corporate boardrooms, and technologists. They incorporate the legacies of 1950s Cold War machine logics and military science, the 1960s counterculture, and Silicon Valley's origin myths, as well as 1990s libertarian techno-utopianism and free-market idealism, and the underlying currents of Romanticism enduring throughout.[41] As these diverse forces have all been variously mapped onto our collective understanding of cloud technologies and platform infrastructure over time, they have also helped shape the values and parameters of governance and policy. That is to say, this ideological cocktail as it has evolved has delivered a regulatory landscape with a very complicated historical relationship to the core influences of centralized, hierarchical control; the principles of decentralization and individual freedom; the fiction of the "free" market; systemic corporate welfare; and even the spirit of "techno-democratic optimism"[42] that largely persists despite current conditions.

Various scholars have addressed these paradoxes when pointing to the crucial role of what Barbrook and Cameron have identified as the "Californian ideology" shaping Silicon Valley tech culture.[43] This ideology was a relentless form of techno-utopianist orthodoxy emerging from a combination of countercultural ideals and institutional distrust with a faith in free market economics that produced Big Tech's signature branding, what Barbrook and Cameron described as a "contradictory mix of technological determinism and libertarian individualism."[44] Their seminal essay was a critique of the widely accepted ethos that held up information technologies as forces that "empower the individual, enhance personal freedom, and radically reduce the power of the nation-state" while remaining willfully ignorant of the massive public subsidies and state interventions that helped build them. As Barbrook and Cameron noted, "the West Coast hi-tech industrial complex has been feasting off the fattest pork barrel in history for decades."[45] That caretaking relationship between the state and Big Tech has continued, along with the wide spectrum of ideological contradictions inherent in the governance of all cloud infrastructure.

The trajectory of platform governance is also a triumph of monopoly capitalism, thanks in part to what Thomas Streeter has described as "a general transformation in the dominant, governing ideas of American society in the early 1980s, when a radical belief in markets and an accompanying suspicion of all forms of governmental regulation—beliefs that were once thought to be fringe—would become common sense among many in positions of power, with global effects."[46] Ronald Reagan's joke that "the nine most terrifying words in the English language are: 'I'm from the Government, and I'm here to help'" still gets approving laughter from audiences and lawmakers today. The protracted naturalization of these ideological traditions has subsequently served to solidify the tenets of "self-regulation," market consolidation, and unaccountable privatized power as preordained realities of modern life that citizens are just meant to expect and endure.

Digital media platforms were initially hailed as agents of transformative global connection, tools of convenience, and gateways to information as they began life in the late 1990s (Amazon, Google) and early 2000s (Facebook, Twitter). As their business models became more dependent on the extraction and sale of user data, this ecosystem became increasingly destructive to the rights of its users. Nevertheless, the inflated myths of "California ideology" continued to fuel Silicon Valley's self-promotion machinery. Mark Zuckerberg's letter to potential investors in advance of what became one of the largest IPOs in history is a prime example.[47] When making his pitch in 2012 for Facebook's future as a public company, Zuckerberg wrote that "we hope to rewire the way people spread and consume information" and "change how people relate to their governments and institutions." According to Zuckerberg, giving people "the power to share" would help them "transform many of our core institutions and industries." Moreover, he described the company as practically a communist enterprise, declaring, "Simply put: we don't build services to make money; we make money to build better services.... We don't wake up in the morning with the primary goal of making money, but we understand that the best way to achieve our mission is to build a strong and valuable company."[48]

Fred Turner has argued that platforms like Facebook have been the ones to help usher in numerous threats to democracy, despite the fact that they were designed and promoted with the promises of preserving it. Turner reminds us that "the same technologies that were meant to level the political playing field have brought troll farms and Russian bots to corrupt our elections.... The same networked methods of organizing that so many thought would bring down malevolent states have not only failed to do so—think of the Arab Spring—but have instead empowered autocrats to more closely monitor protest and dissent."[49] Turner contends that we need to understand history in order to resist these despotic trends, which are sure to persist if other forces of governance do not rise up to stop them. Indeed, this prediction has been borne out repeatedly, including at the January 6, 2021, insurrection and riot at the US Capitol. After this bloody event, organized on social media and staged by conservative extremists to prevent the certification of the 2020 presidential vote, Facebook staffers expressed outrage at the role their platform played in enabling and inciting what happened that day, claiming "We are not a neutral entity"[50] and "History will not judge us kindly."[51] Many of them were still incensed that the platform disbanded its Civic Integrity Team—the unit in charge of protecting elections and identifying the most active political groups that violated rules regarding hate speech and calls for violence—just one month prior to the insurrection.

When platforms themselves are the chief agents of their own governance, the values that contribute to their profit and influence become inextricable from those embedded in cloud policy. Consequently, the most serious harms posed by this sector arise mainly in three key areas: privacy and surveillance, freedom of expression, and market competition/antitrust. Each one is in urgent need of a formal and coordinated policy overhaul. The siloed approaches to addressing the privacy, free speech, and antitrust violations taking place have been ineffective thus far, and there are currently no policies that begin to identify the critical matter of their interrelationship in the platform ecosystem, let alone relative to pipelines and data. Addressing these interdependent dimensions of cloud infrastructure would put the necessary focus on the rights and democratic protections rooted within that are now being eroded, often with our unwitting consent.

Privacy and Surveillance

There is no right to privacy named outright in the US Constitution. It was vaguely enshrined in the Fourth Amendment as a protection against unreasonable searches and seizures by the government, and in the Fifth Amendment as the right against self-incrimination and the forced disclosure of private information about oneself. It has been vigorously debated, affirmed, denied, and reframed since the late nineteenth century by the legal community, the courts, Congress, activists, and now platform companies. Sarah Igo writes that, from the earliest deliberations, "arguments about privacy were really arguments over what it meant to be a modern citizen."[52] Privacy is at the heart of state and corporate power over individuals and is essential to the functioning of a democracy. Its presence or absence has been fundamental to the state's ability to identify people, monitor their behavior and movements, or intrude on their lives and civil rights. As one legal scholar eloquently put it, "privacy, in its various forms, is ultimately about control."[53] In the digital era, such control has been largely transferred to the new tech sovereigns, but the fact remains that without uniform privacy rights, there is no personal autonomy. There is no equality. There is no true freedom.

The right itself was first articulated and famously delineated by Warren and Brandeis in their 1890 *Harvard Law Review* article "The Right to Privacy." This lasting, influential piece formulated privacy as "the right to be let alone,"[54] stemming from the development of new media technology at the time (photography), a Victorian aversion to gossip and "intrusion upon the domestic circle," as well as the desire for protection from the press "overstepping in every direction the obvious bounds of propriety and decency."[55] The Supreme Court inferred constitutional rights to privacy in 1965 in *Griswold v. Connecticut*, which overruled a ban on the use of contraceptives by married couples, thus including the marital bedroom as a sanctioned "zone of privacy." Subsequent landmark rulings further extended privacy rights to unmarried people (*Eisenstadt v. Baird*, 1972), women seeking an abortion (*Roe v. Wade*, 1973),[56] and same-sex partners (*Lawrence v. Texas*, 2003). However, arguments for reproductive and equal rights for all genders and sexualities rooted in privacy (as opposed to bodily autonomy or economic equality) have not held up over time. Indeed, their durability has proven to be fragile at best in the twenty-first century, and some, including the right to an abortion, have already shattered.

In addition to these intimate decisions about reproductive health, the right to privacy has since been afforded by Congress and the courts to additional dimensions of health care. There are also other sector-specific protections for data, such as those related to education, finance, and communication, for example.[57] Children under thirteen have received some protection from the Children's Online Privacy Protection Act of 1998 (COPPA), which put restrictions on digital data collection and marketing while requiring parental consent and more transparency from online services. There is also the Video Privacy Protection Act (VPPA), which stands as a monument to all that is wrong with privacy protections in the digital age. This law was passed in 1988 in reaction to the publication of Supreme Court nominee Robert Bork's video rental history in a newspaper during the debate over his nomination, which was then considered the outer limits of privacy invasion. The VPPA is one of the strongest (and only) legislative protections afforded to consumer privacy against a specific form of data collection. It prevents video rental services from knowingly disclosing your information to a third party. As a result, your ISP and mobile phone can surveil you, the platforms you use can track your digital footprint for profit, your data is bought and sold without your permission, but your rental history of VHS tapes that no longer exist from stores that also no longer exist is strictly off limits. It is notable that in 2012 Netflix successfully lobbied to have the law revised to allow consumers to share their viewing data through social media, in yet another example of the tech sector dictating policy.

Most of these laws were created in an analog society, long before platforms became the main arbiters of privacy. In fact, the cultural momentum to protect this right was strongest when computers and digital databases first emerged in public consciousness during the late 1960s, and the state was viewed as the main potential threat to individual privacy. This pervasive mistrust in part helped lead to the Privacy Act of 1974, signed in the wake of President Nixon's resignation and designed to establish coordinated practices for personal information maintained by the federal government. This was the last time Congress passed comprehensive federal legislation protecting individual privacy. It was enacted largely in response to Watergate and the FBI's COINTELPRO surveillance scandal; according to the Department of

Justice, the law "sought to restore trust in government and to address what at the time was seen as an existential threat to American democracy."[58]

Ironically, congressional interest in protecting public data from the perceived dangers of centralized state collection and storage had the ultimate effect of delivering control over personal information to the private sector. Corporations thus became the main architects of platform infrastructure and the primary repositories of private data, and the chief agents of privacy invasion. As historian Margaret O'Mara explained, this mistake of focusing exclusively on the government and ignoring the threat of private industry at the outset "would ultimately make possible one of modern Silicon Valley's greatest business triumphs: to gather, synthesize, and personalize vast amounts of information and profit richly from it."[59] Public protections from these practices—along with the individual right to privacy in the digital space—would be collateral damage.

In fact, the culture of anti-privacy in America is inseparable from dominant platforms and their dependence on digital surveillance, data monetization, targeted advertising, and algorithmic curation. In *Profit over Privacy* (2021), Matthew Crain explains that these practices of "surveillance advertising" are not new developments, but in fact they are the continuation of trends in technology, marketing, and governance that were well established by the 1990s. He argues that the federal government's decision to privatize and commercialize the Internet and its infrastructure, formally beginning under George H. W. Bush in the late-1980s and ultimately carried out by the Clinton administration in the 1990s, was the most critical in this regard and resulted in privacy becoming "a market transaction like any other."[60] With the private sector in control of this infrastructure, regulation practically nonexistent, and zero public interest protections in place, the platform ecosystem was vulnerable to the worst excesses of advertising-driven, data-thirsty business models—particularly as they developed in the frenzy of the late-1990s tech bubble. It was not long before widespread surveillance was normalized and the rapacious mining and sale of personal data was powering the insatiable engines of the online economy.

It is therefore no surprise that the default settings on dominant platforms such as Google and Facebook are designed to expose—and monetize—our thoughts, our whereabouts, our preferences, and our social networks. Publicity, not privacy, is their fundamental architecture. Recalling Lessig's work, privacy policies are part of the codes that define the core values of our online culture. To take one small example, Facebook's *default* privacy settings went from allowing different types of your personal data to be visible to only you in 2005 to allowing the entire Internet to see them by 2010. Privacy was "disappeared" by design, as it has been across the platform ecosystem. In 2022, executives in Google's Privacy and Data Protection Office actually acknowledged, "Our data infrastructure is not designed for privacy."[61]

In addition to platform settings, their "terms of service" model of governance that forces users to consent to a host of rights infringements and indignities plays another key role. These "agreements," which Frank Pasquale has characterized as "contracts surrendering your rights to the owners of the service,"[62] are collectively serving as the official charter for individual privacy rights online (or lack thereof). They have included acknowledgments that one has no reasonable expectation of privacy when using a service (Google, regarding Gmail, 2013); that one must "not criticize the product or service publicly" (AT&T, 2007); that your online activity can be tracked even after you leave the site for marketing purposes (Facebook, Amazon); that any of your conversations can and will be bugged and the recordings will be kept by the company (Amazon for Alexa, 2019); that law enforcement can access home security footage and other data from their platform and proprietary devices without a warrant (Amazon, Google, 2022); and that the service can sell your uploaded photos to third parties for use in advertising (Instagram, 2012).[63]

As a society, we have all enabled the myriad psychographic invasions of privacy by platforms' dataveillance and extraction models by clicking "I agree." However, users can hardly be blamed for their participation, given the fact that many of these platforms have become essential infrastructures for daily life. As José van Dijck has written, "when it comes to the possibility of opting out, we are confronted not only with techno-economic hurdles, but also with social norms and the ideological imperatives and cultural logics that scaffold them."[64] It is easier said than done to live without Big Tech now that their platforms are involved with mediating everything from our professional and social lives to our health care, education, transportation, and entertainment. Nevertheless, it is important to remember that we are repeatedly making the bargain to exchange surveillance of ourselves and our data for their "free" services. This trade, which defines the digital business model, will continue to extract terms with escalating cultural and political consequences, including the heavy price we pay for giving up our privacy in the deal, knowingly or not.

The role of platform companies in regulating the boundaries of individual privacy would be incomplete without mentioning their participation in the NSA's massive PRISM program—a secret blanket surveillance operation that included spying on American citizens and foreign leaders—exposed by Edward Snowden in 2013. The program has relied on the servers of Apple, Google, Facebook, Yahoo!, and Microsoft (among other platform companies), along with the records of pipeline companies including AT&T and Verizon.[65] Thanks to the participation of platforms in these efforts, NSA analysts were able to examine their stored information; monitor live audio, video, chat, and file transfers taking place; and even receive notifications when their targets logged on to specific platform messaging or email services[66] (see chapter 3 for a detailed discussion of this program).

This insidious American partnership between the private sector and the state at the expense of individual privacy has endured across cloud infrastructure, and it is not limited to domestic abuses. Big Tech's capitulation to authoritarian foreign governments in the form of handing over the data of users who are suspicious in the eyes of the state has been a continuing tragedy. "If you want to do business there you have to comply." This was Yahoo cofounder Jerry Yang's response to learning of Chinese journalist Shi Tao's ten-year sentence to hard labor at a prison factory in 2005 after Yahoo handed over his emails to the Chinese government.[67] As Nicholas Kristof wrote at the time, "Yahoo sold its soul and is a national disgrace."[68] However, Yahoo is far from the only company complying with such requests. Google, Apple, and Amazon's ventures in China have all been used to help the government conduct mass surveillance against its citizens or censor online material with no resistance from the platforms, as they claim they are simply "enforcing a country's laws."[69]

This practice has become so prevalent that Congress (deflecting attention from and ignoring domestic abuses by the same firms) took Big Tech executives to task for their complicity in denying human rights to global users and helping authoritarian governments to punish dissidents. At a 2006 House hearing on Internet Freedom in China, Representative Chris Smith singled out platform companies for handing over data information to Chinese officials:

These are not victimless crimes. We must stand with the oppressed, not the oppressors. "Should business enable the continuation of repressive dictatorships by partnering with a corrupt and cruel secret police and by cooperating with laws that violate basic human rights?" I believe that two of the most essential pillars that prop up totalitarian regimes are the secret police and propaganda. Yet for the sake of market share and profits, leading U.S. companies like Google, Yahoo, Cisco and Microsoft have compromised both the integrity of their product and their duties as responsible corporate citizens. They have aided and abetted the Chinese regime to prop up both of these pillars, propagating the message of the dictatorship unabated and supporting the secret police in a myriad of ways, including surveillance and invasion of privacy, in order to effectuate the massive crackdown on its citizens.[70]

As with all other congressional hearings related to Big Tech, no legislation followed to curb the behaviors being called out by lawmakers, and such events remain largely spectacles of regulatory theater. Moreover, there is still no comprehensive federal law to protect digital privacy in America despite consistent urgings from the FTC for Congress to pass comprehensive consumer privacy protections going back to 2000,[71] constant and massive data breaches across all sectors of platform activity,[72] and numerous failed attempts at establishing industry and tech standards over the past three decades. This includes the Obama administration's 2015 Consumer Privacy Bill of Rights, and the promising Do Not Track legislation (2011) and Do Not Track browser setting, which would have provided a standard opt-out across websites with a simple click, thus restoring some modicum of privacy on the Internet.[73]

In 2022, Big Tech did begin to see some financial consequences for their privacy violations. For example, Google agreed to a settlement with forty states attorneys general of $391.5 million in a privacy case related to their location tracking feature, which continued to track users even when it was turned off. This ended a four-year investigation into Google's practices from 2014 to 2020 in which they were accused of violating consumer protection laws. Google reached one more settlement with the state of Arizona for $85 million just a day earlier. While this is undeniably a large sum of money, Google's annual revenue in 2021 was $256 billion. At the same time, discovery documents in Google Chrome's privacy lawsuit disclosed that the company regularly abandons its assurances to users related to personal information, privacy, and user data. Executives in Google's Privacy and Data Protection Office were quoted as saying, "There is no coherent strategy [on] privacy at Google," "We have gaps in how our system works and what we promise to people," and "At Google, we still seem to believe in that fantasy that users agreed to this."[74]

Global privacy protections remain inconsistent at best, but Europe has done a far superior job at legislating digital privacy rights for its citizens than the US. In 2014, for example, the Warren and Brandeis "right to be let alone" was modernized and reborn in the European Union as the Right to Be Forgotten. The EU also formally instituted the General Data Protection Regulation (GDPR) in 2018 to protect data security, transfer, and privacy of EU citizens. In the GDPR, the Right to Be Forgotten (Article 17) was enshrined as "The Right to Erasure." This right was won in the 2014 "Google Spain case,"[75] in which a Spanish citizen successfully sued Google in his efforts to remove personal data regarding the forced sale of his house from their search results. In this case, the Court of Justice of the European Union (CJEU) issued a decision that required Google to allow EU users to remove unwanted personal information (in the form of links to web pages) from search results. The court further determined that Google acts as a *data controller* as opposed to a mere custodian in providing its Google Search service. With this ruling, search engines that are supported by advertising became responsible for the content that they link to, and as such, these platforms were no longer legally regarded as "neutral intermediaries" in the EU.[76]

The revised "Right to Erasure" extended the ruling far beyond search engine indexes to include all corporate data controllers (with certain exceptions including "exercising the right of freedom of expression and information," "archiving purposes in the public interest," and "scientific or historical research purposes"[77]). In her definitive book on the topic, Meg Leta Jones writes about how collective memory is outrunning state control, noting that "policy makers around the globe are being pressed to figure out a systematic response to the threat of digital memory—and it is a complex threat involving uncertain technological advancements, disrupted norms, and divergent values."[78] This is expanding regulatory lag in the platform arena, particularly for the parts of the world where this right does not exist. Leta Jones also argues that the issue should be framed as one about information stewardship as opposed to information permanence, which opens up opportunities for reconsidering interoperable privacy, accountability, and preservation standards in the digital age outside the borders of the EU.

This ruling ultimately forces lawmakers and citizens everywhere to reconsider the cultural value of the right to individual privacy, and the role of platforms in determining the boundaries of that right. Should it be individuals or corporations that get to decide which embarrassing secrets—or innocent private details—one is allowed to erase from the digital public record that Google Search has become? This issue of discoverability has become particularly urgent in the current era of ubiquitous tracking and sharing of data across global platforms. It brings us back to Alan F. Westin's view of privacy, which he defined in his 1971 seminal book *Privacy and Freedom* as "the claim of individuals, groups, or institutions to determine for themselves when, how, and to what extent information about them is communicated to others."[79]

Many still object to the right to reinvent and manipulate the historical record online, and some even view it as censorship and a violation of the right to free expression for the press.[80] Librarians and lawyers have been among those ardently opposed to the idea of removing information from public record. Nevertheless, this growing demand for what Leta Jones calls the culturally specific and dynamic concepts of "digital redemption" or "digital reinvention" has been recognized in the right to be forgotten. It is a right that represents the tightrope every society must walk when balancing restraints and protections regarding "reputation, identity, cultural history, corporate power, expression, access, and exceptions."[81] The ability to delineate what side of the line that information falls on is also connected to freedom of expression, as well as privacy rights, and is emblematic of how deeply entwined these dimensions of platform governance are.

These negotiations are taking place during a complicated point in cloud policy's history. The continued struggle over the right to be "let alone" is being confronted with the invasive business models of digital platforms, as well as new generational norms of utilizing social media to willingly surrender one's privacy in order to "broadcast yourself" as the early motto of YouTube implored. In one profoundly dystopian example, Amazon announced in 2022 that it was utilizing the footage captured from Ring doorbells (its corporate subsidiary) for a television show entitled *Ring Nation*, produced by MGM Studios, which Amazon also owns. The company promoted the show as showcasing "incredible, hilarious, and uplifting must-see viral moments," but it was plainly obvious even to casual observers that this is simply another way that for-profit surveillance is being normalized, commodified, and repackaged as entertainment by Big Tech corporations. It leaves little else to do but wonder where rock bottom is.

The collective outrage necessary to create change has yet to bubble up to a point of inflection. But this is in fact a terrifying hour of reckoning for America, as Shoshana Zuboff writes: "The intolerable truth of our current condition is that America and most other liberal democracies have, so far, ceded the ownership and operation of all things digital to the political economics of private surveillance capital, which now vies with democracy over the fundamental rights and principles that will define our social order in this century."[82] It is a problem borne of policy, culture, and ideology that cannot be solved with laws alone, or laws about privacy alone. Resolving this self-inflicted crisis will require an approach to governance that addresses the interdependence of digital civil rights and the outsized power that platforms have over our lives. It will also demand a reconciliation between privacy's declining cultural valuation among younger generations with the damages being imposed on our society by the platforms they have known all their lives.

Freedom of Expression

Despite their status as "neutral intermediaries," platforms are key vectors of privacy *and* free speech rights, and in the US they have become chief arbiters of both. Jack Balkin has written about the infrastructure of twenty-first-century free expression as "merging with the infrastructure of speech regulation and the infrastructure of public and private surveillance," articulating a web of technological and legal connections between these areas of governance in which platforms play the dominant role.[83] There have been decades of struggles in the courts to define and understand platforms vis-à-vis the First Amendment.[84] Many of the inherited concepts and safeguards as applied to nineteenth- and twentieth-century communication have grown inadequate for the digital media environment of the new millennium. For example, the proliferation of fake news, harassment, filter bubbles, and bots across platforms have served as evidence that the reliance on "counterspeech," based on the idea that the best answer for harmful speech is more speech, has become impoverished or even obsolete.[85] A small handful of state policies are attending to freedom of expression online, but all roads to privacy and free speech travel the same terrain, and they run directly through the dominant digital platforms. In less than two decades, these companies have amassed extraordinary unchecked power in the realm of speech rights.

Section 230 The most significant law related to online speech in the US was enacted just as the commercial Internet was taking off. It is found in Section 230 of the Communications Decency Act (CDA), an amendment to the 1996 Telecommunications Act, which says in part, "No provider or user of an interactive computer service shall be treated as the publisher or speaker of any information provided by another information content provider."[86] This sentence, famously called "the twenty-six words that created the internet" (Kosseff 2019) offers infrastructure providers—both platforms and pipelines[87]—protections against what their users say or "publish" online and removes legal liability from online intermediaries for the third-party content that they distribute.[88] This protection granted to pipeline (telecomm) and platform (Big Tech) providers has shielded them from responsibility for their users' posts, videos, comments, or activities even as they moderate such "speech."

The origin of Section 230 lies in a New York court case decided just nine months before the 1996 Telecommunications Act was signed into law. The case involved Prodigy, an early online service popular in the 1990s, and a Long Island brokerage firm called Stratton Oakmont (most famous for the greed and criminal escapades of its founder Jordan Belfort, also known as "the Wolf of Wall Street"). When an anonymous user of Prodigy's Money Talk "bulletin board"—an information exchange service accessible to subscribers and curated by Prodigy—posted in 1994 that Stratton Oakmont was a "cult of brokers who either lie for a living or get fired" and the firm was involved in "major criminal fraud," Stratton Oakmont sued Prodigy on various grounds, including libel.[89] They went after the company instead of the individual user, arguing that Prodigy was a publisher of the material and thus responsible for what was posted. Of key importance was the fact that Prodigy moderated its content by using software to screen posts "on the basis of offensiveness and bad taste."[90] It also maintained community guidelines for posts, which bulletin board leaders were required to enforce. By utilizing technology and labor for such purposes, the court ruled that Prodigy was indeed exercising editorial control over its content and was therefore serving as a publisher, rather than simply a distributor. In its ruling, the court also cited an earlier case, *Cubby, Inc. v. CompuServe* (1991), in which a New York court found that one of Prodigy's competitors could *not* be held liable as a publisher for content posted by its users. The reasoning was that CompuServe made *no* effort to review any of the content it hosted, and therefore "had neither knowledge nor reason to know" of the allegedly defamatory statements it was accused of posting.[91] In that case, it was ruled that CompuServe was simply a distributor, not a publisher, of the content in question. As the court later pointed out in *Stratton Oakmont, Inc. v. Prodigy Services Co.*, "Prodigy's conscious choice, to gain the benefits of editorial control, has opened it up to a greater liability than CompuServe and other computer networks that make no such choice."[92]

This caught the attention of Congressmen Ron Wyden (D-OR) and Chris Cox (R-CA) at a time when debates over the telecommunications reform bill had been ongoing for two years. At this pivotal juncture for the new medium of the Internet, they were concerned that these decisions—the first to address online speech liabilities—would discourage investment from the private companies that were to become its main engines. After all, in the wake of the *Stratton Oakmont* ruling, any efforts at content moderation would result in serious legal jeopardy for the companies involved. They realized this would lead to unhealthy environments, for both online speech and the new digital economic sector. Therefore, they joined forces to "incentivize private efforts aimed at combatting 'offensive' material."[93] Engaging in a very rare moment of bipartisanship in 1995, Cox and Wyden together wrote what would become known as Section 230, along with input from civil society organizations and tech companies.

Section 230 was buried deep in the contentious CDA, which itself was written in part to regulate/criminalize the distribution of porn and other "patently offensive" material on the internet to minors.[94] In fact, that goal was the singular impulse in the 1996 law behind more interventionist regulatory action. Section 230 received very little attention at the time, as the focus of the Telecommunications Act was on the pipelines—the cables and telephone wires delivering the Internet. And yet, it is responsible for protecting contemporary platform infrastructure from some of its most threatening legal liabilities. Section 230 has been deemed instrumental to an online environment that supports free expression, despite its many imperfections. Without it, most informed observers agree that the Internet as we now know it would never have come into being. However, it also exacts steep social costs in the form of widespread harassment, disinformation, hate speech, and targeted abuse, among other online scourges. We have yet to fully reconcile the terms of this bargain.

In addition to providing liability protection for posting third-party content ("No provider or user of an interactive computer service shall be treated as the publisher or speaker of any information provided by another information content provider"), Section 230 also provided liability protection for attempts to remove it, overruling *Stratton Oakmont*. As such, it states that no provider shall be held liable for good-faith restriction of access to material considered to be "obscene, lewd, lascivious, filthy, excessively violent, harassing, or otherwise objectionable, whether or not such material is constitutionally protected."[95] This is often called the "Good Samaritan" or "takedown" provision and it is the legal foundation for content moderation. It allows platforms to simultaneously maintain their distinction as "neutral conduits," while also making evaluative decisions and judgments about whether to take down the posts, videos, images, and other content that they distribute for not meeting their community standards. As Tarleton Gillespie has explained, thanks to Section 230, "choosing to delete some content does not suddenly turn the intermediary into a 'publisher,'" which would then "create legal jeopardy for intermediaries that chose to moderate in good faith, by making them more liable for it than if they had simply turned a blind eye."[96] The two contrasting provisions of Section 230—platforms are not liable for putting material up *or* for taking it down—represent the balancing act involved in maintaining an environment valuing free expression while also minimizing censorship.

There has been much debate over the years about Section 230: whether it is currently beneficial or harmful to the character of the internet; whether it has been interpreted too broadly; whether it affords platforms undue "power without responsibility"[97]; whether its broad immunity eliminates the incentives for platforms to prevent online bullying, criminal activity, or stem the spread of disinformation and hate speech—or removes the cost for not doing so. However, as Citron and Franks have noted, "Congress cannot fix what it does not understand."[98] The confusion and lack of knowledge about Section 230 among legislators, along with the broad disagreement over the law's potential modifications and effects, have led to congressional inertia, reinforcing the status quo and the default embrace of "the devil you know."

The contours of Section 230 are deeply rooted in and connected to pipeline policy. As with the telephone and even the telegraph, the carrier of speech is not liable for the content of that speech. Congress wrote in the original conference report that the Communications Decency Act was intended to "modernize the existing protections against obscene lewd, indecent or harassing uses of a telephone."[99] Such protections, however, were meant and enacted for the infrastructure providers, not their users. Accordingly, platforms inherited the critical separation of content and conduit, along with the attendant liability protections, from the foundation of pipeline principles that were enshrined in Section 230 of the CDA.

Section 230 has been called a "uniquely American law"[100] as the primacy of free speech is ingrained in the country's origin story and valued to the extent that it often takes precedence over almost all other rights, including privacy. Section 230 also offers a level of immunity unavailable to online intermediaries in most other countries.[101] The designation of platforms as neutral intermediaries no longer holds in the EU after the Right to Be Forgotten was won in 2014. And yet, Section 230's legal invulnerability has not lived up to the hype. Julie Cohen has argued that, while Congress may have proclaimed the lack of liability for platforms "would foster and preserve the emerging network as a vibrant marketplace of ideas," such narratives and justifications "are premised on assumptions about the affordances of media infrastructures that no longer hold. Platform-based, massively-intermediated processes of search and social networking are inherently processes of market manipulation."[102] This is undeniable, as business models based on unaccountable data harvesting, targeted advertising, and user surveillance have created an environment where the "free market" and "consumer choice" are nothing more than fictive marketing constructions. Google executives have themselves admitted that their "system as design[ed] doesn't really give the user choice."[103]

Content moderation is a primary feature of social media platforms in order to deal with the proliferation of disinformation, hate speech, and online bullying, as well as violent and disturbing graphic images. Gillespie (2018) has detailed the various self-regulation practices that platforms use to combat such material, whether through their terms of service agreements, community guidelines and community flagging, AI detection software and filters, or human content moderators, none of which are without major flaws. Sarah Roberts has written extensively about the human element of content moderation—the globally outsourced workforce conducting the physically and psychologically grueling labor of screening the worst humanity has to offer in order to enforce platform policies.[104] All of these mechanisms contribute to delineating the acceptable boundaries of speech, creating the values of the digital public sphere, and determining what content should be removed and why. We are reminded by Zuboff, however, that these platforms where such moderation takes place are not so much a public sphere any longer as they are private domains "governed by machine operations and their economic imperatives, incapable of, and uninterested in, distinguishing truth from lies or renewal from destruction."[105]

It should therefore not be surprising that the dominant platforms are still the main hosts and amplifiers of all the above-mentioned types of harmful speech, including campaigns of propaganda, fraud, and violent extremism directly targeting the democratic process. These include the fake news and disinformation that permeated social media during the 2016 US election via accounts set up by the Russian state's Internet Research Agency; the 2018 Cambridge Analytica scandal in which Facebook data of eighty-seven million users were sold to a Trump campaign consulting firm trying to manipulate voters[106]; and the planning and promotion related to the January 6, 2021 attack on the US Capitol by Trump supporters attempting to stop the counting of electoral votes to certify the 2020 presidential election.[107] In light of this ever-present dynamic online, platforms have become key infrastructure for the sociotechnical systems identified by Philip Howard as "lie machines," spreading disinformation, propaganda, and a "global economy of political lies" in the service of ad revenue generated by engagement.[108]

Facebook and Twitter both began fact-checking and labeling Donald Trump's lies on their platforms in 2020 as the COVID-19 pandemic dragged on and the presidential election drew near. They ultimately banned Trump after he used their platforms to praise the rioters at the January 6 insurrection and perpetuate his fraudulent claims about the 2020 election results. Posts containing incorrect information about COVID-19 were also flagged for violating misinformation policies (but not removed), along with false claims about COVID-19 vaccines. YouTube indefinitely suspended Trump's account as well. Across social media, political ads underwent new scrutiny or were prohibited altogether (as on Twitter), and algorithms were purportedly "tweaked" to be more attuned to disinformation. Facebook also established an independent Oversight Board in 2020. These and other policy changes were part of an industry-wide effort to install self-regulatory mechanisms to avoid potentially harsher ones imposed on them by the state—a media industry tactic as old as the Motion Picture Producers and Distributors of America's 1927 list of "Don'ts" and "Be Carefuls," which codified images, subjects, and words that Hollywood films should voluntarily avoid in order to forestall federal regulation that could potentially be much worse.[109]

Congressional momentum has been building to enact Section 230 reform and impose accountability measures for the dominant Big Tech platforms.

a

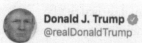

Donald J. Trump ✓
@realDonaldTrump

These are the things and events that happen when a sacred landslide election victory is so unceremoniously & viciously stripped away from great patriots who have been badly & unfairly treated for so long. Go home with love & in peace. Remember this day forever!

⚠ This claim of election fraud is disputed, and this Tweet can't be replied to, Retweeted, or liked due to a risk of violence

6:01 PM · Jan 6, 2021 · Twitter for iPhone

b

Donald J. Trump ✓
@realDonaldTrump

700,000 ballots were not allowed to be viewed in Philadelphia and Pittsburgh which means, based on our great Constitution, we win the State of Pennsylvania!

⚠ Election officials have certified Joe Biden as the winner of the U.S. Presidential election

7:28 PM · Nov 13, 2020 · Twitter for iPhone

81.1K Retweets **23.6K** Quote Tweets **411.7K** Likes

Figure 2.1
(a): January 6, 2021, tweet by Donald Trump. (b): November 4, 2020, tweet by Donald Trump. (c): November 2, 2020, Facebook post by Donald Trump. (d): November 7, 2020, tweet by Donald Trump.

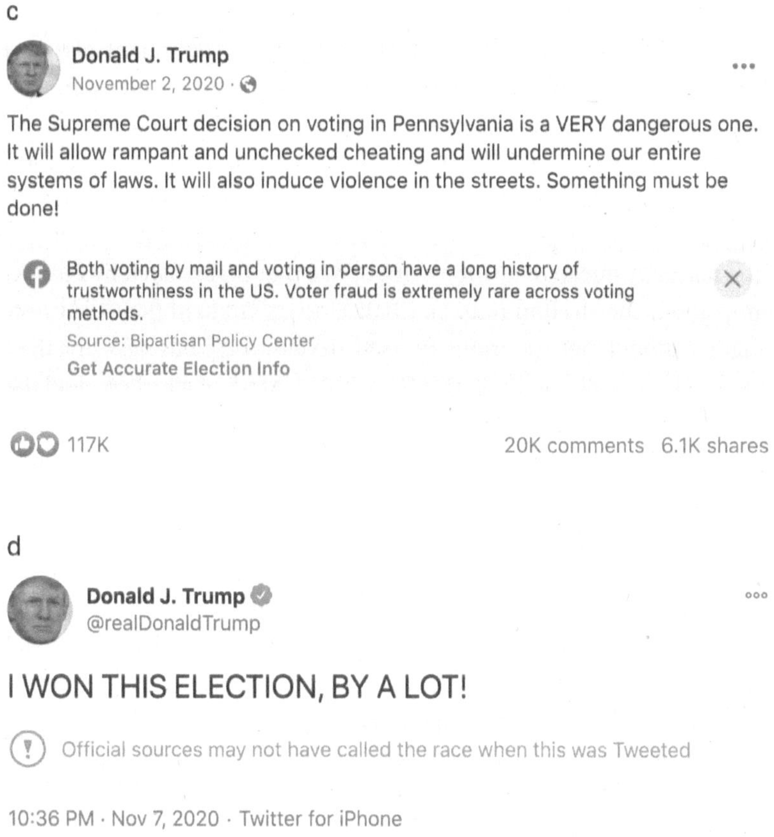

c

Donald J. Trump
November 2, 2020 ·

The Supreme Court decision on voting in Pennsylvania is a VERY dangerous one. It will allow rampant and unchecked cheating and will undermine our entire systems of laws. It will also induce violence in the streets. Something must be done!

 Both voting by mail and voting in person have a long history of trustworthiness in the US. Voter fraud is extremely rare across voting methods.
Source: Bipartisan Policy Center
Get Accurate Election Info

👍❤️ 117K   20K comments 6.1K shares

d

Donald J. Trump ✓
@realDonaldTrump

I WON THIS ELECTION, BY A LOT!

⚠️ Official sources may not have called the race when this was Tweeted

10:36 PM · Nov 7, 2020 · Twitter for iPhone

Figure 2.1
(continued)

Representative Sheldon Whitehouse is one of many who has argued that Section 230 has not kept pace with the operating practices of social media giants, noting that "its unprecedented immunity protection has grown, as courts allowed Section 230 to shield a range of illicit corporate behavior unimaginable at the time of its passage."[110] The advocacy organization Ranking Digital Rights has further emphasized that, in addition to the affordances of Section 230, lawmakers should also be addressing the fundamental issue of the "underlying technical infrastructure and business models that have created an online media ecosystem that is optimized for the convenience of advertisers rather than the needs of democracy."[111]

On the heels of the 2020 election, the US Senate convened hearings to ostensibly address Section 230 and the behavior of dominant platforms. However, this event did little more than expose the profound dysfunction in Congress, with Republicans admonishing the CEOs of Facebook and Twitter for allegedly censoring conservative viewpoints on their platforms, and Democrats blaming their colleagues for "working the refs" and purposely distracting from the core issues related to Section 230 reform (such as the rampant spread of disinformation, like that perpetuated by President Trump about the election results). Chair Lindsay Graham (R-SC) suggested Congress should "let the industry itself develop best business practices." Senator Richard Blumenthal (D-CT) scolded Mark Zuckerberg and Jack Dorsey for building "terrifying tools of persuasion and manipulation" and catastrophically failing the American public at their "civic and moral responsibility to ensure these instruments of influence do not irreparably harm our country." But Blumenthal's true ire was saved for his Republican colleagues, as he disparaged the hearing as an unserious "political sideshow—a public tar and feathering" by GOP senators who chose to ignore threats to democracy in favor of cynical political theater.[112] Again, talks stalled out as Congress turned on itself, and true Section 230 reform was back to being a mere thought experiment.

The Supreme Court heard two cases about the limits of Section 230's immunity in 2023: *Twitter, Inc. v. Taamneh* and *Gonzalez v. Google*[113], neither of which did anything to assist those reform efforts beyond perhaps putting more pressure back onto Congress to write new legislation. The cases involved allegations by families of terrorism victims that Big Tech platforms were in some way responsible for the attacks that killed their loved ones—either through failing to police their own harmful content (*Twitter, Inc. v. Taamneh*) or through acting as a recruiting platform for ISIS by virtue of their recommendation algorithm (*Gonzalez v. Google*). The court ruled on behalf of the defendants, the tech companies, in both cases. In essence, the justices sidestepped a direct confrontation with the limits of Section 230's immunity and declined to narrow the scope of its liability. With that, the onus was returned to legislators to grapple with the growing calls to address Section 230, particularly given the expanding role of artificial intelligence in content generation and the myriad attendant legal questions that has raised.

The EU Digital Services Act Package

Outside the US, countries have begun to expect more accountability from digital platforms. In March 2022, the EU passed the Digital Markets Act (DMA) and Digital Services Act (DSA) in one of the world's most comprehensive legislative attempts to rein in the power of Big Tech. The Digital Markets Act is concerned with competition in digital markets and applies to "gatekeeper platforms" with a market value of more than 75 billion euros, such as the dominant platforms run by Alphabet (Google and YouTube), Amazon, Apple, Meta (Facebook), and Microsoft. European Commission President Ursula von der Leyen explained that, at its core, the DMA package embraces the philosophy "the greater the size, the greater the responsibilities of online platforms."[114] Among the DMA's many significant effects, Apple was forced to open up its app store and allow other sellers and billing systems into its ecosystem; Amazon is not allowed to use data collected from outside sellers in order to offer competing products; and Meta and Google are prohibited from using targeted ads without consent.[115]

The Digital Services Act is focused on consumer protection and user empowerment, platform accountability, and transparency. It applies to online intermediaries including ISPs, cloud hosting services, large search engines, app stores, social media, e-retailers, and platforms with a reach of more than 10 percent of the 450 million consumers in the EU. The DSA updates the 2000 European Union's e-Commerce Directive and its position of "near-absolute broad immunity for online intermediaries," setting new standards of platform accountability for illegal and "harmful content."[116] Together, the DMA/DSA package combines efforts to protect freedom of expression, competition, and privacy while also combating the spread of disinformation, recognizing that these rights and protections are indelibly entwined. Violators face penalties of up to 20 percent of their global revenue for repeat offenses. For some, this could reach into the tens of billions of dollars.[117] EU regulators frequently speak of legislation limiting the powers of Big Tech as upholding "the respect of human rights, freedom, democracy, equality and the rule of law." They demonstrate that it is possible to have values beyond profit inscribed in the vision for platform governance, and to have a comprehensive regulatory approach to curbing platform power in society.

It was in this spirit that Germany enacted the Network Enforcement Act (NetzDG) in 2018. Also known as the "Facebook Act," NetzDG holds social media platforms with at least two million German users partially liable for the content they carry. Jens Pohlmann notes that NetzDG represents the first time that a Western liberal democracy has enacted a law against unlawful speech that undermines the doctrine of US intermediary liability law (Section 230), as under this act platforms are partially held liable for user-generated speech.[118] It is an anti-hate-speech law that is still quite controversial and carries high fines for platforms that fail to remove illegal content from their sites. It was the only legislation to make public transparency reporting mandatory for major platforms until Europe's DSA instituted its own rules on algorithmic transparency in 2022.[119] Terry Flew argues that NetzDG marked "the clear existence of new forms of platform governance that are led by nation states and imposed upon digital platform companies."[120] The GDPR, as well as the DMA package, are similarly state-led, although they are imposed by regional bodies. The US has yet to follow suit.

Finally, it is important to recognize the role of civil society and advocacy organizations in this arena. These groups advocate to policymakers, to the courts, and to Big Tech companies on behalf of the public, while also mounting legal challenges, educating consumers, and continuing to fight for some issues over many years, even decades in some cases.[121] They also form coalitions, such as those found in the collectives of academics, journalists, and advocacy groups from all over the world that wrote the Manila Principles on Intermediary Liability (2015) and the Santa Clara Principles on Transparency and Accountability in Content Moderation (2018).[122] Together, these documents offer examples of robust public interest frameworks for best practices and standards in content moderation, platform accountability, and transparency. One of the authors of the revised Santa Clara Principles, the Ranking Digital Rights (RDR) organization, also generates a Corporate Accountability Index that ranks digital platforms and telecommunications companies based on their policies and commitments to freedom of expression and privacy.[123] Their 2020 report, headlined "Companies Are Improving in Principle, but Failing in Practice," reported that zero of the twenty-six companies they rank (which include all of the dominant global platforms) "came even close to earning a passing grade on our international human rights-based standards of transparency and accountability."[124]

The work of these groups is critical for the future of equity in cloud policy. They inject a perspective into the process that often escapes the narrow focus of most policymakers, such as the outsized importance of digital privacy and secure communication to marginalized communities, human rights defenders, and journalists, for example. They balance "incrementalism with long-term vision," as president and CEO of the Center for Democracy & Technology Alexandra Reeve Givens has described the balance of navigating the tension between the small wins and the larger goals for the future.[125] And they do so by maintaining a steadfast, almost impossible measure of optimism regardless of the setbacks and challenges they face. Ultimately, they light the way toward more just and democratic policy for all cloud infrastructure.

Antitrust and Market Competition

There is truly nothing new under the sun when it comes to monopolists in America; platforms are simply the most recent iteration. The negotiation of the threat posed by industrial consolidation to democratic forms of governance is a story that includes chapters set in the American Revolution and its aftermath. Thomas Jefferson proposed a Bill of Rights with twelve constitutional amendments, including one that would ban commercial monopolies like the transnational British East India Company that triggered the Boston Tea Party. That effort notably failed, and the public continues to pay the price centuries later. The railroads of the industrial revolution forced one of the country's earliest and most dramatic reckonings with monopoly power. In a pioneering journalistic exposé of the Standard Oil Trust in the late nineteenth century, H. D. Lloyd wrote, "Our treatment of 'the railroad problem' will show the quality and calibre of our political sense. It will go far in foreshadowing the future lines of our social and political growth. It may indicate whether the American democracy, like all the democratic experiments which have preceded it, is to become extinct because the people had not wit enough or virtue enough to make the common good supreme." He concluded, "The time has come to face the fact that the forces of capital and industry have outgrown the forces of our government."[126]

These words, written almost 150 years ago, still haunt cloud policy today. Platforms are now shining examples of the political perils Lloyd foretold. Google currently has a 90 percent share of the global search market and two-thirds of the world conducts its browsing on Google Chrome, allowing the company to dominate digital information access and the vast online advertising market that goes with it. In the US, the Amazon webstore takes in one of every two dollars spent online, and the company regularly uses data from

Figure 2.2
Joseph Keppler, "The Bosses of the Senate," 1889.

its third-party sellers to unfairly undercut competition and benefit its own private-label and retail businesses.[127] Apple now has over 50 percent of the global smartphone market and runs the world's largest mobile app store, with an 80 percent market share, taking a 30 percent cut of all transactions. All the while, platform CEOs (and members of Congress whom they fund and support) repeatedly invoke the specter of "chilling innovation" at the mere mention of regulatory investigations, knowing full well their own acquisitions, investment strategies, algorithms, and business practices strangle market and technology innovation by design.[128] While the European Union has levied major fines and structural remedies against Big Tech for their anticompetitive behavior, as of this writing US agencies and legislators have yet to impose anything close to such consequences, all while the dominant platforms have expanded their market control to the point that they are now apparently too big to curtail.

In many ways they are reminiscent of the "economic royalists" carving up industrial dynasties that FDR referred to in his 1936 renomination acceptance speech. The US had just endured decades of struggle against monopolies and trusts that controlled the railroads, banking, tobacco, steel, and oil industries and utilized their money and power to influence policy. As Roosevelt spoke to an audience about the freedoms won from eighteenth-century monarchs, he warned the public of new corporate kingdoms and industrial dictatorships being built on the concentration of wealth and market dominance. He lamented the will to power of these "privileged princes" and their dangerous incursion into all aspects of life as they "reached out for control over Government itself" and succeeded. Roosevelt also addressed the human cost of this new social order, as industrial magnates grew rich on the backs of the people, and yet, for the workers, "opportunity was limited by monopoly. . . . The political equality we once had won was meaningless in the face of economic inequality." In the end, Roosevelt recognized this fight against economic tyranny as "a war for the survival of democracy."[129] I invoke this history to emphasize that not only does the inequity of the platform economy have ample precedent—comparisons of Big Tech to the "robber barons" and trusts of the Gilded Age are now commonplace—but that it also comes with centuries of warning about the links between political and economic equality, and the critical role played by antitrust enforcement to the health of a democratic society.

Antitrust enforcement was functioning at its peak from the Progressive Era through the 1960s. This was a time when the importance of market competition was ideologically related to the preservation of social equality, and neoliberalism had not yet ascended in the realms of policy and governance. President Kennedy's antitrust chief, Lee Loevinger, had worked briefly under Thurman Arnold, who had busted trusts in the New Deal era. He said in his initial meeting with Robert Kennedy about the job, "I believe in antitrust almost as a secular religion."[130] In testimony before Congress, Loevinger said that "the problems with which the antitrust laws are concerned—the problems of distribution of power within society—are second only to the questions of survival in the face of threats of nuclear weapons in importance for our generation."[131]

From the mid-twentieth century until the late 1970s, the modern courts and regulatory agencies were guided primarily by tenets of what was known as the Harvard School of antitrust. This approach was skeptical of mergers and rather intolerant of concentrated markets, and its guiding economic theories "assumed that firms with market power would act in an anticompetitive manner."[132] During this time, anti-monopoly law was used to block mergers of corporations that had just 5 or 6 percent of market share[133]—for perspective, when Facebook bought Instagram in 2012, it was said to control 95 percent of social media in the US.[134] The Chicago School began to take hold in the 1980s and has kept a firm grip on regulators and the courts ever since, eventually enabling the current state of the platform market. The Chicago School transformed the approach to antitrust enforcement, in large part because of a pivotal law journal article written by Robert Bork in 1966. In it, Bork argued that that the sole intent of federal antitrust law was actually to promote economic efficiency and the "maximization of wealth" through what he called "consumer welfare."[135] By this, he meant lower prices and increased output of goods and services, or, as he termed it, "consumer want satisfaction." Economic and legal experts have addressed how damaging the ideology underlying the "consumer welfare" standard and its near-fanatical embrace by regulators has been for citizens, for market competition, and for democracy. Tim Wu for one has written about how Bork's "radically narrow reading of the Sherman Act threw out the broader concerns that had long animated the Act and its enforcement, . . . that antitrust represented a democratic choice of economic structure and a check on the political and economic power of the monopolies."[136] Matt Stoller has argued that, according to Bork, "Congress intended the Sherman Act . . . not as a means of protecting democracy, or markets, or the rights of citizens to produce and exchange free from interference by a monopolist. The only thing antitrust was meant to do was get consumers more stuff."[137]

Coinciding with the rise of the Chicago School, the US elected President Ronald Reagan, who appointed antitrust critic William Baxter as his Antitrust Division chief at the Department of Justice. The administration issued new merger guidelines in 1982 that represented a dramatic change from the previous set put forth under President Lyndon Johnson in 1968. Waves of industrial concentration followed, along with conservative federal judge appointments, beginning an era of renewed tolerance in the political climate for mergers and monopolies. Gone was the use of antitrust as a tool to restrain corporate misbehavior and safeguard competition. Instead, America became "the land of the big and the home of the consolidated."[138] It should come as no surprise that, as antitrust enforcement has declined, income inequality has increased—to the point that the top 1 percent now owns nearly one-third of the wealth in the United States, and this share continues to grow, trending back to levels of wealth concentration not seen since the late nineteenth century.

President Biden's FTC Chairwoman Lina Khan rose to prominence writing about the modern collapse of antitrust—specifically how the consumer welfare standard "fails to capture the architecture of market power in the twenty-first century marketplace."[139] Khan focused on Amazon as the paradigmatic example of this failure because of its dominant role in multiple markets, including acting as the distribution infrastructure for its rivals. To this point, Khan has compared the company to the railroads, noting, "The thousands of retailers and independent businesses that must ride Amazon's rails to reach market are increasingly dependent on their biggest competitor." However, even she admits that "Amazon is not the problem—the state of the law is the problem, and Amazon depicts that in an elegant way."[140] Matt Stoller also noted that Amazon has participated in a type of vertical integration that even the railroads were ultimately prohibited from because the company "exploded in a legal environment crafted by Bork, where vertical integration was a signal not of monopolization but efficiency."[141]

In an attempt to address these issues, the House Judiciary Antitrust Subcommittee investigated Amazon, Apple, Facebook, and Google for thirteen months in 2020–2021, obtaining more than 1.3 million documents in the process. These platform companies were accused of a wide range of anticompetitive abuses "concerning the extent to which they have exploited, entrenched, and expanded their power over digital markets" in retail, social networking, advertising, search, and apps.[142] The committee questioned the platform's CEOs: Jeff Bezos, Tim Cook, Mark Zuckerberg, and Sundar Pichai of Alphabet, two of whom are the world's richest humans. These four collectively run businesses that were worth over $5 trillion at the time, yet they spent much of their testimony arguing that they were not all that powerful. Gigi Sohn, former senior adviser at the FCC, wrote that this felt like the tech platforms' "Big Tobacco moment,"[143] recalling the 1994 congressional hearings when executives from the seven largest tobacco companies testified that they did not think cigarettes were addictive.

In their 450-page report, the committee argued that these platforms serve as gatekeepers over key channels of distribution and control the "infrastructure of the digital age" along with access to markets; utilize surveillance and their "data advantage" to limit, buy out, cut off, and destroy their competition; and abuse their position as intermediaries to entrench and expand their dominance through self-preferencing, predatory pricing, and/or exclusionary conduct.[144] Facebook's $1 billion purchase of Instagram in 2012 was one example used to demonstrate these rapacious practices. Facebook recognized the nascent threat Instagram was beginning to pose externally, so it bought out the company. Facebook then further engaged in anticompetitive practices characterized by one former employee as "collusion, but within an internal monopoly" in order to "position Facebook and Instagram to not compete with each other" or cannibalize one another.[145] Amazon's role in online retail was another frequently cited example, particularly with respect to its treatment of third-party sellers on its platform.

The committee characterized the CEOs' answers as "often evasive and non-responsive, raising fresh questions about whether they believe they are beyond the reach of democratic oversight."[146] It is hard to believe that such questions still exist, given Big Tech's near-unanimous adherence to the gospel of Peter Thiel, one of Silicon Valley's most successful entrepreneurs, who has proudly proclaimed "competition is for losers" and "a relic of history," while "monopoly is the condition of every successful business."[147] Since 1998, Google, Amazon, Apple, and Facebook have collectively purchased more than five hundred companies without a single acquisition blocked by US regulatory agencies.[148] Google alone has acquired 270 companies between 2001 and 2021, including Android, Nest, YouTube, and Waze. Between 2011 and 2021, Microsoft made more than one hundred acquisitions, with companies such as Activision Blizzard, Skype, Nokia Devices, LinkedIn and GitHub now part of their corporate holdings. Amazon has gone on a similar spending spree and now owns MGM Studios, iRobot, Ring, Twitch Interactive, and Whole Foods, among more than a hundred other purchases. Facebook (Meta) has acquired nearly a hundred companies, including numerous startups, AI companies, and larger potential competitors such as Oculus VR, Instagram, and WhatsApp.[149] These purchases have shown no signs of slowing down.[150] This is not the behavior of an industry concerned about the reach of democratic oversight. The committee's concluding commentary on these developments was stunningly obvious: "It is unclear whether the antitrust agencies are presently equipped to block anticompetitive mergers in digital markets. The record of the Federal Trade Commission and the Justice Department in this area shows significant missteps and repeat enforcement failures. While both agencies are currently pursuing reviews of pending transactions, it is not yet clear whether they have developed the analytical tools to challenge anticompetitive deals in digital markets."[151] This was government-speak for what was already apparent: antitrust is much too slow as a remedy, and is running decades behind market conditions.

To the committee's credit, the report's final recommendations were substantial, albeit ignored. They included structurally separating dominant platforms (i.e., "breaking them up"); requiring rules for nondiscrimination, interoperability, and data portability; discouraging the acquisition of potential rivals and nascent competitors by dominant platforms; and instituting specific measures to strengthen antitrust laws and revive antitrust enforcement. The committee also addressed the "core conflict of interest" faced by the dominant platforms in their function as critical intermediaries that compete with rivals using their services, noting that the collection and exploitation of surveillance data from their competitors to enhance their own dominant position is a threat to the digital economy. Therefore, the committee recommended congressional legislation that prohibits dominant platforms from competing in the markets dependent on their infrastructure, and limiting the markets in which dominant platforms can engage. This was based on precedents found in late nineteenth-century and mid-twentieth-century legislation related to railroads and finance, respectively. They also referenced the broadcast television industry that was once subject to the "fin-syn rules" (the 1970 Financial Interest and Syndication Rules) preventing networks from competing in both the production and syndication markets simultaneously in order to curb their anticompetitive practices and dominant control of the medium.[152]

Thus far, Big Tech's success record against antitrust regulations in the US has been an epic failure for consumers (see appendix). The dominant platforms have been investigated by the Department of Justice, the Federal Trade Commission, nearly every single state attorney general, the US Congress, the Securities and Exchange Commission, and a long list of European regulators. As of 2023, Puerto Rico, the District of Columbia, and Guam, plus every US state except Alabama, was suing Google over its search business. Facebook was fined $5.1 billion for charges related to the Cambridge Analytica scandal in 2019 and is being investigated by the FTC along with Amazon. The *U.S. v. Google* case launched in 2020 was the first antitrust case brought against Big Tech since the Department of Justice went after Microsoft in 1998.[153]

However, *none* of the antitrust investigations against Big Tech in the past five years have yet to result in a single action against the dominant platforms—except those brought in the EU, where they have been fined billions of euros for anticompetitive practices. Moreover, at least five antitrust bills have been written in the US to regulate these companies and their anticompetitive practices in 2021–2022 alone, and zero were put forward for a vote.[154]

The intractable partisan dysfunction in Congress was intensified by the millions of dollars that tech companies spent lobbying against these bills while they were still in committee. Much of this money funded cynical disinformation campaigns playing on the ignorance of voters, claiming that the bills would lead to greater dependence on China, weaken US technology, and threaten personal and national security. Behind many of these ads, op-eds and "commissioned studies" is the Facebook-backed company American Edge, which is funded by the social media platform to attack antitrust legislation in Washington, DC, and disguise Facebook's participation in such efforts. Often the company funneled money to other sympathetic groups to create the appearance of grassroots opposition to antitrust regulation.[155] In addition to being common practice in industries like pharmaceuticals, telecommunications, and tobacco, this strategy is one more that connects Big Tech to the trusts of the Gilded Age, as the railroads were frequent practitioners of such propaganda campaigns. Former FCC Chairman Tom Wheeler has argued, "As the railroads fought regulation, there was no emotion-generating tactic that was too low. The railroads even played to racism by suggesting that regulation would prevent them from segregating African Americans in Jim Crow railcars."[156] While the railroads' campaigns ultimately failed and led to legislation that provided nearly a century of protection against monopolistic behavior in industry, a contemporary revamped approach to antitrust is long past due. The digital platform economy has proven that the current foundation is well beyond its expiration date.

President Biden began to turn attention and resources toward reining in Big Tech with anti-monopoly hires such as Lina Khan to run the FTC, Columbia Law professor Tim Wu as an adviser on technology and competition policy, and Jonathan Kanter as assistant AG for antitrust at the Department of Justice. The FTC further signaled in 2021 that they wanted to move into regulating "discriminatory AI" with a post on their website that included the warning about algorithmic performance: "Hold yourself accountable—or be ready for the FTC to do it for you."[157] Biden also notably signed an expansive executive order in July 2021 related to competition in numerous industries, including health care, finance, farming, airlines, and telecommunications. In it, the president urged the FTC to write its own rules protecting consumers from data surveillance, linking the issues of privacy and competition in a federal vision for regulation for the first time, as opposed to isolating them in separate orders. This holistic approach was supported by a former head of advertising at Google who spoke to these interrelated dimensions of policy when he said that "competition in tech is needed to ensure people are able to have private online experiences, because large companies like Google will never truly care about user privacy."[158]

The 2021 executive order was also a rare formal recognition that personal data is a commodity that fuels the platform economy at the public's expense, and the control over that data—which effectively denies individual privacy—is directly linked to competition in that space. Labor scholar Nelson Lichtenstein said of Biden's order that it recalls "the great antimonopoly tradition that has animated social and economic reform almost since the nation's founding. This tradition worries less about technocratic questions such as whether concentrations of corporate power will lead to lower consumer prices and more about broader social and political concerns about the destructive effects that big business can have on our nation."[159] In the platform ecosystem, antitrust and competition are intrinsically linked not only to individual privacy but also to freedom of speech, access to information, and the maintenance of a surveillance culture. These broader concerns of private monopoly power in the arena of civil liberties grow increasingly dire as dominant platforms cement their control over "the necessaries of life."

The Survival of News

The social costs of these systemic failures of antitrust additionally extend to the public's ability to access quality, diverse sources of news and information, one of the hallmarks of a democratic society. Roughly two-thirds of advertising expenditures are now digital, leaving the news industry in freefall. Facebook/Meta and Google control roughly 50 percent of the US ad revenue and the market infrastructure itself. In 2021, Google's parent firm Alphabet made $257.6 billion, 92.5 percent of which came from its advertising businesses.[160] These companies also own intermediaries such as Google's DoubleClick for Publishers, Ad Exchange (AdX), and DV360 that help negotiate advertising sales and set prices.[161] When Google purchased DoubleClick in 2008 for $3.1 billion after a bidding war with Microsoft, it bought the biggest ad server in the market in a merger approved by the FTC with no conditions. The agency noted there was no evidence to support theories of potential competitive harm and they had no concerns about the effects on competition.[162] The commission's inability to connect the anticompetitive dots between Google's advertising business and its own websites that buy and compete for ad dollars was laughable, and remains a savage indictment of the "consumer welfare" standard that US antitrust enforcement has followed since the 1980s. Almost immediately, Google predictably began restricting its competitors' access to information and privileging its own properties in the market, and by 2020, Google was selling 85 percent of the advertising it brokered to itself.[163]

This funneling of revenue that was once the lifeblood of the newspaper industry to the coffers of two Big Tech platforms has led to catastrophic layoffs, consolidation, and the creation of "news deserts" across the US. In the 1930s, newspapers experienced a similar existential threat from another form of "new media." With the arrival of widespread commercial radio broadcasting, the monopoly position in information gathering and distribution that they had enjoyed since the colonial era was under attack. In what would become known as the "Press-Radio War," broadcasters began to siphon off a dramatic portion of the advertising revenue that was once the sole domain of print.[164] At first, most newspapers fervently fought against the threat from broadcast journalism. However, Robert McChesney has argued that attempts to engage the general public in a debate about commercial media reform and its importance to the public interest were "a resounding failure."[165] The press lost its war against radio but nevertheless endured, forced to share economic, cultural, and political power with a new rival medium. The current threat from digital platforms has proven even more destructive to the survival of news itself. It has devastated local journalism and left most communities in the dark about the information that most directly impacts their lives, and without a check on accountability for civic leaders and elected officials. Since 2005, roughly 2,200 local papers have shut down, and the number of newspaper journalists fell by more than half between 2008 and 2020.[166] Presently, more than 2,000 of the 3,143 counties in America—*nearly two-thirds*—have no daily newspaper, a critical democratic safeguard that is no longer part of the societal fabric for millions of citizens.[167]

To compound the problem, many of these citizens turn to social media platforms instead, with about half of Americans getting at least some of their news on social media, and nearly one-third from Facebook.[168] Unfortunately, on such platforms, algorithms reward "engagement"—what Zuboff has called "a euphemism used to conceal illicit extraction operations"[169]—which is most effectively generated from content that is salacious, hyperpartisan, emotional, and most often politically extremist and false.[170] On Twitter, lies were found to travel faster than the truth.[171] When Facebook overhauled its algorithm in 2018 to increase engagement, it was shown to "reward outrage and lies" and misinformation was "inordinately prevalent among reshares."[172] In fact, Facebook researchers discovered that "publishers and political parties were reorienting their posts toward outrage and sensationalism. That tactic produced high levels of comments and reactions that translated into success on Facebook."[173] Success in this case means more advertising revenue, essentially eliminating the incentive for fact-checking or moderating content in a way that negatively impacts profit.

Whistleblower Frances Haugen, a former data scientist for Facebook who filed complaints with federal law enforcement and testified before the US Senate in 2021, revealed how the platform's algorithm amplified misinformation and undermined public welfare. She shared thousands of documents with lawmakers, telling them, "I saw that Facebook repeatedly encountered conflicts between its own profits and our safety. *Facebook consistently resolved those conflicts in favor of its own profits*. The result has been a system that amplifies division, extremism, and polarization—and [is] undermining societies around the world." She continued, "Facebook became a $1 trillion company by *paying for its profits with our safety*," and stressed that "*Facebook's closed design means it has no oversight—even from its own Oversight Board, which is as blind as the public*. Only Facebook knows how it personalizes your feed for you. It hides behind walls that keep the eyes of researchers and regulators from understanding the true dynamics of the system." Haugen implored Congress to act quickly, emphasizing "Facebook chooses profit over safety every day—and without action, this will continue."[174] Algorithms built to leverage attention for advertising revenue are one of the more dangerous and destructive dimensions of social media's economic and cultural power. As of this writing, they remain totally unregulated and, as Haugen testified, "accountable to no one."

The growing spread of dis- and misinformation on dominant platforms such as Facebook, YouTube, and Twitter is part of a perfect storm. Along with the diminishing trust in news organizations and the concurrent collapse of the news industry's business model, it is a trifecta that has delivered yet another crisis for democracy borne of cloud policy. To be sure, the loss of trust in news organizations is not entirely unwarranted. The mainstream news is deeply flawed; it is regularly incapable of articulating complex systemic issues, often due to its own commercial biases and lack of resources, among other problems.[175] Moreover, half of all daily newspapers are now controlled by financial firms, such as "vulture hedge fund" Alden Global Capital, which owns roughly two hundred papers including the *Chicago Tribune*, the *Baltimore Sun*, and the *New York Daily News*. Alden is widely known for gutting newsroom staffs and treating newspapers like "any other commodity in an extractive business."[176] The ascendance of platform power alongside the evisceration of the news industry is a double blow for a democratic citizenry.

It is notable that the 2020 congressional report "Investigation of Competition in Digital Markets" included a formal recommendation to "Create an Even Playing Field for the Free and Diverse Press."[177] The report further related the lack of access to and availability of trustworthy news sources to the outsized power of Google and Facebook in digital advertising, which has left many news publishers at their mercy. To support them, the committee recommended a (temporary) safe harbor for news publishers and broadcasters to collectively negotiate with dominant platforms. This recommendation evokes the retransmission consent rules enacted as part of the 1992 Cable Act, which gave broadcast stations the right to negotiate compensation from cable companies that carried their signal. The rules were designed to mitigate the threat posed by the newly emerging cable industry to the viability of broadcasting, and correct "a system under which broadcasters in effect subsidize the establishment of their chief competitors."[178] Dominant platforms have replicated many of these same anticompetitive conditions in their role as distributors of news content, generating advertising revenue for themselves but none for the journalists or news organizations.

While the practice of requiring platforms to compensate news producers has yet to be translated to action in the US, Facebook and Google have fought against similar international initiatives and lost. As Terry Flew has written, in 2021 the companies "made explicit forms of power that had long been tacit in the media environment"[179] and threatened to shut down in

Australia if the government passed legislation that required them to negotiate payment with news media companies for their content. Australian public interest advocates noted, "When a private corporation tries to use its monopoly power to threaten and bully a sovereign nation, it's a surefire sign that regulation is long overdue."[180] After a federal investigation, Australia passed such a law in 2021, and now the News Media and Digital Platforms Mandatory Bargaining Code forces all platforms to pay publishers for their news. Following numerous related battles between Big Tech and news publishers in Europe, the European Union also adopted a new digital Copyright Directive in 2021 that affords news publishers the right to receive compensation from platforms that use their content. At the very least this begins the process of paying news organizations for their work that was previously just stolen by dominant platforms. It also redirects some of the advertising revenue that formerly supported professional journalism back to news producers.

Without legal directives such as those implemented in Australia and Europe, or a revised version of retransmission consent for the digital economy, the economic power of dominant platforms will continue to threaten the survival of news in the US. As a case in point, in 2022 Facebook decided not to renew contracts worth over $100 million with news publishers for featuring their content.[181] Taking this issue on directly, the Charleston, West Virginia, *Gazette-Mail* (circulation 28,000) filed an antitrust lawsuit against Google and Facebook for monopolizing the digital advertising market and the revenue that supports local news. The publisher argued that because of the vertical integration, market domination, and unlawful, anticompetitive practices of these tech giants, "the freedom of the press is not at stake; the press itself is at stake."[182] They further pointed to an alleged secret agreement between Google and Facebook (codenamed "Jedi Blue") in which the two former rivals in digital advertising were accused of unlawfully conspiring in a quid pro quo agreement to rig the market in their favor, illegally undermining competition and cementing their worldwide dominance in the industry.[183] What began with a tiny local paper taking on the two Goliaths of digital advertising has grown to include the actions of two hundred papers across the country seeking to receive back payment and future compensation along the lines of that offered to publishers in Australia. As these conflicts last and multiply, they are compounded by the current states of antitrust *and* the news industry. Understanding their interdependence is key to restoring them both.

Alternative Visions

Imagining life beyond our current platform ecosystem in which democracy and the public good are at the mercy of "an internet kleptocracy that profits from disinformation, polarization, and rage"[184] will depend on holistic policy thinking, and a multifaceted approach. Piecemeal solutions will not suffice. At this point nothing short of a regulatory revolution will shift platforms off their scorched earth practices. It will require taking governance regimes down to the studs and rebuilding from the bottom up. Remedies must include both structural and behavioral interventions. The business models of Big Tech's monopolistic platforms have contributed to the erosion of privacy, civil debate, and a healthy, diverse news industry. Reinvention further demands a dramatic reorientation of policy values guiding this infrastructure of the platformized public sphere, such as that offered by Nicholas Suzor's "digital constitutionalism," which would delineate new norms and expectations for this ecosystem along with enforcement measures for protecting the rights of its global users.[185] Inspiration for this project of value recovery can be found in the policy histories of broadcast, national parks, and even the formative pipelines of the cloud.

Legislative proposals for the formation of a new agency designed to regulate this sector of the economy have emerged. Among them is the Digital Platform Commission Act of 2022, which recommended establishing the first federal body to oversee and regulate digital platforms. In the bill's introduction, Senator Michael Bennet (D-CO) notes that "digital platforms remain largely unregulated and are left to write their own rules without meaningful democratic input or accountability." The bill articulates the ongoing and expanding threat to the public interest as platforms increase their control over access for civic engagement and economic and educational opportunities, as well as to government and public safety services. It included goals of preventing harmful levels of concentrated private power over critical digital infrastructure and preserving a competitive marketplace of ideas with a diversity of views at all levels of governance. The bill further pointed to the growing need for consumer protections and algorithmic transparency. The unregulated policies and operations of digital platforms leading to their role in abetting the collapse of local journalism, the dissemination of disinformation and hate speech, and the destruction of privacy enacted through data mining without informed consent were additional reasons offered in the argument for the creation of this agency.[186]

Recognizing that oversight of digital platforms "cannot simply be a replay of what worked in the industrial era," former FCC Chairman Tom Wheeler along with law and policy experts Phil Verveer and Gene Kimmelman have also advocated for the establishment of a "digital platform agency"—a regulatory agency with "digital DNA"—to comprehensively address the "inadequate public policy tools available to protect consumers and promote competition" in the platform space.[187] In conversations about what led to their idea for a digital platform agency, Wheeler has explained that policymakers tend to define today and tomorrow in terms of what they knew yesterday; indeed, he notes that the FCC's muscle memory dates back to the late nineteenth and early twentieth centuries. With this proposal, he and his coauthors are offering "a new brush to represent a new era" as embodied in this more agile, born-digital agency.[188]

To enact such reform, it will be essential to renew our view of how platforms can and should function in our society. In that spirit, Victor Pickard has put forth the idea of forging "a new 'social contract' between the platforms, regulators, and society writ large" that "clarifies the normative understanding that media firms' purpose is not merely to accumulate profit but also to support democracy."[189] Pickard looks back to earlier policy battles in broadcasting during the 1940s and reminds us of the lessons that endure from attempted reforms of that era: such remedies must be *structural*. Only by rejecting extractive commercial business models and their underlying capitalist logics will reinvention be possible. Philip Napoli has similarly proposed the idea of regarding dominant platforms as "public trustees"—a model traditionally reserved for the protection of natural resources but one that could instead be part of a regulatory overhaul for cloud policy. Much like broadcasters were designated to be public trustees of the airwaves in their capacity as stewards of a monopoly slice of the broadcast spectrum, Napoli suggests we might view the aggregate user data compiled by dominant platforms as a similar, collectively owned public resource. In exchange, platforms would assume responsibilities and public interest obligations, particularly with respect to content regulation.[190] Napoli's proposal takes the well-regarded idea of Balkin (2016) to treat platforms as "information fiduciaries" (one that carries an attendant legal obligation to be trustworthy, fair, and responsible to their end users[191]) and builds out those commitments to a more robust array of concerns.

Anne Applebaum and Peter Pomerantsev look to environmental protection as a model for regulating cloud infrastructure. They argue that, "to improve the ecology around a river, it isn't enough to simply regulate companies' pollution. Nor will it help to just break up the polluting companies. You need to think about how the river is used by citizens—what sort of residential buildings are constructed along the banks, what is transported up and down the river—and the fish that swim in the water."[192] They advocate collective solutions to these societal problems of tech governance for technology that is now "as integral to our lives and our economies as rivers once were to the emergence of early civilizations."[193]

Most regulatory alternatives have been offered in a more sector-specific manner, such as those focused on competition and antitrust remedies. The predominant calls for breaking companies up, such as forcing Facebook to sell Instagram, or requiring Google to divest YouTube, have gained some traction, but they have the slowest and most politically challenging path forward; recall that the AT&T breakup took ten years, and after thirteen years the antitrust case against IBM was abandoned. The last time the government (somewhat) successfully sued a tech company for antitrust violations was the 1998 case against Microsoft, and regulatory tolerance for corporate size and scale has only grown since. Further, many of those sympathetic to intervention nevertheless feel that *more* parties involved with dataveillance is not necessarily better, and that such a remedy does not address the foundational problem of platform's extractive business models.

There is also the more specialized tool of "structural separations," as advocated by current FTC Chairwoman Lina Khan, for platforms controlling a disproportionate share of online commerce, communications, and access to markets, while also functioning as economic and political gatekeepers that can thwart competition and stifle innovation.[194] Khan has called structural separations "common carriage's forgotten cousin," as they both aim to eliminate discriminatory behaviors of critical networks and infrastructure. Structural separations are limits that apply to businesses that operate in a market and also provide infrastructural service to their competitors, such as Amazon selling its own private label products in its Marketplace that hosts (and collects data from) millions of third-party sellers. Apple's app store is another prominent example—the company is currently under scrutiny in the EU after complaints from Spotify regarding Apple's discriminatory treatment and fees imposed on Spotify's app users.[195] Structural separations can prohibit companies from entering certain markets. They can also prevent dominant intermediaries from directly competing with the businesses reliant on their service, thus eliminating conflicts of interest and the threat of self-preferencing.[196] These restrictions have been applied to numerous industries in the past: railroads, which were not allowed to own the cargo they carried; telecommunications, including AT&T's 1956 consent decree preventing the company from entering any business beyond "common carrier communications" and the subsequent pipeline policy of "maximum separation" between data processing and communication services instituted in 1971 by the Computer I Inquiries; and the 1970 Financial Interest and Syndication Rules ("fin-syn") in broadcasting that prohibited television networks from producing or owning the syndicated programming that they aired, until the repeal of the rules in 1995.

While such approaches to antitrust enforcement were once common, they have since become historical curiosities. However, the the 2020 congressional report "Investigation of Competition in Digital Markets" did recommend structural separations for the dominant platforms, with the committee concluding,

> Through using market power in one area to advantage a separate line of business, dominant firms undermine competition on the merits. By functioning as critical intermediaries that are also integrated across lines of business, the dominant platforms face a core conflict of interest. The surveillance data they collect through their intermediary role, meanwhile, lets them exploit that conflict with unrivaled precision. Their ability both to use their dominance in one market as negotiating leverage in another, and to subsidize entry to capture unrelated markets, have the effect of spreading concentration from one market into others, threatening greater and greater portions of the digital economy. To address this underlying conflict of interest, Subcommittee staff recommends that Congress consider legislation that draws on two mainstay tools of the antimonopoly toolkit: structural separation and line of business restrictions."[197]

These proposals are facing a predictable torrent of pushback from decades of entrenched politicized standards and regulatory practices, as well as Big Tech lobbyists pouring millions into their efforts to influence legislators and future policy. Reclaiming individual privacy is confronted with similar hurdles. Solutions begin with federal privacy legislation that protects individuals as opposed to Big Tech platforms. Many involve data regulation as well, including the idea of individual control and/or ownership of one's own data, limits on the amount and types of data that can be collected, and compensation for the use and sale of personal data. So far, no jurisdiction has considered any of those as a legal solution, nor successfully articulated property rights for user data. Many have called for all platform default settings to be opt-in consent only. Keeping users perpetually uninformed about how their privacy is being invaded is another key strategy for platform companies, so widespread digital media literacy is also essential for creating change.

Some of that awareness can be taught in schools, but it also depends on widespread news and investigative reporting about these issues. Reinvigorating public media is crucial to that cause. Timothy Karr and Craig Aaron of Free Press have proposed creating a tax on targeted advertising to fund a public-interest media system "that places civic engagement and truth-seeking over alienation and propaganda."[198] They have further called on Congress to create a multibillion-dollar public interest media endowment in order to fund independent, noncommercial news outlets, protected from political interference and concentrating on local news, investigative reporting, and the creation of alternative platforms emphasizing diverse and underserved communities. Similar campaigns for taxes on dominant digital platforms to promote journalism have been launched in Brazil.[199] Public funding for journalism, particularly for local outlets, was also among the expert recommendations for Congress in the 2019 hearings, Online Platforms and Market Power.[200]

What unites these alternative visions for platform governance are the beliefs that the status quo is unacceptable, Big Tech's power has grown beyond reasonable measures, and more regulatory intervention is needed. Band-aids in the form of impermanent and politicized applications of antitrust law or agency regulation are insufficient; change can come only from universal structural reform. The project of imagining alternatives ultimately invites the articulation of more equitable cultural values for the evolving digital public sphere, and for the platforms serving as its primary host. This is a true test, as Mariana Mazzucato, et al. have noted: "Creating an environment that rewards genuine value creation and punishes value extraction is the fundamental economic challenge of our time."[201] If nothing else, the path forward must inject public values into the regulatory calculus, as cloud policy in the service of platform royalty is destroying the remaining promise of this critical infrastructure.

3 Data

> What we are opposed to, and what we are trying to perhaps resist, is the totally computerized man. . . . For what you are trying to seek as complete knowledge of a person, at the expense of this man's privacy, does in effect place him in a position of jeopardy at some future point.
> —Representative Cornelius Gallagher, House Hearings on Invasion of Privacy, 1965

> Our data wanders far and wide. Our data wanders endlessly.
> —Edward Snowden

The Church Committee was a Senate select committee convened in 1975 to investigate "a wide range of intelligence abuses by federal agencies, including the CIA, FBI, Internal Revenue Service, and National Security Agency."[1] It was chaired by Frank Church (D-ID) in the Senate. The investigation was prompted by the whistleblowing revelations of a former captain in army intelligence, Christopher Pyle, who delivered a long and detailed exposé about the army's vast domestic intelligence operations in 1970. "Intentionally or not," Pyle wrote, "the Army has gone far beyond the limits of its needs and authority in collecting domestic political information. It has created an activity which, by its existence alone, jeopardizes individual rights, democratic political processes, and even the national security it seeks to protect."[2] At the time, James Reston wrote about these developments and the exposure of Nixon's "enemies list" as the result of the fact that "the scientific capacity to use the arts of wartime espionage on private citizens has greatly expanded while the political capacity to control all this has actually declined."[3]

Along with other concurrent investigations, including the Pike Committee in the House led by Otis Pike (D-NY),[4] the Rockefeller Commission investigating "large-scale spying on American citizens by the CIA,"[5] and the Watergate hearings themselves, the Church Committee exposed widespread and long-standing intelligence abuses and government fraud in the service of political agendas. The reports detailed more than three decades of wiretapping, bugging, tax investigations, mail surveillance, and harassment campaigns directed at US citizens and peaceful protesters who were viewed as adversaries of those in power. The targets of this abuse included reporters, Supreme Court justices, congressional representatives, and student activists. So-called subversives such as Dr. Martin Luther King Jr., antiwar activists, and citizens involved in the "Women's Liberation Movement" were also on the list.[6] All of this was conducted "in the name of collecting intelligence about threats to national security."[7]

In the report, Senator Church lamented that "we have seen segments of our Government, in their attitudes and action, adopt tactics unworthy of a democracy, and occasionally reminiscent of the tactics of totalitarian regimes."[8] It detailed an exhaustive litany of surveillance against American citizens, even turning information about lawful activities into such a massive quantity of data that official pressure mounted to "use it against the target."[9] The report was a scathing indictment of the US intelligence community at large and their practices in the postwar era, which often resembled that of our foreign enemies' secret police forces. "Too many people have been spied upon by too many Government agencies and to[o] much information has been collected. The Government has often undertaken the secret surveillance of citizens on the basis of their political beliefs, even when those beliefs posed no threat of violence or illegal acts on behalf of a hostile foreign power."[10]

Moreover, in a chilling interview on *Meet the Press* in the summer of 1975, Church amplified his warnings, pointing to "a future in which technological advances 'could be turned around on the American people' and used to facilitate a system of government surveillance."[11] If the US continued down this path, he cautioned, "No American would have any privacy left, such is the capability to monitor everything: telephone conversations, telegrams, it doesn't matter. There would be no place to hide. . . . I know the capacity that is there to make tyranny total in America and we must see to it that this agency and all agencies that possess this technology operate within the law and under proper supervision so that we never cross over that abyss. That is the abyss from which there is no return."[12] This stands as one of the most prophetic statements in cloud policy's history.

As the Church Committee's lessons demonstrate, data has inherited significant legacies of policy from the analog era, just as platforms and pipelines have. These legacies were shaped by wars, geopolitics, technocultures, and generations of government corruption. Mostly invisible on the surface, they are nevertheless alive and well in this layer of infrastructure through which privacy rights and other digital civil liberties are further filtered and controlled. Along with pipelines and platforms, data shares the digital-era dichotomy of being subjected to dual regulatory regimes: formal policy such as legislation, case law, agency rulings, and global accords for data governance; and a growing host of "informal" policy such as end-user license agreements (EULAs), terms of service (TOS) agreements, and other rules embedded in code and design created by Big Tech providers. Accordingly, this chapter focuses in part on the tensions between public and private forces involved in the collection, access, and control of data in the cloud, following the discussion in chapter 2 of data extraction and monetization in the platform ecosystem. Here we also look to the ideological influences embedded in data's many historical pathways to illuminate the core values that have formed its contemporary policy frameworks.

(Im)materiality and (In)visibility

Data is a resource that has become the main currency of the platform economy, and the most significant immaterial infrastructure of the twenty-first century. The United Kingdom National Infrastructure Commission acknowledged in 2017 that "data is now as much a critical component of our infrastructure as bricks and mortar . . . and needs maintenance in the same way that physical infrastructure needs maintenance. It must be updated, housed and made secure."[13] It is housed in the material spaces of data centers, where it is stored, processed, and redistributed. The "black-boxing" of these infrastructures is rather extreme, rendering their operational details and most other information about the industry effectively invisible. As the joke goes, "The first rule of data centers is: Don't talk about data centers."[14] Very little is publicized about these complexes, "where the cloud touches the ground," for security purposes and other strategic reasons. One notable exception to this practice occurred with Google's 2012 public relations push to promote their data centers as visible, accessible, and environmentally friendly. Inviting the public to "come inside" and "see where the Internet lives," the

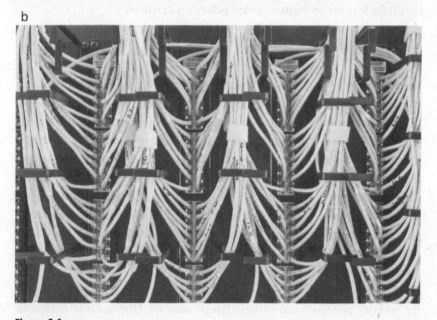

Figure 3.1
(a): Tape library backups, Google data center, Berkeley County, SC. (b): Ethernet switches in Google data center Berkeley County, SC. (c): Water pipes at Google data center Douglas County, GA. (d): Copper coils in Google data center in Singapore.
Credit: Google

Data

c

d

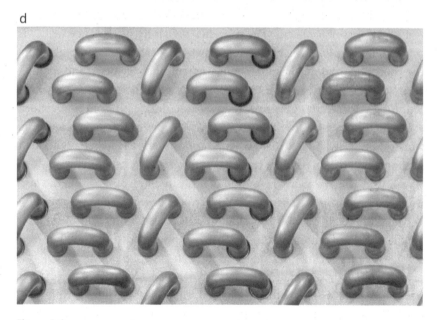

Figure 3.1
(continued)

images of technology on the site were devoted to "revealing" Google's data centers with colorful shots of computers, wires, routers, switches, pipes, and hard drives that arguably make this infrastructure much *less* visible when it is decontextualized (see figure 3.1).[15]

While the technology appears almost as abstract art, the buildings themselves (figure 3.2) are styled in a way to showcase their natural environment and surroundings: the photographs highlight the stunning sunrise or sunset, the spectacular mountains or ice fields in their midst, or even a "family of deer" near the hulking data center in Council Bluffs, Iowa, blissfully unaware of the tonnage of steel and cement right behind them. The point was to "show off" the data centers, but in reality, they were rendered practically inconsequential by the vistas and natural beauty stealing the camera's focus. These images perfectly represent Parks's argument about the politics of infrastructural visibility in her work on "antenna trees" designed to obscure cell towers. "By disguising infrastructure as part of the natural environment," she explains, "concealment strategies keep citizens naïve and uninformed about the network technologies they subsidize and use each day."[16] Johnson and Hogan have made a similar case for the political peril of obscuring the reality of data storage itself. "The risk when data is invisible," they write, "is of a failure of collective citizen engagement in decision-making regarding the conceptualization, meaning, emplacement, management, and maintenance of these infrastructures."[17] Nicole Starosielski (referring back to Lewis Mumford's work characterizing sewage, electrical, and communications infrastructure as "the invisible city"[18]) has also argued, in her pioneering work on undersea cables, that "whether infrastructure is materially hidden or simply ignored, invisibility has been naturalized as its dominant mode of visuality."[19] This strategic convention of infrastructural obfuscation has also extended deep into the realm of cloud technologies and cloud policy.

The chronic "invisibility" of cloud policy is also a function of its legal complexity. Data centers and the data they house exist within and between countless policy regimes, including an intricate mix of domestic and global, private and public, and state, regional, and national. Much of data regulation has also been obscured by the unheralded triumph of privatized "informal" policies and corporate self-governance over countervailing forces supporting the public interest. These issues remain largely shadowed, even as scholars, journalists, and public interest advocates work to expose how the regulatory sausage is made. Consequently, most of the stakes of data regulation hide

Figure 3.2
(a): Google data center in Council Bluffs, IA. (b): Google data center in Berkeley County, SC. (c): Google data center in The Dalles, OR. (d): Google data center in Hamina, Finland. Credit: Google

c

d

Figure 3.2
(continued)

in plain sight. Some high-profile events such as Edward Snowden's whistleblowing in 2013 (discussed below) and the Cambridge Analytica scandal of 2018 have put a spotlight on some of data policy's key vulnerabilities. However, the legal and regulatory debates about data control and protections are still largely impenetrable for the general public and elicit scant attention in popular discourse despite their connection to everyday life. They are nevertheless at the very core of what constitutes the public's interest, and rehabilitating this element of cloud policy requires rendering the import of data policy and its many consequences broadly intelligible.

Locating Control

Currently, there are no legal or regulatory frameworks for treating data in the cloud as property, with attendant rights to ownership. There are ongoing, infinitely complex discussions of digital property rights that far exceed the scope of this book,[20] but a defining feature of cloud policy is that you do not own your personal data in the cloud. The Big Tech companies distributing it, storing it, mining it, or selling it don't own it either. Katharina Pistor has argued that these companies have benefited from the legal ambiguity surrounding data ownership, and instead they treat the data they capture "as *res nullius*, or wild animals: things that belong to no one but can be claimed by whoever catches them first."[21] Indeed, the locus of control over such data has been primarily enacted through dynamics and domains beyond the construct of ownership, the most prominent being custodianship, and the privacy policies and terms of service agreements of digital platforms.

Data jurisdiction is another primary legal arena where the battle over access and control has been waged. It contains many unresolved conflicts related to determining data's physical location, and the relevant authorities over and protections available to that data. At the outset, the economics and functionality of the cloud are at odds with most data protection law. As one former FCC official explained, "data protection law is largely based on an understanding that you know where your data is located within particular borders, whereas the economics of the cloud is dependent on data being able to flow across borders in a fairly seamless way."[22] The contents of the cloud have been dispersed all over the world, but the jurisdiction and governance of data remains largely determined by national policy regimes that are often irreconcilable with one another.

Legal scholars and lawyers have been grappling for decades with the myriad jurisdictional challenges created by the global cloud. Some have argued that these challenges are not novel or unprecedented, as there are other components of the regulated global economy that move across borders without a material presence, such as money, pollution, stock, and debt.[23] However, the case *for* "data exceptionalism" has also been made by claiming that conventional legal tenets and precepts cannot easily apply to digital data because of its immateriality, and the resulting complexities in assigning territoriality. In 1996, for example, legal scholars David Johnson and David Post wrote that the global computer-based network of communications "cut across territorial borders, creating a new realm of human activity and undermining the feasibility—and legitimacy—of laws based on geographic boundaries." Instead, new boundaries created by screens and passwords now define "a distinct Cyberspace that needs and can create its own law and legal institutions." Moreover, they argued, it is separate from and threatens established legal doctrine tied to territorial jurisdictions and the related power of those governments.[24] That same year, John Perry Barlow released his polemical cyberlibertarian manifesto, characterizing cyberspace as a utopian, sovereign domain that should be free of government control, declaring it a "global space" being built "to be naturally independent of the tyrannies you seek to impose on us."[25] A year later, in *Reno v. ACLU*, the US Supreme Court described cyberspace as "a unique medium . . . located in no particular geographical location but available to anyone, anywhere in the world, with access to the Internet."[26]

More recently, Jennifer Daskal has argued, "Whereas territoriality depends on the ability to define the relevant 'here' and 'there,' data is everywhere and anywhere and calls into question which 'here' and 'there' matter."[27] Zachary Clopton points to some of the many related questions created by assigning territoriality for cloud-based data when he asks, "Does territoriality refer to the location of the source, the recipient, the storage, or the government access? Certainty may be important and clear rules may be desirable, but territoriality does not achieve these ends."[28] Paul Schwartz has added that "data in the cloud raise different issues than . . . data in a filing cabinet. Most crucially, one size does not fit all when current law assesses legal access to global clouds."[29] There is still no settled global accord for how data should be regulated, and who or what entity has the right to make that determination. Data originates in one country, passes through and is stored in others, often simultaneously existing in several international territories as well as both federal and state jurisdictions in the US. As a result, this data travels through just as many policy regimes of privacy laws, data processing and protection laws, and other regulations affecting the rights to, responsibilities for, and control over that data while it is being distributed and stored in the vast infrastructure of the cloud.

This multi-jurisdictional nature of cloud computing has created a policy landscape littered with holes and impending crises. This uncoordinated legal maze of transnational data flows has created quite a test for regulators and the courts. The lack of universal legal standards for a global digital ecosystem has also created an incredibly precarious policy environment for the data of the future. The US federal government published its first cloud computing strategy in 2011 under President Barack Obama. The key and unresolved "issues to consider" at the end of that document highlighted the policy problems that many global stewards of cloud governance continue to face:

- Data sovereignty, data in motion, and data access: How do countries strike the proper balance between privacy, security, and intellectual property of national data?
- Are there needs for international cloud computing legal, regulatory, or governance frameworks?
- Cloud computing codes of conducts for national governments, industry, and non-governmental organizations
- Data interoperability and portability in domestic and international settings
- Ensuring global harmonization of cloud computing standards[30]

Most of the relevant questions surrounding data regulation and jurisdiction have been left to the courts—state, federal, and regional bodies such as the European Court of Justice have all contributed important rulings. In the United States, jurisdiction has largely been based on the geopolitical determinant of state borders, and data centers have achieved some standing as the legally defined location of data in the cloud. As legal expert Andrew Keane Woods has explained, "Cloud-based data resides on servers—essentially large hard drives—and wherever those servers sit, they are subject to territorial assertions of jurisdiction.... Jurisdiction over cloud-based data has nearly everything to do with territoriality."[31] However, whether data is determined to exist in one place (that of one data center) or multiple locations (wherever the data are collected, processed, and stored, usually in fragments at a variety of centers)—either simultaneously or sequentially—still remains in flux. For example, if a data server replicates one's information for safekeeping, multiple countries could have concurrent, overlapping jurisdiction.[32] Moreover, providers like Google or Microsoft can "shard" or partition email messages "and store the resulting 'slices of data' in servers in California, Ireland, and Japan,"[33] thus potentially subjecting a single email to the data laws and regulations of multiple countries.

As a result of these and other unsettled legal questions relating to data policy, we are, as one legal scholar has put it, "sailing into the future on a sinking ship."[34] In fact, this has become the case, quite literally, as companies begin to sink their data into the sea to offset cooling and storage costs. Microsoft's Project Natick has had an undersea data center in the Northern Isles, off the coast of Scotland, since June, 2018. The project, which has the tagline "50% of us live near the coast. Why doesn't our data?" uses renewable electricity generated from wind, solar, and tidal power, and takes full advantage of the cold-water temperatures in the North Sea. It contains 864 servers with almost twenty-eight petabytes of data, or "enough storage for about 5 million movies."[35] In keeping with Big Tech's custom of putting complex infrastructural technologies behind more palatable, user-friendly veneers, the project utilizes an underwater camera that livestreams on their website and also captures still shots from a "favorite visitor" and schools of fish swimming among the data (see figure 3.3). This focuses the public's attention on the adorable sea creatures now coexisting with this hulking structure, as opposed to the many environmental quandaries represented by this data center's location, such as its contribution to rising ocean temperatures, the creation of noise and other forms of pollution that disturb marine life, and the eventual acceleration of climate change should these technologies be fully scaled up.[36]

And yet, location matters more than ever. Where corporations decide to place their data centers is determined by a variety of factors beyond legislation and rules governing data. Considerations include the sophistication of local infrastructure, local tax codes, and, crucially, proximity to both affordable electricity and other energy resources. Julia Velkova has argued that such determinations are heavily affected by "current [international] policy efforts to incentivize new data center projects with the promise of corporate tax reductions, cheap land and electricity cost packages, eased access to high-voltage electricity grids, and low-latency fibre connectivity. Policies like these

Figure 3.3
Project Natick FishCam.
Credit: Microsoft Research

have already converted the Nordic countries into central nodes in the global cloud infrastructure by hosting the data centers of Microsoft, Amazon, Apple, Google, Facebook, Yandex and global collocation providers like Equinix or Interxion."[37] Many Big Tech companies including Google, Facebook, Twitter, Amazon, and Dropbox have chosen to house much of their data in Dublin, Ireland (primarily in the Grand Canal Dock area now known as "Silicon Docks"), not only for the well-educated, English-speaking workforce and extremely low corporate taxes, but also for Dublin's chilly weather, which significantly lowers the energy costs to keep their data centers cool.[38]

Electricity costs are significant, as data centers are one of the fastest growing consumers of energy (to both power and cool the servers). It is estimated that they are responsible for up to 3 percent of global electricity consumption, eclipsing the demand of entire countries.[39] To contend with such massive energy requirements, half of which is needed for cooling, locating data centers in colder climates has become a common strategy. Consequently, the Nordic countries have become one burgeoning data center destination for US companies to build data centers. Facebook's facility in Luleå, Sweden, for example, relies on the average daily temperature of 2°C (35.6°F) to do most of the work to cool the server halls, with dams on the nearby Luleå river generating renewable electricity to supply the rest of the facility's power needs.[40] The ocean's built-in cooling system is free of charge, and a perfect location for a data center from an energy standpoint. Project Natick is capitalizing on these many benefits with their unconventional, watery home for data servers and is likely a harbinger of the cloud's future footprint underseas.

The power demands of data centers have also created an increasingly interdependent relationship between the infrastructures for data and energy. These complex geopolitics have even led to data centers becoming energy sources in and of themselves, as the heat generated by servers becomes repurposed into what Julia Velkova has described as "data furnaces." These data furnaces provide environmentally sustainable energy to European cities, rerouting "waste heat" to apartments and water supplies in places like Helsinki, Stockholm, and Paris, among others.[41] Even the cave network below the nineteenth-century Orthodox Uspenski Cathedral in downtown Helsinki, once intended as a World War II bomb shelter for protecting city officials in the event of a Russian attack, is now a data center that is also providing heat to hundreds of homes and apartments in the city.[42]

Figure 3.4
The Uspenski Cathedral.
Credit: Timo Noko under CC BY 2.0. https://creativecommons.org/licenses/by/2.0/deed.en. No modifications were made.

Further, the allure of cheaper energy to supply data centers often creates jurisdiction shopping for global data hosting. This is quite problematic, in that the decision to use the most hospitable location in terms of economic incentives, climate, and other benefits also has serious consequences for the privacy and security of the data being housed. After all, the places with the most affordable energy or the best tax breaks do not necessarily also have the most favorable laws to protect the privacy of user data or safeguard its security. The growth of Guizhou, a remote mountainous province in southwest China, is one prominent example. It is now known as the country's "cloud computing capital," hosting data centers set up by companies including Apple, Qualcomm, and Oracle.[43] Among its attractions are the "karst landforms" and their underground caves, the abundant hydropower, cool temperatures, and other natural geographical advantages providing the affordable, environmentally friendly cooling for data center infrastructure.

However, the Chinese government insists on numerous concessions in exchange for operating inside the country, which include submitting to governmental regulations that allow for the state surveillance of user data and activity. Such surveillance has frequently resulted in harsh prison sentences for ordinary citizens as well as for political dissidents, members of religious and ethnic minorities, journalists, and human rights activists. This is one of many factors that has led to China's designation as "the world's worst environment for internet freedom" for eight consecutive years and counting by Freedom House, a global human rights advocacy organization.[44]

These dynamics only approximate a small window into the connections between data location and governance on the development of global cloud formation. To fully appreciate the impact on the constellation of rights involved, it is imperative to contextualize this web of relationships within the longer history of remote data storage and related policy legacies. Particularly, it is crucial to note the ways in which this history has defined the relationship of personal data to the state, the corporation, and the privacy of an individual. Only then does this dimension of cloud policy become truly legible as a long-standing locus of political power and social control.

"The Abyss from Which There Is No Return"

The history of formal data regulation in the US is a web of legislation, lawsuits, judicial orders and decisions, agency rulings, and the actions of intelligence and law enforcement communities. The relevant oversight bodies have long suffered from regulatory hangover to varying degrees, as technological development and cultural practices continually outpace the vision and approach of policymakers. This long-standing systemic disconnect has created a policy landscape that is littered with insufficient privacy protections, insecure data, and irreconcilable conflicts of law. Add in the successive administrations that have used privately owned public infrastructure to spy on American citizens decade after decade, and we are left with an almost unimaginably dysfunctional component of cloud policy that has played an outsized role in the progressive erosion of digital civil liberties. Indeed, the infrastructures for mass surveillance, entertainment, and communication have been one in the same since the army's Black Chamber relied on Western Union to surveil diplomatic and military messages after World War I. With that in mind, it is imperative to understand these sociotechnical histories as media histories, even though they are most often contextualized as belonging to the military, law enforcement, and national security.

The National Data Center

Cultural fears about the state's ability to track its citizens have circulated at least since the 1930s when the New Deal ushered in Social Security and a panic ensued over being assigned an identification number that would follow one all the way to the grave.[45] These fears continued through the 1950s with the Red Scare, loyalty oaths, and the anti-Communist crusades of the House Committee on Un-American Activities. However, Congress did not devote much attention to the privacy of individual citizens until the 1960s, when concerns reached new heights, thanks in part to technological advances. Portable recording technologies and computing began to sound alarms, as their capabilities elicited new threats to privacy rights. Such worries were amplified by the Supreme Court, as Chief Justice Earl Warren stated in a 1963 opinion regarding recording devices and entrapment: "The fantastic advances in the field of electronic communication constitute a great danger to the privacy of the individual."[46] In addition, a wave of writing by scholars and journalists at this time, focused on technology, privacy, and personal autonomy, helped inform public debate. In many ways this work anticipated current anxieties about the price of life under Big Tech. Vance Packard's *The Naked Society* (1964), Alan F. Westin's *Privacy and Freedom* (1967), and Arthur Miller's *The Assault on Privacy* (1971) were among the most influential in this genre. Miller understood then that the time would soon come when "our primary source of knowledge will be electronic information nodes or communications centers located in our homes, schools, and offices that are connected to international, national, regional, and local computer-based data networks."[47] Westin evoked many present-day issues in his wide-ranging, foundational book, paying great attention to "data surveillance" and how new technologies were affecting norms of privacy in order to recuperate this "cornerstone of the American system of liberty."[48] He viewed privacy and freedom as inextricably linked, defining privacy as "the claim of individuals . . . to determine for themselves when, how, and to what extent information about them is communicated to others."[49] *Privacy and Freedom* is still useful today for thinking about the malleable parameters of privacy, and its power in defining an individual's relationship to the state.

This was the context in which President Johnson proposed a federally controlled data center called the National Data Bank in 1965 as part of the Great Society project. The data center was imagined as a tool for efficiency and organization that would consolidate federal databases at the dawn of computerized record-keeping. However, concerns about technology and privacy were becoming widespread enough that a congressional Special Subcommittee on the Invasion of Privacy was established in the House of Representatives. Four separate hearings were held in the House and Senate between 1966 and 1967 to discuss the threats to privacy posed by the computer and the government control of data. They were dominated by overwhelming expressions of concern about the sanctity of individual privacy and civil liberties. The government's power combined with the yet-unknown capabilities of digital technology were positioned as the main potential threat. The determination that the public needed to be protected from the centralized state collection of data above all else, without sufficient attention to the dangers lurking elsewhere, was a defining moment for cloud policy that has only grown more consequential over the decades that followed.

The chair of the Subcommittee on the Invasion of Privacy running the House hearings, Representative Cornelius "Neil" Gallagher (D-NJ), introduced the investigation of the National Data Center in July 1966 by saying, "The possible future storage and regrouping of such personal information . . . strikes at the core of our Judeo-Christian concept of 'forgive and forget,' because the computer neither forgives nor forgets."[50] Representative Frank Horton (R-NY) warned that "the magnitude of the problem we now confront is akin to the changes wrought in our national life with the dawning of the nuclear age. . . . It is not enough to say 'It can't happen here'; our grandfathers said that about television."[51] One of the original network architects of the Internet, Paul Baran, alluded to threats posed by the future cloud in his expert-witness testimony, noting that "a multiplicity of large, remote-access computer systems, if interconnected, can pose the danger of loss of the individual's right to privacy—as we know it today."[52] Author Vance Packard called attention to the "suffocating sense of surveillance" engendered by a centralized government database, noting the "hazard of permitting so much power to rest in the hands of the people in a position to push computer buttons, . . . [because] we all to some extent fall under the control of the machine's managers."[53]

Gallagher had a remarkably prescient grasp of technological threats to individual privacy, which was likely a result of being persecuted and having his own privacy violated for many years by J. Edgar Hoover and the FBI.[54] Gallagher's speech to the American Bar Association in 1967, titled "Technology and Freedom," was quite striking in its predictive accuracy. It included the following, partially adapted from the statement of Professor Arthur Miller at the Senate hearings that same year:

> Although the technology of computerization has raised new horizons of progress, it also brings with it grave dangers.... The computer, with its insatiable appetite for information, its image of infallibility, its inability to forget anything that has been put into it, may become the heart of a surveillance system that will turn society into a transparent world in which our home, our finances, our associations, our mental and physical condition are laid bare to the most casual observer. If information is power, then real power and its inherent threat to the Republic will not rest in some elected officials or Army generals, but in a few overzealous members of a bureaucratic elite.[55]

The final report from the House Committee was clear about the links between data privacy and democracy: "A suffocating sense of surveillance, represented by instantaneously retrievable, derogatory or noncontextual data, is not an atmosphere in which freedom can long survive.... This report, therefore, charges the Federal Government as well as the computer community with a dual responsibility.... They must ... guarantee Americans that the tonic of high speed information handling does not contain a toxic which will kill privacy."[56] The committee further noted that the dangers of unauthorized access to information was great, and "a grave threat to the constitutional guarantees exists in the National Data Bank concept," leading to their emphasis on prioritizing privacy in the center's eventual design and implementation. However, the committee's ultimate recommendation was to stop working on the National Data Bank until privacy protections were fully explored and guaranteed "to the greatest extent possible to the citizens whose personal records would form its information base."[57] Once again linking privacy to politics, the authors emphasized, "While computerized data bases hold great promise, they must contain procedures which can assure the continuation of freedom of thought and action that is such a vital part of the American tradition. The collection and processing of statistical data should not and need not be gained by sacrificing the guiding principles of our democracy."[58]

At the same time, the reporting in the popular press was highly alarmist. One representative article in *Look* magazine titled "The Computer Data

Figure 3.5
From the November 1967 cover of the *Atlantic*.
Credit: Drawing by Edward Sorel.

Bank: Will It Kill Your Freedom?" posed various questions that could easily be answered by "any snooper with a computer," such as, "Did your sister have an illegitimate baby when she was 15? Did you fail math in junior high? Are you divorced or living in a common-law relationship? Do you pay your bills promptly? Are you willing to talk to salesmen? Have you been treated for a venereal disease? Are you visiting a psychiatrist? Were you ever arrested?"[59] Chairman Gallagher was quoted in the same article, warning, "Computer data banks are at the same stage of development as the early railroads and the first telephone companies, which took a number of years to link themselves together in a nationwide network. Welfare departments, credit bureaus, hospitals, police departments and dozens of other institutions are putting their files into hundreds of relatively small data centers. No matter what you call them, they're still data centers, and they can be linked." The public uproar in response to all these developments led to National Data Bank discussions and debate being shut down by 1970.

Unfortunately, it would be a pyrrhic victory. The focus on protecting public data from the perceived dangers of centralized state collection and storage blinded legislators to the problems created by the solution: putting data in the hands of private companies. Corporations ultimately filled the vacuum created by the National Data Bank's failure, and became the chief custodians of US citizens' private data. As O'Mara has argued, these decisions actually created the very problem they were trying to prevent. "The privacy warriors of the 1960s would have been astounded by what the tech industry has become. They would be more amazed to realize that the policy choices they made back then—to demand data transparency rather than limit data collection, and to legislate the behavior of government but not private industry—enabled today's tech giants to become as large and powerful as they are."[60] The congressional attempt to defend US citizens from experiencing "big brother" and the world as imagined in Orwell's *1984*, which were mentioned relentlessly during the hearings, ended up creating exactly what they were trying to avoid, albeit serving a different master. This is not to suggest that government control over public data is preferable, but instead to emphasize that private control without regulatory oversight has proven to be undeniably disastrous for individual and collective privacy, and a signature failure of contemporary cloud policy. To his credit, Senator Long (D-MO) who presided over the Senate hearings in 1966 and 1967 did warn that if the proposals for a National Data Bank "concerned themselves only with Government interests, and if individual, private interests were ignored, we might be creating a form of Frankenstein monster,"[61] but his words went unheeded.

The cultural tensions around surveillance lingered, as evident in the 1973 report, *Records, Computers, and the Rights of Citizens*, put out by the Secretary's Advisory Committee of the Department of Health, Education, and Welfare (see figure 3.6). This report was about computerized record-keeping, privacy safeguards, and the issue of the social security number. It is a stunning document that catalogs record-keeping practices going back to the Stone Age through the advent of automated systems. It included similar work on computerized record-keeping and privacy being done in Canada, Great Britain, and Sweden. The report also newly identified citizens as "data subjects," emphasizing privacy safeguards and the individual's loss of control

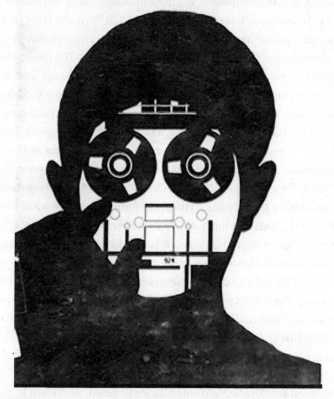

Figure 3.6
The report of the Secretary of Health, Education, and Welfare's Advisory Committee on Automated Personal Data Systems, 1973.

over the use of their personal data. In so doing, this 1973 report predicted many of the problems created by Big Tech business models, arguing that concerns about computerized records usually center on privacy, particularly as "privacy is considered to entail control by an individual over the uses made of information about him. In many circumstances in modern life an individual must either surrender some of that control or forego the services that an organization provides. Although there is nothing inherently unfair in trading some measure of privacy for a benefit, both parties to the exchange should participate in setting the terms."[62]

The report also recommended a federal "Code of Fair Information Practice," which contained principles for transparency, autonomy over one's personal data, and safeguard requirements regarding data usage by third parties. None of these recommendations were adopted at the time. However, they went on to inform future agency recommendations and early privacy legislation such as the Privacy Act of 1974, which was enacted in the wake of President Nixon's resignation. And yet, as O'Mara has pointed out, much of this legislation concentrated on the right to know about what information that federal databases held, but none of it "addressed the question of whether this information should have been gathered in the first place."[63] The 1974 act did not stop data collection, it merely revealed how much of it was taking place on the federal level. According to Sarah Igo, despite being "designed to empower citizens vis-à-vis the record keepers, the law would wind up stoking fears that the United States had become a full-fledged surveillance society in which individuals were outmatched from the outset."[64]

These widespread concerns at the dawn of the computerized era led to yet another round of hearings in the Senate in 1975. This time the focus was on "surveillance technology," as news had emerged about the Pentagon's surveillance of Vietnam War protestors, and journalists exposed the Johnson and Nixon administrations for utilizing a computerized, networked domestic spy operation that linked "the CIA, the Defense Intelligence Agency, the National Security Agency, more than 20 universities, and a dozen research centers, like the Rand Corporation."[65] Echoing the foreboding words of Frank Church issued just a month earlier, Chairman John Tunney (R-CA) opened the hearings saying, "Technological developments are arriving so rapidly and are changing the nature of our society so fundamentally that we are in danger of losing the capacity to shape our own destiny." He further stated that "control over the technology of surveillance conveys effective control over our privacy, our freedom, and our dignity—in short, control over the most meaningful aspects of our lives as free human beings."[66] MIT President Jerome Wiesner testified that the surveillance problem had become a crisis because "information technology puts vastly more power into the hands of government and private interests that have the resources to use it" and "to the degree that the Constitution meant for power to be in the hands of the 'governed,' widespread collection of personal information poses a threat to the Constitution itself."[67] Ultimately, Weiser argued that there is "serious danger of creating an 'information tyranny' in the innocent pursuit of a more efficient society."[68] The committee echoed his tone, raising alarm that the "continued ignorance of surveillance technology—its size and structure as a separate industry, the justifications for its growth, its impact on society—could prove to be an Orwellian catastrophe for our privacy and our freedoms."[69] As it turned out, all of these fears were well-founded. These proceedings contained vital warnings and lessons for the future of cloud policy that have since been lost to history.

The PATRIOT Act and FISA

What becomes clear from reading these hearings, reports, and news stories from the 1960s and 1970s is that cloud policy's history is in part the history of the US as a modern surveillance state. One of the most important components of its modern architecture is the Uniting and Strengthening America by Providing Appropriate Tools Required to Intercept and Obstruct Terrorism Act (better known as the USA PATRIOT Act[70]). The PATRIOT Act was signed into law on October 26, 2001, just forty-five days after the attacks on the World Trade Center and one week after its introduction to Congress, with almost no debate. The bill was more than three hundred pages long and sweeping in its scope, covering issues ranging from border security and money laundering to criminal law procedures and intelligence collection. It also created numerous changes related to data security, privacy, and surveillance, including the broad expansion of collection procedures and territoriality limits related to federal court warrants for digital data.

Most significantly, the PATRIOT Act greatly enhanced the surveillance powers of the government while simultaneously reducing democratic checks on those powers, including judicial oversight, transparency, and public accountability. Section 215 of the PATRIOT Act was particularly egregious in its overreach, and many civil society organizations and privacy experts have claimed it violated the Constitution, specifically the First, Fourth, and

Fifth Amendments. It gave the NSA permission to collect metadata on every American's phone records, and allowed the FBI to force any person or institution, including doctors, libraries, universities, and Internet service providers, to turn over records on their clients or customers.[71] The PATRIOT Act also authorized the US government's ability to compel the handover of data held by US companies, *regardless of where the data is located geographically*—that is, regardless of where it is being stored in the world. That meant that, when served with a warrant, companies could be required to surrender data stored remotely in the cloud even if the servers in question were located outside US jurisdiction, without any notice to those whose data was involved. The PATRIOT Act went so far in its reach that, as policy scholar Andrea Renda put it, "even contract provisions specifying that data will be governed by foreign law can be ignored by the US government."[72] Its scope and reach led the PATRIOT Act to be characterized as "one of the only laws that affects the entire cloud computing industry."[73]

The extensive nature of surveillance powers in the PATRIOT Act are commonly thought to originate in the 9/11 attacks on the World Trade Center and the ensuing war on terror. While that is true in terms of timing, in fact the PATRIOT Act was also an opportunity to revive an extensive and historical neoconservative wish list of changes to privacy and surveillance laws in the US. It had been building for many decades, ever since the Church Committee's scathing report on domestic intelligence abuses and surveillance campaigns was released in the wake of the Watergate scandal. In its 1976 report, the Church Committee issued ninety-six recommendations designed to prevent future abuses. It took many years to fully enact them, but eventually Congress, with the support of Presidents Ford and Carter (who issued executive orders barring the CIA from operating inside the United States), created new Senate and House intelligence committees to monitor the work of the CIA, NSA, and other intelligence agencies.[74]

Further reforms were enacted after the Church Committee report, including additional levels of judicial review and legislative oversight designed to curtail FBI and NSA abuses in the name of national security. The most notable was the Foreign Intelligence Surveillance Act (FISA) of 1978. This act was passed after much debate in Congress that ultimately acknowledged the need for "statutory procedure authorizing the use of electronic surveillance in the United States for foreign intelligence purposes" in light of "almost 50 years of [somewhat inconclusive] case law dealing with the subject of warrantless electronic surveillance, and . . . the practice of warrantless foreign intelligence surveillance sanctioned and engaged in by nine administrations."[75] Initially, FISA *added* privacy protections for individuals, requiring new procedures for warrants that tapped electronic communication (while making some exceptions for gathering foreign intelligence) and restoring civil liberties that had been systemically abused by the government for decades. Its evolution is quite ironic and tragic, given that FISA has become one of the key laws that has helped to degrade cloud policy and diminish privacy rights all over the world, due in large part to subsequent amendments by the PATRIOT Act and other legislation.

The timeline of FISA's evolution from enhancing the privacy of US citizens to invading it with impunity was relatively brief in the context of policy's long arc. In less than thirty years, FISA was transformed from a law helping to end domestic spying to a tool of mass surveillance targeting US citizens. It went from stringently regulating the collection of "foreign intelligence" while also making it much more difficult to spy on US citizens[76] to conscripting infrastructure providers such as phone companies, social media platforms, and Internet service providers (ISPs) in dragnet surveillance efforts. It also "for the first time created a power of mass-surveillance specifically targeted at the data of non-US persons located outside the US,"[77] utilizing many of these same global infrastructures of cloud computing.

FISA had some initial help beginning in 1981 from President Ronald Reagan's Executive Order 12333. This order has allowed the NSA to conduct extensive surveillance abroad, including signals intelligence (i.e., intelligence derived from the interception of electronic signals and systems—communications, radar, weapons), and also maintain domestic data collection and retention operations.[78] But it is FISAs numerous amendments—four and counting since 1978 as of this writing[79]—that have progressively moved it away from the original legal intention of limiting the powers of the government to spy on its citizens and closer to being an instrument of what it was first designed to thwart: the unnecessary and unlawful surveillance of US citizens by their own government. The initial and substantive amendments by the PATRIOT Act and the 2008 reauthorization of FISA were the most consequential in this regard.

These changes to FISA greatly expanded the pool of legal surveillance targets and the time period that the government is allowed to conduct certain surveillance activities. They also further lowered the bar for legality and oversight and infringed on the constitutional rights of US citizens. For example, Section 702 of the FISA Amendments Act of 2008 created the authority for the government to "conduct targeted surveillance of foreign persons located outside the United States, with the compelled assistance of electronic communication service providers."[80] This surveillance was also warrantless. Such operations (which have enlisted the help of Big Tech companies) have additionally "swept up" the data of many Americans who are emailing or communicating with a non-US citizen outside the United States. This has been termed "incidental collection" by the intelligence community, but is better known by privacy advocates and lawyers as a violation of the Fourth Amendment.[81]

FISA also created the Foreign Intelligence Surveillance Court (FISC or FISA Court) to review national security–related eavesdropping (mostly wiretapping) requests made by the Justice Department, and administer search warrants. It was supposed to be another level of judicial review installed by Congress to *prevent* the government from abusing its surveillance powers in the future as it had in the past. If the government agent wants to surveil a US citizen or resident as part of a national security investigation, they need approval from a federal judge who sits on the FISA Court. Unfortunately, the FISA Court has been widely criticized in recent years for its politicization and lack of serious oversight.

One of the major issues is that FISC is a "secret court," with hearings and records of proceedings off-limits to the public. The government is the only party present, without any adversarial presence—that is, there are no witnesses or legal advocates for anyone under suspicion. The court's decisions cannot be appealed or made public in any way. Moreover, every one of the eleven FISC judges, who sit for seven-year terms, is appointed by the chief justice of the Supreme Court without any outside confirmation process. Of the eighty-four judges on the court since 1979, less than 10 percent of appointments—eight in total—have been women.[82] Currently, Chief Justice Roberts is solely responsible for the present composition of the court of ten Republicans and one Democrat that has the incredible power of deciding whom the government can spy on, and how much surveillance of US citizens by their own government is acceptable.

The FISA court's long-standing acquiescence to the DOJ and FBI has led to it being labeled a "rubber stamp"[83] or "something like an administrative agency"[84] without any checks or meaningful judicial review. From 2001 to

2012 alone, FISA judges approved 20,909 warrants and rejected a grand total of ten.[85] Looking at an even longer time period that pre-dated the PATRIOT Act, from the court's inception in 1979 to 2015, FISA judges approved 38,269 applications for surveillance orders and rejected seventeen.[86] In the very rare instance when the court does turn down a request, the Department of Justice also has the power to subpoena data from service providers such as AT&T or Comcast for the calls or emails of anyone they want to surveil, and to use gag orders so that those providers can't tell a customer that their data has been turned over to the authorities. This is done through a tool called National Security Letters (NSLs), the use and powers of which have grown exponentially since the passage of the PATRIOT Act.[87] They also have no judicial oversight. The total number of NSL requests prior to the PATRIOT Act was roughly in the "hundreds" between 1978 and 2001, exploding to 143,074 from 2003 to 2005.[88] The use of NSLs was also found to have been massively and systemically abused by the FBI in a 2010 audit by the Department of Justice's Office of the Inspector General.[89]

Moreover, since the FISA Amendments Act of 2008, the court has been required to approve entire surveillance systems and not just individual warrants as they previously did; this method ultimately led directly to the NSA's PRISM program, which FISC supervised—a secret, blanket domestic and international surveillance operation revealed by Edward Snowden in 2013 (addressed in detail below). Thus, the FISA court, along with the laws enacted by the PATRIOT Act, FISA itself, and their many subsequent amendments were in many ways responsible for PRISM and the assault on individual privacy rights that went with it. This led to the eventual unraveling of the policy regime for data security utilized by the United States and the European Union. Before moving on to that geopolitical debacle, it is important to return to key legislation regarding electronic privacy that emerged in the pre-Internet era and continues to haunt cloud policy today: the Electronic Communications Privacy Act (ECPA) of 1986.

The ECPA and the CLOUD Act

The ECPA[90] was passed during Ronald Reagan's second term in office, when telephones were anchored to a wall and messages were either written in longhand on scraps of paper or left on the ever-unreliable answering machine. It was enacted in response to an acknowledged regulatory lag, with the Senate Judiciary Committee commenting, upon its introduction, that current electronic privacy law was "hopelessly out of date."[91] The report stated that existing legislation "has not kept pace with the development of communication and computer technology. Nor has it kept pace with changes in the structure of the telecommunications industry."[92] At the same time, the report noted that the same "tremendous advances in telecommunications and computer technologies have carried with them comparable technological advances in surveillance devices and techniques."[93] The ECPA was drafted, according to one of its coauthors Senator Patrick Leahy, "to ensure that all Americans would enjoy the same privacy protections in their online communications as they did in the offline world, while ensuring that law enforcement had access to information needed to combat crime."[94] It is currently the law that primarily determines privacy rights for stored online communications in the US, and it applies to the behavior of private corporations such as Google, Facebook, and AT&T, as well as to government and law enforcement agencies. It has never been updated. Unsurprisingly, this law that was written five years before the first web page was created, based on archaic understandings of digital media and technocultures, is now a woefully insufficient tool to protect personal data in the era of wireless communication and cloud-based storage.

The law was created to revise federal wiretapping and eavesdropping rules that were written in 1968, which mainly focused on telephone lines and their use by organized crime figures. The ECPA was designed to extend protections against unconstitutional government wiretaps to digital conversations and was therefore applied to computer-based communications, including email and other digital data. It was supposed to be enhancing the privacy protections for US citizens. However, as Senator Leahy acknowledged in 2010 during a failed attempt to amend the bill, "At the time, ECPA was a cutting-edge piece of legislation. But, the many advances in communication technologies since have outpaced the privacy protections that Congress put in place."[95] In another case of technology moving faster than those regulating it, the ECPA—much like FISA—evolved into a statute that actually endangers the privacy rights of individuals instead of protecting them. However, instead of amendments being the impetus for the change in values, in the case of the ECPA it is the *lack* of any substantive updates that is responsible for the law shifting so far from its original intentions. At this point, the ECPA is a main reason that much of the personal, private data in the US remains extremely vulnerable.

Most of these vulnerabilities are because of the "third-party problems" created by the law and based on extraordinarily resilient legal constructs developed in an entirely different media ecosystem. The "third party doctrine" grew out of two Supreme Court cases in the late 1970s[96] and holds that if one voluntarily provides information to a third party, they have "no legitimate expectation of privacy" from warrantless government access to that information. Moreover, the Fourth Amendment offers no protection. These problems were not only left unresolved by the ECPA, it could be argued they were made even worse, thanks to the Stored Communications Act (SCA), which is Title II of the ECPA.[97] The SCA regulates access to "stored wire and electronic communications and transactional records."[98] Cloud data is in storage for most of its existence. With this provision, the SCA holds the key to regulating the privacy of communications and data held/stored by service providers (such as mobile phone companies or ISPs), all of whom qualify as "third parties." Of critical importance is that the SCA considers data stored with third parties for 180 days to be "at rest" and technically abandoned. According to the SCA, a warrant is not necessary for the government to access such data; instead, they simply need to issue a subpoena.

Of course, in 1986 when the ECPA was written, online or third-party document storage was incredibly expensive, and it was conceivable that such data was "abandoned" if it had not been used in six months. After all, the price of storing data remotely was roughly $700,000 per gigabyte in 1981—it is now less than $0.03 per gigabyte to store today.[99] Moreover, if you had email in the 1980s, it was not stored on servers for long periods of time. Instead, it was downloaded to the user's own computer as opposed to being kept in the as-yet-undeveloped cloud. Now that today's email services offer practically unlimited storage, and free options abound, most people store as much data as they want on cloud servers and rarely think about how long it has been sitting there. Common practice for workplace and personal computing has evolved to depend on long-term remote storage for email. However, the parameters for classifying data "at rest" are still defined by the SCA, and have thus created tremendous vulnerabilities for anyone in the US storing data in the cloud today, which is to say, almost everyone using a computer. As the Electronic Frontier Foundation (EFF) has said, "the eighties were good for a lot of things—but not sustainable email privacy law."[100]

The SCA addresses three categories of data. The first is data considered to be "at rest," inactive and stored in servers. It is differentiated from the second category, data "in motion" (traveling through a network or temporarily residing in a computer's memory), and the third, data "in use" (archived data under constant change, also in servers). Since 1986, government access to data "at rest" does not require a warrant, just a subpoena. Consequently, any email or document that has been stored for over six months is less private than even postal mail or phone messages. In fact, it has the same level of protection as one's garbage on trash day. Since the trash collector is also a third party, "no reasonable expectation of privacy exists once trash has been placed in a public area for collection," according to a US Third Court of Appeals decision from 1981. Similar to digital data that has been inert for six months, "the placing of trash in garbage cans at a time and place for anticipated collection by public employees for hauling to a public dump signifies abandonment."[101] The legal "abandonment" of data as viewed by the ECPA has left our digital data almost as vulnerable as junk on a street corner.

While public interest and privacy advocacy groups such as the EFF, the Electronic Privacy Information Center (EPIC), and Public Knowledge have long been calling attention to this issue, the private sector of the digital economy has finally taken notice. A coalition called Digital Due Process has been one source of pushback to this dated regime. The group includes civil rights and civil society groups such as the ACLU and the Center for Democracy & Technology. It also includes some of the biggest corporations affected by the ECPA, such as Dropbox, Facebook, Microsoft, and Google, that recognize the many dangers to their businesses posed by archaic data policy. The coalition supports changes to the law in response to changes in technology and usage patterns, and advocates for modernizing the ECPA to reflect the more complex conditions and privacy requirements brought on by digital technologies and cloud computing.[102] As the coalition has noted, "A single e-mail is subject to multiple different legal standards in its life-cycle, from the moment it is being typed to the moment it is opened by the recipient to the time it is stored with the e-mail service provider."[103] Despite decades of calls for change by a broad spectrum of voices, there has yet to be any reform of these categories and the outmoded language of the ECPA continues to erode civil liberties in the cloud.

Microsoft's then General Counsel Brad Smith (at this writing in 2023 he is the company's vice chairman and president) testified before the Senate Judiciary Committee in 2010 that "it is important to situate ECPA reform in the context of a broader policy agenda that should be advanced to ensure that the full benefits of cloud computing are realized. Such an agenda would encompass not only user privacy interests in relation to parties other than the government (such as the cloud provider itself and private third parties), but also other interests that are inextricably linked with privacy, including security, transparency, and national sovereignty."[104] A 2014 White House working group on big data that consulted Big Tech companies, academics, advertising agencies, legal experts, civil rights groups, and intelligence agencies made yet another unified call to update the ECPA.[105] Among their recommendations were, "Amend the Electronic Communications Privacy Act to ensure the standard of protection for online content is consistent with that afforded in the physical world."[106] This has yet to happen. As a result, a bureaucratic understanding of digital communication envisioned in 1986 continues to dictate electronic privacy law in the US.

What's more, the ECPA (and, in turn, the SCA) were not written with a global communications landscape in mind. They were written with a strictly domestic mindset that "assumed a U.S. Internet with U.S. servers and U.S. users" as legal expert Orin Kerr has argued, despite the fact that more than 90 percent of Internet users are outside the US.[107] Importantly, the ECPA did not apply extraterritorially for the first few decades of its existence. This prevented the US government from reaching into other countries' borders and cloud infrastructure to access data and communications for law enforcement purposes.

This issue of extraterritorial rights was the subject of a long-running case between the United States Department of Justice and Microsoft that hinged on the issue of data location, and whether the US could legally access data stored in other countries.[108] This case began in 2013 and brought data sovereignty, corporate versus state policies, and the tragically outdated ECPA into the spotlight as it made its way through the courts. It began when Microsoft resisted a court order to surrender emails held on their servers that were related to a narcotics case. The company said the emails were stored in Dublin, Ireland, and could not be extracted by using the warrant under the SCA. This refusal sparked a five-year battle over the ability of the US government to force US corporations to surrender their cloud data housed in another country.[109] It went through many levels of appeal, and after it became clear that the case was headed to the Supreme Court, Congress finally held hearings on international conflicts of law and implications for cross-border data requests and data stored abroad in 2016–2017.[110] In the end, and after the

Supreme Court heard oral arguments, Congress passed the Clarifying Lawful Overseas Use of Data (CLOUD) Act which resolved the extraterritoriality issue and rendered the case moot.[111]

The Clarifying Lawful Overseas Use of Data (CLOUD) Act was signed into law on March 23, 2018. It states that providers must surrender information and communication records requested by US law enforcement "regardless of whether such communication, record, or other information is located within or outside of the United States."[112] With that, the US government amended the ECPA and the SCA, and gave itself the legal ability to reach into cloud servers and extract data housed by Big Tech, no matter where in the world that data was being stored. This has thrown global privacy rights into a crisis: as one legal scholar has explained, "as long as U.S. companies dominate Internet services, there is no way for a[nother] country to keep its citizens' data from the U.S. government."[113]

Snowden, Safe Harbour, and the Privacy Shield

In June 2013, the *Guardian* and the *Washington Post* revealed threats posed by the US government to the privacy of their own citizens that were much worse than anything embedded in the SCA.[114] In fact, this news sounded like it came straight out of the anxieties of the Gallagher hearings in 1966, or the Church Committee's stark warnings that followed a decade later. The news stories exposing the NSA's mass surveillance program, the most notorious of which became PRISM, detailed a vast secret operation of spying on US citizens, foreign leaders, and suspected terrorists of all nationalities being carried out with cooperation and collusion of US-based tech companies. Based on documents and intelligence leaked by whistleblower Edward Snowden, the world was suddenly made aware that, for at least five years, "the NSA, whose lawful mission is foreign intelligence, [has been] reaching deep inside the machinery of American companies that host hundreds of millions of American-held accounts on American soil."[115] Snowden characterized this program, authorized under FISA and supervised by FISC, as "the most significant change in the history of American espionage—the change from targeted surveillance of individuals to the mass surveillance of entire populations."[116]

Moreover, this was taking place with the assistance of Silicon Valley's biggest players—Microsoft, Facebook, Apple, Google, Yahoo, and others feeding data to the NSA (see figure 3.7). If they decided not to cooperate, the government simply forced their participation as it did by threatening to fine Yahoo!

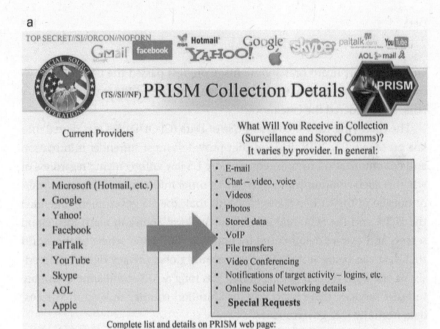
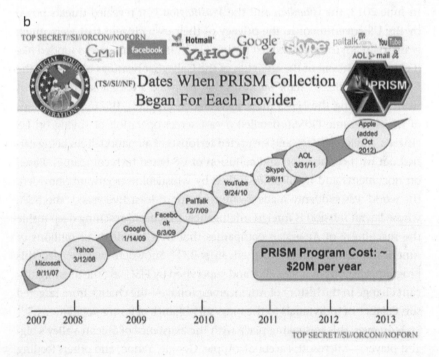

Figure 3.7
NSA PRISM collection details.

$250,000 a day if the company failed to comply with their "requests."[117] It was further revealed that Verizon had been supplying the NSA with the telephone records of millions of US customers as ordered by a FISA Court warrant.[118] AT&T, MCI, and Sprint had also been handing over telephone and Internet records to the government on a previously unimaginable scale—over one billion cellphone records a day from AT&T alone, and filtering messages for the NSA to read.[119] State surveillance has been a prominent use of telecommunications infrastructure throughout its history. And although the NSA's predecessor was once defunded for reading military and civilian telegrams and mail in 1929 because such behavior was thought to be unethical, unprincipled, and ungentlemanly,[120] apparently enough time had passed that egregious invasions of privacy by the state were permissible once again.

The *New York Times* had first written about the NSA's massive domestic warrantless surveillance program in 2005. The paper published multiple stories detailing how the George W. Bush administration had authorized the NSA to spy on US citizens since 2002 in the wake of the 9/11 attacks with essentially no oversight.[121] The government maintained that this sea change of allowing the NSA to operate domestically was necessary to fight the war on terror and prevent future attacks inside the United States. It required partnerships between telecommunications companies and law enforcement taking place without public awareness. The extent of the cooperation—indeed the long-term collusion—between private infrastructure companies and the NSA was ultimately revealed by Snowden when he exposed the extent of the government's domestic surveillance operation in 2013. There is no *official* public record of such partnerships, the data they captured, or the information produced, which makes the scope and scale of offenses impossible to know. What is known has been pieced together by whistleblowers, journalists, academics, and civil society groups. The egregious violations of privacy detail a surveillance state supported by the very phone companies and Internet service providers that we pay every month for their services and entrust our data to for safekeeping without much thought. Prominent examples during this era of cloud formation include

- Whistleblower and former AT&T technician Mark Klein revealed in 2006 that AT&T was providing the NSA full access to its customers' phone calls and routing their Internet traffic to special data-mining equipment located in a secret room in its San Francisco tech hub.[122] This information was included in a class action lawsuit against AT&T that same year led by the Electronic Frontier Foundation, suing the company for violating federal and state laws and allowing government agents to wiretap AT&T customers' phone and Internet communication without consent or warrants.[123] This case was eventually combined with more than fifty other lawsuits filed against telecommunications companies for similar wiretapping abuses. It was dismissed in 2009 because of retroactive immunity awarded to the telecommunications companies by the FISA Amendments Act in 2008.[124]

- *USA Today* reported in 2006 that the NSA had secured the cooperation of the major telecommunications companies including AT&T, MCI, Verizon, BellSouth, and Sprint in its warrantless wiretapping operations. The companies granted the government access to their systems without court orders, providing the NSA the ability to eavesdrop and monitor the calls of tens of millions of Americans as part of the war on terror.[125]
- In 2009, a graduate student (who went on to work as the principal technologist for the ACLU and then to a senior advisory capacity on privacy issues in Congress) exposed Sprint/Nextel's Electronic Surveillance Department for their mass provision of customer data to law enforcement. Christopher Soghoian, then at Indiana University, made public an audio recording of the Electronic Surveillance Department manager "describing how his company has provided GPS location data about its wireless customers to law enforcement over 8 million times."[126] Moreover, the Electronic Surveillance Department maintained a special portal for police to log into without having to obtain a warrant if they wanted to access geolocation information on a particular customer.

It was not long before lawsuits were filed against the NSA, including by the ACLU[127] and the Electronic Frontier Foundation.[128] The EFF's case, *Jewel v. NSA* (2008) was a class action lawsuit on behalf of AT&T customers against the NSA, its former director, and many government officials including President George W. Bush, Vice President Dick Cheney, and Attorney General John Ashcroft. It was intended to stop the NSA's massive warrantless surveillance program revealed by whistleblowers and journalists and hold those who authorized it accountable. The *Jewel* case was one of the longest running efforts to stop government surveillance, wending its way through the courts for almost fifteen years until the Supreme Court refused to hear it in 2022, allowing lower courts' rulings that relied on the "state secrets" privilege to stand. Despite thousands of pages of public evidence and testimony and investigations, the privilege basically allows the government to block the release of any information in a lawsuit that could be considered harmful to national security. In briefs, the EFF said that the courts have "created a broad national-security exception to the Constitution that allows all Americans to be spied upon by their government while denying them any viable means of challenging that spying."[129]

There have also been numerous hearings and "expressions of concern" by government officials and agencies along the way, and even multiple inspectors general publicly criticizing the value and foundational legal reasoning for the NSA's domestic surveillance program,[130] but the program only grew over time, as Edward Snowden eventually exposed. Most striking is that this program that endangered civil liberties for almost two decades continued through at least two more administrations. Three types of surveillance were initiated by the US government in the aftermath of 9/11: a mass telephone records collection program, a mass Internet metadata collection program, and what became known as the Upstream program, which tapped into the Internet backbone with the help of the major telecommunications companies.[131] While the telephone records and Internet metadata collection programs are both said to have formally ended (with the phone program being declared illegal by a US Court of Appeals in 2020), it was revealed years later that the NSA had found a functional equivalent for the Internet surveillance program, and simply began to access the same data through foreign collection efforts.[132] Moreover, as of this writing in 2023, the Upstream program still continues its warrantless surveillance of millions of US citizens. Section 702 of FISA, which authorizes it, will be up for renewal again in 2024.

The reliance on private surveillance infrastructure has been a hallmark of cloud policy throughout its history. Bruce Schneier has addressed this "alliance of interests," noting, "The NSA didn't build a massive Internet eavesdropping system from scratch. It noticed that the corporate world was already building one, and tapped into it."[133] It is not unrelated that intelligence gathering in the US has also become overwhelmingly privatized—70 percent of the US intelligence budget ($62.8 billion in 2020) has been going to private contractors—such as Booz Allen Hamilton, which employed Snowden—since at least 2007.[134] This has created an outsourced culture of corporatized intelligence with very little accountability, particularly to the taxpayers footing the bill. It has also been described as a "transformation of the Cold War intelligence bureaucracy into something new and different that is literally dominated by contractor interests."[135] The difficulty of exposing government corruption by using proscribed whistleblowing channels has been well documented.[136] The increasingly privatized operations of intelligence gathering—whether internally by hiring contract labor or externally by creating partnerships with tech companies—has made it even more challenging (and unlikely) to call out fraud or abuse taking place. Furthermore, while Congress did extend the protections for whistleblowers to private contractors in 2016, that still would not have helped Snowden in 2013, because of the classified nature of the materials he leaked. As professor and author Jill Lepore explained, "Snowden signed an oath not to disclose government secrets, and neither the Whistleblower Protection Act nor its many revisions and amendments extend its protections to people who disclose classified intelligence. . . . If you steal classified documents, you can't be a whistle-blower."[137]

Nevertheless, once the information became public, it created an international crisis of trust in the US with respect to data privacy. Initially, the revelation of PRISM and the cooperation of US tech companies led to many businesses and governments outside the US turning away from US services and their cloud provision, looking elsewhere to house data outside the reach of the NSA and its surveillance network. It was estimated that the US tech industry lost up to $180 billion thanks to global distrust toward US cloud infrastructure providers in the immediate years following the Snowden leaks.[138] One of the more significant such consequences for cloud policy and international data regulation was the implosion of the "Safe Harbour" agreement between the US and the EU. This agreement allowed for the transfer of personal data between the United States and Europe and was the cornerstone of transatlantic e-commerce. It was necessary because the European Union has been much more aggressive about protecting data and the privacy of its citizens than most other countries or regions have been, particularly the United States.[139] Safe Harbour created a bilateral commitment to a watered-down version of European privacy laws so that the US and EU could continue to do business over the Internet. Companies like Facebook, Microsoft, Amazon, Google, and others relied on it because, without Safe Harbour, they would not be participating in the transatlantic digital economy.

Enter Max Schrems, then an Austrian law student taking a semester abroad at Santa Clara University in California. Schrems was shocked when a Facebook privacy lawyer visiting his privacy law seminar seemed tragically uninformed about European data protection laws, and he decided to write his final paper for the class on "how Facebook was flunking privacy in Europe."[140] When doing his research, Schrems discovered that the company kept "dossiers on individual users [that] are hundreds of pages long, and include information users thought had been deleted."[141] He got involved in privacy advocacy, and subsequently filed twenty-two complaints with Ireland's Data Protection Commissioner in 2011 over Facebook's privacy policy and their handling of user data and posts. Schrems then began a long and winding lawsuit against the Irish Data Protection commissioner, as Facebook's European data flows through its Irish subsidiary, Facebook Ireland. Incredibly the lawsuit eventually went all the way to the Europe's highest court, the European Court of Justice (CJEU). In what became known as "The Facebook Case," or now "Schrems I," the Safe Harbour agreement was struck down in October 2015, in part because of what Edward Snowden revealed about the privacy invasions of millions of Europeans by US intelligence services, and in part by the FISA Amendments Act of 2008.[142] In the end, the CJEU determined that FISA and programs like PRISM rendered European citizens insufficiently protected from US government surveillance and, as such, US law did not afford European citizens "adequate privacy protections" for their personal data that was being stored and processed by US-owned servers. With that, data transfer between the EU and US was thrown into legal chaos. What began as a graduate student's seminar paper eventually brought down one of the most important regulatory regimes for the international trade and global exchange of personal data.

    This issue is an extension of a conflict that has persisted at least since 1991, when the EU began trying to impose transborder data flow restrictions and the US attempted to block such policies.[143] The struggles have continued well into the digital era of international cloud storage. Ever since the landmark Schrems I decision, as the EFF has noted, "multinational companies, the US government, and the European Commission sought to paper over the giant gaps between US spying practices and the EU's fundamental values."[144] As the US government chose to neither end the PRISM program nor pursue stronger privacy laws that would allow for a long-term global accord, they simply reinstalled a modified Safe Harbour with a new name: the EU-US Privacy Shield. However, without a fundamental change in how the US treated data security, this was never going to last—and it didn't. The Privacy Shield of

2016 proved too fragile, as predicted by many civil society organizations and legal scholars who viewed this framework as simply putting lipstick on a pig.

Max Schrems, now a lawyer and activist, led the charge once again. In the case that became known as Schrems II, the CJEU decided, in July 2020, that the Privacy Shield was invalid because of the nature of US surveillance programs.[145] Once the bombshell decision was handed down, there was no longer any accord for reduced expectations of privacy protections, and this left the US in the position of finally having to seek legislative remedies. As the EFF described it, "five years after the original iceberg of Schrems 1, Schrems 2 has pushed the Titanic fully beneath the waves." The US lost its privileged arrangement and would have expectations of privacy protections that were the same as any other country going forward. After the Schrems II ruling came out, following the same legal reasoning as Schrems I, Schrems himself took a victory lap with the press:

> The Court clarified for a second time now that there is a clash of EU privacy law and US surveillance law. As the EU will not change its fundamental rights to please the NSA, the only way to overcome this clash is for the US to introduce solid privacy rights for all people—including foreigners. Surveillance reform thereby becomes crucial for the business interests of Silicon Valley. This judgment is . . . the consequence of US surveillance laws. You can't blame the Court to say the unavoidable—when shit hits the fan, you can't blame the fan.[146]

Perhaps the third time will be the charm: the Trans-Atlantic Data Privacy Framework is the new agreement, as of 2022, regarding protection measures for EU-US data flows. Supposedly, it will offer new legal protections for personal data access and use by US national security entities and address the surveillance roadblocks for European regulators contained in the Privacy Shield. Whether it does remains to be seen.

Digital data's policy pathways have taken us through the terrain of state surveillance, espionage, energy and the environment, workplace information cultures, and international trade, among others. Together, they track a constellation of issues and concerns that have enabled a decentralized and privatized global cloud infrastructure that is inconsistently regulated across territories, and in which users' data and privacy rights are often left unprotected. American data policies have begun to fragment even further, as states begin to fill the federal void with their own legislation. California enacted their own comprehensive consumer data privacy law in 2018—the California Consumer Privacy Act (CCPA). This gives California residents greater rights and control over their personal data collected by businesses and compels stronger transparency requirements than federal laws. Four other states have since followed with their own laws for their residents: Colorado, Connecticut, Utah, and Virginia. These laws are set to take effect throughout 2023. As state policies splinter and global agreements falter, there is one area of influence for data policy that continues to grow more powerful: Silicon Valley.

Private Control/Public Data

Amazon Web Services (AWS) is the most successful cloud infrastructure company in the world. In 2006 Amazon began aggressively marketing their cloud computing infrastructure (which included data distribution, storage, and processing). As of 2020, AWS controlled a third of the $100 billion cloud market, more than the combined share of its three largest competitors, Microsoft Azure, (18 pecent), Google Cloud (8 percent), and IBM Cloud (6 percent).[147] Their sales continue to rise, as do their profits. NASA, Netflix, the Federal Reserve, the Department of Defense[148] and the State Department, the CIA, Twitter, Facebook, Dow Jones, and NASDAQ are among their many clients. By the time Google finally offered their own cloud service, Google Cloud Platform, which rolled out in 2012, Amazon had become a massive utility in and of itself. In April 2015, Jeff Bezos revealed that Amazon Web Services had brought in $4.6 billion in revenue the previous year and was on track to out-earn his retail business soon.[149] In 2021, AWS reported over $62 billion in earnings, with no signs of slowing down.[150] Their economic success and industry dominance has translated directly into political power, especially in the realm of data governance.

As services like AWS and Azure become ubiquitous and the public sector's dependence on the private cloud increases, the responsibilities for determining the rules and regulations for data protection have shifted as well. The US has no comprehensive, unified formal vision for data regulation and protection on par with that of the EU. Instead, the terms dictating data privacy and security have increasingly become the province of Big Tech corporations and, more specifically, their terms of service (TOS), privacy policies, and end user license agreements (EULAs). These byzantine documents have stepped into the void created by the lack of appropriate regulation and policies set by the state and now "function as a form of privately made law" at the expense of consumers.[151] In their study of digital ownership and legal property rights, Perzanowski and Schultz explain that in order to participate in digital culture—whether streaming or downloading media, reading electronic copies of books, or even using one's personal email—we must accept licensing and usage terms that are "private regulations that redefine consumer rights . . . dictated by private actors driven by their own self-interest."[152] They argue that "we've replaced courts and due process with code and license terms."[153] These informal policies and processes are now the primary drivers establishing the terms for data governance.

Many of these terms are quite egregious, although most users will never know just how intrusive the policies are that they "agree" to. While EULAs and TOS agreements have undoubtedly become foundational texts for media scholars, they are not light reads. One *New York Times* study analyzed 150 corporate privacy policies and determined "only Immanuel Kant's famously difficult 'Critique of Pure Reason' registers a more challenging readability score than Facebook's privacy policy."[154] These agreements are written in legalese that is vague, impenetrable, and totally inscrutable to most readers. They are also anything but concise, as Perzanowski and Schultz have noted: iTunes' agreement is longer than *MacBeth*, and PayPal's is longer than *Hamlet*.[155] Over the course of two decades, Google's privacy policy expanded along with its increasingly elaborate data collection practices, going from a two-minute read in 1999 to a thirty-minute slog by 2018.[156] Accordingly, the terms of governance in the digital economy now extract entirely unreasonable promises from users regarding their private data in exchange for access to the service. Worst of all is the overall trend, according to Federal Trade Commission Chairwoman Lina Khan, "to eliminate courts as a means for ordinary Americans to uphold their rights against companies."[157] The TOS fine print includes protection from lawsuits and class actions. Users are nevertheless forced to agree in order to use the service, surrendering their own rights and thus creating immunity and dramatically limiting legal accountability for platform corporations.

Most of Silicon Valley's TOS agreements now also contain jurisdiction requirements for any lawsuit that might arise, superseding any kind of territorial-based determination. Amazon's AWS, for example, has locations across the US, as well as in Amsterdam, Athens, Beijing, Berlin, Brussels, Bogotá, Bucharest, Budapest, Cape Town, Dubai, Dublin, Frankfurt, Hanoi, Hong Kong, Hyderabad, Johannesburg, Kolkata, Lagos, London, Lisbon, Manchester, Manila, Milan, Montreal, Nairobi, Osaka, Paris, Prague, Rio de

Janeiro, São Paulo, Seoul, Shanghai, Shenzen, Singapore, Stockholm, Sydney, Taipei, Tel Aviv, Tokyo, Vienna, Warsaw, Zagreb, and Zurich, among many other cities.[158] Despite their presence all over the globe, the AWS End User License Agreement (EULA) and private contracts stipulate choice of law provisions—the set of rules used to select which jurisdiction's laws to apply in a lawsuit, including the question of what courts can determine or even hear disputes about which laws apply. This means Amazon requires that most cloud computing disputes—there are currently eleven countries with exceptions—must be resolved in Washington State, site of their corporate headquarters, even if the user (or the data/server in question) is located in a different part of the world.[159] As with almost all things related to data jurisdiction, location is key; courts in Amazon's home state are far more desirable to AWS for resolving disputes than anywhere else. Therefore, when you click "I Agree," you are also agreeing to the terms of resolution for any future disputes that might arise between you and their company. This has effectively rewritten the map according to global cloud territories governed by private terms of service agreements, creating an extralegal landscape wherein Big Tech is writing the rules that global governments must follow.

The Apple-FBI standoff in 2016 over breaking the encryption of an iPhone is a dramatic example of the private sector's stealthy but formidable power in the cloud policy space. In February of that year, the FBI ordered Apple to unlock the phone of the deceased suspect in the San Bernardino mass shooting that took place in December 2015. The government wanted Apple to create a type of "back door" in to the phone through a custom operating system that could disable its security features. Apple refused, stating that if this "unlocked" version of its operating system were to be leaked, stolen, or misused, it could potentially compromise the security of hundreds of millions of their customers. Of course, it would also destroy consumer confidence in the brand. Tim Cook, Apple CEO, published an open letter about the need for encryption, the FBI's threat to data security, and the dangerous precedent of the government's "chilling" demands, which would "undermine the very freedoms and liberty our government is meant to protect."[160]

The FBI sued Apple, and thus began a very public battle over data policy and the complicated relationship between the federal government, law enforcement, and Big Tech companies in the privacy ecosystem. Even the United Nations came out in support of Apple, noting that encryption is "fundamental to the exercise of freedom of opinion and expression in the digital age." Then-FBI Director James Comey responded publicly with the FBI's view, that "there is no place outside of judicial reach.... There is no such thing as absolute privacy in America."[161] The FBI's position on US citizens' (lack of) privacy rights, in addition to the public spectacle of the standoff with Apple, also pointed to one of cloud policy's main conflicts of power and trust. As the public evaluated whether they believed Apple (and by extension, Silicon Valley) or their government to be the more trustworthy stewards of their data, they were also witnessing the latest clash between formal and informal policy, and public versus privatized governance, even if unwittingly.

That particular round ended when the FBI dropped the case after they managed to access the data on the phone using professional hackers, but it did not begin in 2016. The US government and Apple had been locking horns since the debut of their encrypted operating system in 2014, which had such powerful security that Apple maintained they could not bypass a phone's security and access the data even if they wanted to. According to Tim Cook, Apple's leadership and legal team "had been meeting regularly with heads of the FBI, the Justice Department, and the attorneys general in both Washington and Cupertino."[162] The FBI wanted "access to phones on a mass basis,"[163] and Apple firmly resisted in the name of privacy and their obligations to their customers. Ultimately, Apple walked away without having to comply with the government's demands, winning that fight in the ongoing data governance war. As Susan Landau has explained in her extensive analysis of this event, even though the problem of this particular iPhone in this specific case was resolved, "the FBI's fear of 'going dark'—of losing the ability to listen in or collect data when this information was encrypted—remained."[164] Their reliance on Big Tech and the insistence on granting law enforcement "exceptional access," to encrypted data, she argues, is *not* "the FBI weighing the demands of security versus privacy. Rather, it is pitting questions about the efficiency and effectiveness of law enforcement against our personal, business, and national security."[165] This reframing by Landau emphasizes the longer arc of how privacy has been damaged by the dependence of law enforcement on the data controlled by private tech companies.

In fact, the ACLU reported that the government has been forcing these companies to help unlock their customers' devices in dozens of cases spread across twenty-three states since 2008, and that instead of invoking such power only in "extraordinary" cases, its use has "actually become quite ordinary."[166] The government's power in these cases is derived from the All Writs Act of 1789, which "allows federal judges the power to issue court orders" and basically gives them the ability to compel people to do things that are within the boundaries of the law.[167] Signed by George Washington himself, the act was designed to build in the necessary flexibility for a new democratic government to function and deal with unanticipated challenges, without allowing so much latitude that the founders' very recent problem with tyranny would again be able to rear its head. It is rather doubtful, however, that it was written in the spirit that would allow the FBI to compel a private company to design new software that allowed the government to surveil individual devices at will.

It is important to note here that, in 1994, the US Congress passed the Communications Assistance for Law Enforcement Act (CALEA), which specifically disallows the government from telling "manufacturers how to design or configure a phone or software used by that phone—including security software used by that phone."[168] The law did require telephone companies to alter their networks in order to make wiretapping of digital calls easier for law enforcement, and it was later expanded in 2005 by the FCC to apply to Internet service providers (ISPs) and VOIP services such as Skype.[169] However, CALEA still ensured that the government could not compel Apple or any other phone manufacturer to create a "back door" into its encrypted software, and that was the law for twenty-two years. That is why their lawyers turned to the All Writs Act despite the very clear language in CALEA. As CALEA was rolling out, the NSA also lost a formative struggle in the 1990s "crypto-wars" over the same issue. Referred to as "the first holy war of the information highway,"[170] this fight involved the "Clipper Chip," a device that gave the government a "back door" or digital "key" to encrypted telephone conversations. By adding the chip to AT&Ts landline telephones, it provided the government with surveillance capabilities that the NSA was lacking in the early 1990s. However, it was an unqualified failure thanks to widespread objections from telecommunications and technology industries, as well as privacy activists, technologists, civil liberties organizations, and even some prominent politicians. The NSA had to basically admit defeat, and General Michael Hayden, who went on to head both the NSA and the CIA, lamented that the NSA "lost" this crypto-war: "We didn't get the Clipper Chip, we didn't get the back door."[171]

Instead, the NSA eventually walked right through the front door, welcomed in by the private sector. However, soon after Snowden's bombshells, some of the very same companies that had been integral to the PRISM program decided that public exposure of their participation was bad for their brand, so they sued the US government over their inability to comment on FISA requests. At first it was just Microsoft and Google that were unhappy about being silenced from discussing their roles, and with the government's "continued unwillingness" to publish more information about anything related to FISA. They filed suit in June 2013, citing First Amendment concerns over the freedom to reveal details about how much data they were sharing. The true issue at hand was the harm to their bottom lines, as they also demanded the right to "defend [their] corporate reputations battered by Edward Snowden's revelations."[172] Microsoft and Google were eventually joined by other tech companies including Facebook, Yahoo!, LinkedIn, and Twitter. In 2014, they came to a settlement when the government relaxed rules "restricting what details companies can disclose about Foreign Intelligence Surveillance Act (FISA) court orders they receive for user information," allowing the companies to show a more "limited involvement" in the widespread surveillance program that led to tremendous financial losses and global distrust in their brands.[173]

Microsoft sued the government again over more gag orders in 2016, this time joined by dozens of tech companies (including Apple and Mozilla), civil rights groups (ACLU, the EFF), and news media from NPR to Fox News, as well as airlines, law enforcement and even biomedical companies as signatories. This suit was against the Department of Justice for compelling tech companies to turn over customers' data without their knowledge, a provision enabled by the ECPA. Such gag orders are customarily used in national security investigations, but Microsoft argued that "its customers have a right to know when the government obtains a warrant to read their emails," and the company had a right to tell them. Further, Microsoft claimed in its filing that the government "has exploited the transition to cloud computing as a means of expanding its power to conduct secret investigations."[174] Once the Justice Department agreed to "limit" its use of gag orders, ending their indefinite timeline and secrecy demands in 2017, the suit was dropped. To avoid a broader court ruling that might determine the practice to be unconstitutional, the DOJ struck a deal; nevertheless, it is not a victory for consumers enshrined in legislation or covered by any First or Fourth Amendment protections.

Much of this public fanfare over privacy and security is ultimately about public relations. Big Tech has historically used these concepts for branding purposes while simultaneously undermining them all over the world. Tim Cook, for example, has said that he views privacy as a "fundamental human right" and regularly calls for federal privacy laws in the US similar to those in Europe. He has also publicly criticized Facebook and Google, equating their services to surveillance as their growing invasions of user privacy and reckless exploitation of user data are revealed.[175] Meanwhile, Apple operates a data center in China and complies with governmental regulations allowing user data to be inspected and investigated by the state without question or exception. As a condition of opening the center, legal ownership and control was handed over to a government-backed company in southern China called Guizhou-Cloud Big Data (GCBD). In so doing, Apple "has constructed a bureaucracy that has become a powerful tool in China's vast censorship operation."[176] The company heavily censors its Chinese App Store, does not allow the use of the Taiwanese flag emoji on iPhones in the region, and shows Taiwan as part of China on their maps. Apple is also required to have all of its encryption technology used in China approved by the government. These concessions in exchange for a presence in the Chinese digital market have virtually guaranteed the Chinese government access to the emails, contacts, and geolocation and personal information of millions of Chinese citizens, thereby endangering the "fundamental human right" of privacy that Apple so publicly claims to value.

The expanding power of Big Tech intermediaries, particularly with respect to data and surveillance, has drawn attention to their lack of accountability and oversight; their role in the destruction of public values; and their threats to democracy and the autonomy of citizens. Taking these issues (and their attendant crises) to the realm of geopolitical power, we are reminded of Alan Rozenshtein's argument that these companies also "challenge the state's monopoly over security, the very locus of traditional conceptions of sovereignty."[177] Katharina Pistor has discussed data "as a tool for governing others on a scale that rivals that of nation states with their law."[178] As such, we are also witnessing the mounting battle between public and private determinants of sovereignty as waged through data and its control. Making sense of this profound shift requires circling back to the relationships between data location and regulation, where we find one last piece of this global puzzle: national clouds.

Data Sovereignty, Data Localization, and National Clouds

The growing trend of "data sovereignty"—subjecting data to the laws of the state or territory in which it is collected and stored—has been rising ever since the Snowden revelations in 2013. Once the global community realized that data passing through US cloud infrastructure would be exposed to the NSA's well-documented regimes of surveillance and privacy violations, governments all over the world sought to establish greater control over their citizens' data. Law and policy scholar Kristina Irion wrote in 2012 that data sovereignty was, at that point, "not an established legal concept but simply shorthand for retention of authority and control over information assets" and conceptually broader than privacy and data security.[179] At its core, the goal of data sovereignty aims to connect data to the territorial jurisdiction and laws of a specific state and protect it from all others. This requires a broad acceptance of the territorial logics of nation-states that impose the geopolitical borders and boundaries on the immaterial resource of data, a necessary precondition for the current stage of commodification and extractive capitalism described by Couldry and Mejias as "data colonialism." As of this writing, more than a hundred countries have some sort of data sovereignty laws in place,[180] leading to growing fears about the fragmenting, divided "splinternet" and the attendant balkanization of global communication, technologies, and markets.[181]

As more governments have embraced the idea of greater control over their citizens' data, the movement around data sovereignty has also led to the adoption of "data localization" and, subsequently, the birth of "national clouds" or "sovereign clouds." Data localization is a strict construction of policies "whereby national governments compel Internet content hosts to store data about Internet users in their country on servers located within the jurisdiction of that national government (localized data hosting)."[182] Depending on the particular government and data context, it could be simply a copy of the data, a particular type of data, or it could mean that all data is restricted from crossing any state border once it is stored. This movement to "nationalize" clouds or "localize" the global Internet is being increasingly used by governments seeking more control over data security. It can be partially attributed to the worldwide mistrust engendered by the fallout from the Snowden revelations and, in other cases, stems from the desire to tightly control information flow and/or citizen behavior in authoritarian countries.

In July 2020, Turkey passed a law that orders social media platforms with over one million daily users (e.g., Facebook, Twitter, YouTube) to open offices in Turkey and remove content deemed offensive by the government or face stiff penalties in terms of fines and bandwidth throttling that would stifle their distribution. They also would be forced to store user data inside

Turkey. This follows increasing media control by the Erdogan regime aimed at stifling dissent and controlling the information available to citizens.[183] Bruce Schneier has addressed the deeper tensions, noting that "cyber sovereignty is often a smoke screen for the desires of political leaders to monitor and control their citizens without interference from foreign governments or corporations. And the fight against cyber sovereignty is often viewed as a smoke screen for the NSA's efforts to gain access to more of the world's communications."[184]

The "national cloud" movement has proceeded unevenly, but it has many legal scholars concerned about the complex data security, privacy, interoperability, and law enforcement issues this dynamic has created. Jennifer Daskal, for one, has written about such developments as creating "a race to the bottom, with every nation unilaterally seeking to access sought-after data, companies increasingly caught between conflicting laws, and privacy rights minimally protected, if at all."[185] Paul M. Schwartz has cautioned that "all clouds are not created equal" and has pointed to numerous technical and geopolitical problems for the future of global access.[186] The balkanization created by data localization also threatens the free exchange of ideas, information, and even protections for human rights. Naturally, Big Tech US-based cloud providers have been waging a war of rhetoric against "national clouds," and yet these same companies have been quite active in creating the very same infrastructural formations as concessions to global regulators. For example, in a 2015 blog post published in the wake of the EU-US Safe Harbour dissolution, Microsoft's President Brad Smith cautioned that a balkanized Internet could mark "a return to the digital dark ages."[187] Less than five years later, the company was proudly touting three national cloud deployments—one for the US government's data, one in Germany, and one in China. These are defined by the company as "physical and logical network-isolated instances of Microsoft enterprise cloud services that are confined within the geographic borders of specific countries and operated by local personnel."[188] Amazon is essentially creating their own version of national clouds all over the world with their thirty-one "geographic regions" that currently house ninety-nine "availability zones," described by AWS as "fully isolated partitions of our infrastructure."[189] This new corporatized division of the globe into "availability zones" of cloud provision has Amazon designing a new geography of private "nations" based on territoriality of infrastructure with an unelected Big Tech company as sovereign.

Data localization has also been adopted to varying degrees by countries from Germany, Canada, and Australia to Russia, Turkey, Brazil, and China.[190] In 2012, even before Snowden's revelations about the NSA came out, a report by the European Parliament on privacy and the cloud recommended that the EU countries build their own cloud computing data centers and locate them on the European continent. The report emphasized, "It is important to reiterate that jurisdiction still matters. Where the infrastructure underpinning cloud computing (i.e., data centres) is located, and the legal framework that cloud service providers are subject to, are key issues."[191] In addition to the concerns over the PATRIOT Act and fears about data insecurity and government overreach, much of the reaction in the EU driving regional bodies and individual countries toward sovereign clouds has been due to the inordinate "hyperscaling" of US tech companies and their occlusion of local services. Big Tech's dominance in the global cloud infrastructure space is unparalleled and only growing: AWS, Microsoft, and Alphabet's Google have a combined worldwide share of 64 percent of the market for data storage in the cloud with no real competitors in sight.[192]

The General Data Protection Regulation (GDPR) was implemented in 2018 in the European Union to protect data security and transfer and the privacy of EU citizens. GDPR data privacy protections afforded to "data subjects" in the EU (which apply only to personal data relating to individuals, not to corporations and other legal entities) include the right to erasure (formerly known as the "right to be forgotten"), the right to rectification of inaccurate or incomplete personal data, the right to object to processing personal data, and the right to access information about the purpose and types of personal data being processed.[193] Fulfilling these rights are the obligation of the data controller and enforced by the state, highlighting the dramatically different standards between the US and EU, as well as the lack of consistency in global privacy protections. Interestingly, the GDPR created greater European demand for US-based cloud providers, thanks to the promises of Amazon, Microsoft, and other US giants in the industry to "keep data far from the prying eyes of American spies by sequestering it in Europe."[194]

In their quest to provide cloud infrastructure to the rest of the globe, including (and perhaps most especially) the world's largest market, Big Tech has made many concessions to authoritarian regimes. Microsoft's Azure was the first to provide cloud services in China in 2014, partnering with a Chinese company (21Vianet) to operate data centers in Beijing and Shanghai. AWS has also partnered with Chinese companies since 2014 to provide cloud services in Beijing and, three years later, in Ningxia. Both regions are "isolated from all other AWS regions" and its global network, to create a restricted national cloud according to strict Chinese regulations and censorship protocols.[195] China's Cybersecurity Law of 2017 mandated that Chinese data must be kept on servers within China's borders, in their own "national cloud." Consequently, Apple's Chinese iCloud operations also moved to the city of Guiyang in 2018.[196] Hosting data in government-controlled data centers and enabling state censorship and privacy violations has become part of the price of doing business in China.

Russia moved in the same direction as China, enacting the country's Sovereign Internet Law, which effectively allows Russia to cut itself off from the World Wide Web and global data servers. The law came into effect November 1, 2019, after being signed into law amid great protests. The *Moscow Times* called it an "Internet Isolation Bill,"[197] and it has been widely criticized for taking the country's Internet infrastructure even further down the path of centralized totalitarian control—an "online iron curtain" of sorts modeled on China's "Great Firewall."[198] Vladimir Putin reportedly enacted the law to "shut Russians off from information contradicting the Kremlin's approved narrative," but it also utilizes an alternative domain name system from the rest of the global Internet. The law gives officials the power to block access to websites (including Facebook and Twitter), platforms, and VPNs, and requires more filtering and surveillance by Russian Internet service providers and state administrators.[199] Despite worldwide human rights concerns, the "sovereign Internet" has removed free speech rights, individual privacy, and information freedom online for Russia's 146 million citizens.

The anti-Putin "Smart Voting" app that was initially made available through Apple's and Google's app stores before the 2021 Russian parliamentary election was another casualty of Big Tech capitulation to autocratic demands. It was designed by the team of imprisoned Russian dissident Aleksei Navalny in order to provide opposition-minded Russians in each of the country's 225 electoral districts with recommendations for candidates to vote for that would "create turbulence in the system."[200] Russian officials pressured the tech giants to remove it, and they eventually complied as the threats escalated.[201] Such responses from the US-based corporate data stewards in

Figure 3.8
Protesters against Russian bill to establish a "sovereign Internet" in March 2019.
Credit: Printed with the permission of Radio Free Europe/Radio Liberty, "Moscow Protests Against Internet Bill," March 10, 2019.

authoritarian countries are simply excused by stating they are obliged to "follow the country's laws," but in doing so, they are also supporting regimes of censorship, human rights abuses, and fascism.

The role of Big Tech companies in data governance and the long arc of data policies compromising the right to privacy across global jurisdictions have contributed to what Edward Snowden has described as "the transformation of the free and fragmented internet into history's first centralized means of global mass surveillance." Moreover, he added that, in the US, "this violation of our fundamental privacy occurred without our knowledge or consent, or even the knowledge and consent of our courts and most lawmakers."[202] With this understanding, where do we turn? What do we do, now that our cloud infrastructures have been exploited in ways that are violating the rights and freedoms of unsuspecting citizens? Thoughts on those questions are what follows.

In the end, history is always our best teacher. If cloud policy has taught us anything, it is that the same legal and cultural struggles will await the next critical infrastructural technology and the one after that—until the issues they represent are widely understood as those necessary to defend civil liberties and the health and vitality of democracy, and they are regulated accordingly. Sadly, most citizens remain unaware, uninterested, or unsure of what to do about our current predicament. This is in part attributable to a lack of public education about the issues and stakes of cloud policy, to impoverished and compromised political leadership, and to the poor quality of media coverage about the regulation of cloud infrastructure. In turn, the breakdown of public values in this policy domain has snowballed at a truly alarming rate—bringing us ever closer to the "abyss from which there is no return" that we were warned about so many years ago.

Epilogue: Preserving the Cloud's Future

We all forget history.
—former FCC Chairman Newton Minow

If people were the deciders, they wouldn't take the deal, but they are not the deciders.
—Google privacy and data protection executive

Another flaw in the human character is that everybody wants to build and nobody wants to do maintenance.
—Kurt Vonnegut

Developments in cloud policy happen at a dizzying speed, and writing about them often feels like chasing a moving target. As I was finishing this book, in 2023, Elon Musk bought Twitter for $44 billion and renamed it "X." Microsoft was fighting to save its $69 billion deal for video game publisher Activision Blizzard, encountering a rising global skepticism of tech mergers. The Biden administration funded a massive expansion of broadband on tribal lands and created an Office of Indigenous Communications and Technology in the Department of the Interior. Google and Facebook began to face increasing penalties for privacy violations, seeing fines in the hundreds of millions of dollars and euros. Section 230 had its first day in the Supreme Court. And the FCC remained deadlocked more than halfway through President Biden's first term, with the eminently qualified fifth commissioner's nomination blocked for sixteen months in Congress and, ultimately, destroyed by a coordinated lobbying and smear campaign led by News Corp and Comcast.[1]

Thankfully, a historical lens slows the pace. It also expands the analytical terrain and affords the necessary panoramic perspective to understand long-term trends without the whiplash that comes from reacting to individual events as they unfold. In the case of cloud policy, that brings a recognition of how infrastructures come to embody a society's values, aspirations, and political power struggles over time. History reveals the long-term ideological progression in cloud networks from being a public resource to becoming a means for private gain. It further shows us that the limits on monopolies, speech rights, access to information, and privacy are interwoven in the fabric of these infrastructures. We can then recognize how the determination of civil liberties has become privatized, and democracy has become compromised, by the evolution of pipeline, platform, and data policies. The historical approach reminds us that we have been here before and that there are lessons to be learned from the triumphs, the failures, and the battles themselves.

Former FCC Chairman Newton Minow has maintained that the words "public interest" are "at the heart of what Congress did in 1934, and they remain at the heart of our tomorrows."[2] If we are to preserve this elegant vision of regulatory purpose, we must work backward to that model of Progressive Era–inspired stewardship to move forward with revitalized cloud policy. This includes rejecting vacant corporate slogans like "don't be evil" as stand-ins for the responsibility of infrastructure providers to support the public interest. It demands reclaiming the commitments to the modern infrastructural ideal that posits infrastructure as a public good, too important to be surrendered to market forces. The connective tissue from the Gilded Age to the twenty-first century includes the tyranny of corporate trusts, but monopolistic control over this ecosystem was not the only option, and the current predicaments of cloud policy are neither irreversible nor inevitable. Nonetheless, for these twenty-first-century infrastructures of democracy to function for the public once again, we cannot afford to look away from the regulatory landscape. Every moment of technological change or "disruption" is important, but so are the long stretches of entrenchment. These seemingly idle phases are when cultural practices become solidified and policy becomes a way of life. The fights over cloud policy are dependent on all of our sustained attention and engagement.

In addition to privatized governance, we have been enduring decades of what Des Freedman has called "*negative policy* . . . a form of nonintervention where media markets and institutions are left to govern themselves without outside interference."[3] The nonintervention begins with a long-held resistance to regulating digital technology that was evident during the legislative debates preceding the 1996 Telecommunications Act. It has maintained an intractable chokehold on contemporary lawmakers in the US ever since. Former FCC Chairman Tom Wheeler recalls being scolded by Congress multiple times for "trying to regulate the internet, . . . as if regulating the internet would break it. It was as if there was something magic about it, and if you messed around with it, you were going to break the magic."[4]

Nevertheless, vibrant activist communities and organizations have kept the pressure on for decades, pushing for more from policymakers. While broadband pipelines are not yet "bound by the law of public service undertakings" such as those imagined for Parkhill's computer utility, the groundswell of recognition that the "free market" has absolutely failed the public in this arena is growing. The only people left to convince are the politicians. Public interest advocates have remained resolutely focused on ensuring the Internet is available to all without undue cost. The history of cloud policy still carries the echoes of Theodore Vail promoting universality in 1910, and the urgency of wiring the land and delivering communication "from every one in every place to every one in every other place." Incredibly, a hundred years later, the provision of broadband has not matched what was accomplished with basic telephony.

Most recently, platforms have become the main target for reform efforts in cloud policy, as demands for public governance options and accountability intensify. Populist sentiments embracing public ownership as a way to control unmitigated corporate power and market forces are finding new traction in the digital age. The argument for treating platforms as public utilities has also been widely debated. Such visions in many ways hearken back to the turn of the century nationalization campaigns aimed at the telegraph and telephone, modeled after the post office. They also recall the EU's Digital Services Act package, which is a shining reminder that it is possible for policy to include consumer protections, corporate accountability, and transparency when regulating platforms. Another immediate concern for platform companies is the future of Section 230. The political calls for its repeal are mounting, but as one headline read, "Lots of Politicians Hate Section 230—but They Can't Agree on Why."[5] Some hate it because it offers too much immunity for the content that platforms host, others hate it because they think it allows for politically biased decisions about what viewpoints platforms take down. In fact, the desire to repeal Section 230 might be the one thing that the intensely polarized US Congress might agree on, but there is no consensus about what to do next.

The landscape for data policy also continues to evolve, now incorporating the turn from third-party to first-party tracking technologies that is under way. In response to widespread cultural concerns about personal privacy and legislation, such as the European Union's GDPR, Apple, Google, and Mozilla have announced plans to phase out support for "third-party cookies." These technologies track users across the Internet in order to deliver targeted ads and are foundational to the nearly $700 billion online advertising industry. Now that Apple and Google will no longer allow third parties to trail users' digital activities on their platforms, including iPhones, the Chrome web browser, and Android phones, a major shift is taking place. Among them, the data collection on users and user behavior will be done by those individual platforms themselves. With the growth of first-party tracking, the next obvious step is for the largest companies to become de facto data analytics firms for the rest of the Internet economy, with smaller firms being relegated to supporting, dependent players. Apple and Google's claim that the change was entirely motivated by the desire to protect user privacy is unconvincing. As watchdog group Ranking Digital Rights explained, these changes might shrink the number of companies that catalog our data, but "they also will help to consolidate our digital dossiers in the hands of a few uniquely powerful platforms, and reduce or even eliminate many of the smaller players in the ecosystem."[6] In keeping with the historical trajectory of cloud policy, Big Tech continues to get bigger, scoring another win for consolidated monopoly power and a loss for the public.

The trend toward more localized policies across the pipeline, platform, and data governance regimes in the US has also begun to accelerate. In many cases, these laws and regulations have begun to override or challenge federal agencies. One example is state legislation regarding net neutrality. In addition to the seven states that have currently enacted net neutrality legislation or adopted resolutions, eleven more introduced net neutrality bills in recent sessions. Moreover, twenty-nine states considered privacy bills in 2022 alone. In contrast to these positive turns in pipeline policy, Texas and Florida have passed their own state laws dictating what types of speech are protected from moderation on digital platforms, setting up what looks to be another trip through the federal court system for Section 230. Additionally, the data localization movement and sovereign clouds continue to expand, to the point that 75 percent of countries have some type of localization requirements.

This has created global complications for cross-border data flows and rising tensions between economic and privacy concerns.

The siloed nature of policy in the cloud space is one of its greatest weaknesses. Policy responses to data privacy and surveillance issues, monopolies, and speech rights need to be in dialogue with one another. They are inextricably intertwined with the vitality of the democratic process, civil liberties, and access to information. We are unable to address such matters if there is not a vision encompassing their interdependent relationships. As Moore and Tambini have noted, "policy is fragmented because the underlying thinking is siloed."[7] This applies to policymakers and the disciplinary isolation of policy-related scholarship as well. Economists, first amendment lawyers, specialists in antitrust and competition, data privacy experts, network engineers, cultural geographers, computer scientists, historians, and media scholars are rarely, if ever, in the same room together talking about policy solutions. Further, they often lack a common language and methodology to productively collaborate. If we are to reclaim the digital public interest, such interdisciplinarity and integrated approaches to cloud policy are elemental.

Activism and the Way Forward

Reforming cloud policy in the public interest will require the public's interest. In order to cultivate "an Internet of the public, by the public, and for the public . . . that advances instead of threatens democracy and the public sphere," as Christian Fuchs and Klaus Unterberger have advocated in their public service media and Internet manifesto, we must all get involved.[8] The advocacy community leading the way includes organizations such as the Center for Democracy & Technology (CDT), the Electronic Privacy Information Center (EPIC), American Civil Liberties Union (ACLU), and Electronic Frontier Foundation (EFF), as well as Public Knowledge, Access Now, and Ranking Digital Rights, among many others. These groups work with legislators, policymakers, Big Tech companies, and civil society to protect and advance the public's interest in matters of cloud policy and well beyond.[9] Part of this labor includes directing attention to the gaping holes in legacy regulatory frameworks compromised by more than a century's worth of regulatory capture and corporate lobbying, toward new possibilities that prioritize social justice and public values.[10] Alexandra Reeve

Givens, president and CEO of the Center for Democracy and Technology, addresses one source of her own inspiration despite the challenging circumstances in which advocacy organizations do their work:

> So many of the social justice issues of our time are closely tied to technology. And there's something that a group like CDT can do about them. It feels particularly true at this point in history.... If you care about civil rights in the 21st century, if you care about voting rights, if you care about reproductive freedom, all of those are tech issues. It feels lucky and very motivating to do this type of work when you realize just how much of our day to day lives and freedoms are embodied by technology, and how much we need groups like this focused on those rights.[11]

Labor activists have also begun their own efforts from within platform companies. Silicon Valley is largely built on the invisible labor of content moderators, identity verification workers, and caption teams—what Gray and Suri have termed "ghost work" or "the humans behind the seemingly automated systems that we all take for granted."[12] The labor is grueling, often traumatizing, isolating, and precarious. These employees do not enjoy any union protections, as contract labor, and they are among the most exploited workers in the digital media economy. However, the dominant platforms have begun to face a growing labor organization movement that is reverberating across many sectors of the US economy. In response they have turned to union busting. The threatening success of workers' efforts at Amazon, Apple, and Google have been met with aggressive company-led resistance tactics including worker surveillance, disinformation campaigns, intimidation, and firings.[13] Big Tech's message is the same one that journalist Ida Tarbell found in her groundbreaking investigations of Standard Oil at the turn of the twentieth century: cross us and you will be crushed. Such values have been the privilege of monopoly corporations and have endured across time, markets, and technologies.

The continued vitality of journalism is critical for activists. A healthy, well-funded industry of investigative journalism is one of the advocacy community's best tools. Without it, citizens remain uninformed and democracy is on the ropes. Victor Pickard has emphasized the importance of this connection, writing that "any society that aspires to be a democracy must ensure the existence of reliable news and information systems."[14] The struggle against "trusts" in the Progressive Era was significantly aided by journalists like Tarbell, Upton Sinclair, Henry Demarest Lloyd, Lincoln Steffens, and Ida B. Wells—all writers known as muckrakers who documented government and corporate abuses and corruption. Tarbell's 1904 two-volume *History of the Standard Oil Company* began as a popular series in McClure's magazine that grew to be nineteen parts. Tarbell had watched her father go bankrupt in 1872 thanks to John D. Rockefeller and Standard Oil, and she went on to write the book that led to the company's breakup at the height of Rockefeller's power. She did this without any tradition of investigative journalism to draw on— she invented it as she went along. All the while, her journalistic work influenced the Supreme Court, Congress, and regulatory agencies.[15] Throughout this book, the connections among monopoly capitalism, platform business models, and the collapse of the news industry have been explored for their collective danger to democracy. What is most alarming in this loop of effects is how much of Tarbell's crucial tradition we have lost. The passion for justice and exposing the truth that drives journalists still survives, but the financial support for the industry has withered. The consequences of losing this democratic safeguard have reverberated across the domain of cloud policy, as we find ourselves in a new Gilded Age for the digital era.

Changing that ill-fated course requires confronting other foundational issues as well, including

- **Lobbying and campaign finance reform**: This is ground zero for eradicating the corporate chokehold over our political system. Confronting these issues necessarily involves revising campaign finance laws and tax laws, and radically increased transparency in all areas of political contributions, lobbying, and "outside spending" including by dark money organizations.[16] At one time, comedian Bill Maher suggested that US politicians should look more like NASCAR drivers, forced to wear the corporate logos of their sponsors sewn onto their suits. Why is this more radical than allowing invisible corporate money to dictate our laws and basic rights?

- **Antitrust policy and thresholds for industry competition**: The approach to antitrust must be completely revised, leaving the antiquated, ineffectual, and monopoly-enhancing standard of consumer welfare in the dustbin of history. Instead of consumer welfare, some economists have begun pressing lawmakers to consider market behaviors in terms of "citizen welfare," which incorporates the impact of competition on a democratic system of government.[17] The impact of competition (or lack thereof) on individual privacy is another key consideration for the health of democracy in this age of surveillance capitalism.

- **Education**: Digital media literacy should be a required component of formal education, beginning in junior high school. Thinking critically about issues such as disinformation, media ownership, and online privacy cannot start in elective university courses. Once young people are interacting on digital devices and social media platforms, they also need to be armed with knowledge about this larger ecosystem that dramatically impacts their lives. It is just as urgent as educating our students about climate change, history, or civics.
- **Public visibility**: These issues must become a regular feature of political discourse, and their news coverage should be as ubiquitous as that of the stock market. We can no longer afford for cloud policy to be unintelligible to the general public. To that end, key cloud policy issues such as Internet pricing and access, competition in the platform economy, public/private surveillance partnerships, and digital speech rights should be part of the platforms for all candidates for state and federal office. Citizens have a right to know where their potential leaders stand on matters of cloud policy, and they should be an expected component of all public debates.
- **Legal frameworks**: It is time to move beyond obsolete legislation and policy written in bygone technological eras that keep us tethered to the past. The persistence of legacy constructs such as the "third-party doctrine," "natural monopoly," and the "consumer welfare standard" contribute to regulatory environment that is willfully toothless and wholly incapable of addressing the realities of contemporary technocultures. They are among the many norms in the cloud policy landscape that are long past due for an upgrade.

Creative Solutions

The answers to the many questions that arise in the pursuit of better policy depend on the questions we see fit to ask. For example, legal scholar Alan F. Westin wrote, in 1967, "Will the tools [of surveillance] be used for man's liberation or his subjugation?" adding, "Can we preserve the opportunities for privacy without which our whole system of civil liberties may become formalistic ritual?"[18] Marietje Schaake, former member of the European Parliament, connected platforms to the demise of democracy when pointing to their governance by tech companies and asking the obvious: "With what oversight and legitimacy?"[19] Former FCC Commissioner Nicholas Johnson, when trying to discern whether regulations contributed or detracted from the contribution of the telephone network to American lives, asked during a visit to Bell Labs, "Do you have an anthropologist working here?"[20] Crafting solutions to any of the problems explored throughout this book could only benefit from questions with a similar spirit.

Solutions also come from historical successes. The design elements that have been key to the Internet's survival through military, civilian, and corporate control—flexibility, neutrality, interoperability, decentralization, and longevity—must also find their way into cloud policy's core values today. As van Schewick and Abbate have each argued, adaptability to unpredictability has to extend beyond network architecture to inflect policy values as well.[21] Of course, some of the challenges facing cloud policy do not have direct historical precedent, such as how to address the potential harms of AI in the form of chatbots or the profit-driven algorithms powering the platform economy, for example. They do underscore the necessity for algorithmic transparency, however, a value that has been proposed (thanks to the efforts of civil society) in legislation in the US, Europe, and elsewhere.[22]

It is also important to emphasize that we cannot engineer our way out of this policy crisis. Shoshana Zuboff has argued that the only solutions to this siege on democracy are political, not technological.[23] Fred Turner has also added, "It's time to let go of the fantasy that engineers can do our politics for us, and that all we need to do to change the world is to voice our desires in the public forums they build."[24] The solutions we seek entail new policy regimes that are undoubtedly dependent on political transformation, but also on larger societal-wide shifts. Policy *is* politics, but it is also culture, economics, and ideology. It is a way of constructing the world. It is also a system of power and control. It is time to inject values that prioritize the public good over the welfare of corporations into that system.

To enact the fundamental changes addressed throughout this book, it is also imperative to renew our understanding of public utilities and their role in society, as well as their future in the cloud policy domain. We can begin with Harold Feld's explanation that "when we designate a service as a utility, that means it has become too important to leave to the benevolence of corporations, the kindness of kings, or the cold indifference of the market. We must guarantee fair access for all under a rule of law."[25] Along these lines, Dan Schiller has argued that "our conception of public utility must be refreshed, reimagined. It needs to be sufficiently capacious to permit us to erect a common roof over all segments of contemporary networking: not only terrestrial, submarine, satellite, and mobile carriers, but also search, e-commerce, and social network companies. To accomplish this will require much political creativity."[26] Creativity is not necessarily a strength of contemporary regulatory politics, at least when it comes to serving the public. Perhaps things would begin to change with anthropologists, historians, and artists at the table. Without them, we are often incapable of seeing these less-visible dimensions of the policy emergency that the cloud represents.

Our cultural understanding of data also needs to adapt to current conditions in which it is a ubiquitous resource and a vector of surveillance. Jathan Sadowski has offered one proposal in this spirit, suggesting that we liken data controls for platforms to those of rent control that limits the amount of money that landlords can demand from tenants. He argues that a policy of data control should "[restrict] the conditions, purposes, and uses of data that corporations extract from people, while also overseeing the flow and exchange of data across different markets and industries. Data controls are crucial for reversing the vast political and economic asymmetries that currently exist in our system while delivering more power over platforms to the public."[27] This call for a core of accountability, transparency, and compliance regimes has been echoed in the work of the advocacy community, as well as by scholars across disciplinary divides.

Noted media scholar and activist Ethan Zuckerman has said that "our ability to imagine alternatives is directly related to the histories we tell."[28] Accordingly, it is crucial that we recognize this history of cloud infrastructure as one of failed policy. It is also the historical success of regulatory capture and compromised politicians supporting the unmitigated growth of corporations at the expense of public welfare. These ideological pathways have been paved by corruption and greed. They have been also supported by what Luzhou Li has called "media policy silence," or "policy practices marked by policy opacity rather than policy visibility, by the absence of formal policy or 'un-decisions' rather than decisions and by policy inertia rather than intervention."[29] Des Freedman views such silences and failures to act as pointing to "the options that are *not* considered, to the questions that are kept *off* the policy agenda, to the players who are *not* invited to the policy table, and to the values that are seen as unrealistic or undesirable by those best able to mobilize their policy-making power."[30] When these absences and exclusionary tactics are highlighted in the telling and retelling of policy history, the need for new frameworks and leadership is urgently apparent. The voices of activists and scholars have become louder in these silences of inaction, and so have their alternative visions for cloud policy that will allow for different histories to be told someday. Appel et al. have argued that "infrastructures are important not just for what they do in the here and now, but for what they signify about the future."[31] The future signified by contemporary cloud policy is undeniably bleak. With this book, I hope to render these destructive politics manifest so that, together, we can begin to redirect our infrastructural destiny onto a more equitable civic path.

We Could Have Been a Contender . . .

Much like Marlon Brando's character Terry Malloy famously lamented in *On the Waterfront*, there have been points throughout the history of cloud policy when we had a chance to do so much better. Indeed, things could have gone very differently at multiple junctures over the past hundred years.

The early twentieth-century battles between the government and AT&T, for example, were a series of missed opportunities for telecommunications policy reform. The FCC's 1939 report chronicling the lack of competition and the need for "actual and not nominal regulation" to protect the public interest went nowhere, as did the antitrust suit that followed ten years later, thanks to internal conflicts among government agencies. The first rewrite of the Communications Act of 1934 was a chance to establish policy for the digital era that comes once in a generation. Unfortunately, the Telecommunications Act of 1996 was a deregulatory assault on the public interest, allowing for unprecedented levels of industry concentration and cross-ownership. Six corporations soon controlled most major media properties and distribution channels. The Telecommunications Act also failed to ensure public access to the Internet, classifying ISPs as "information providers" and ultimately derailing the FCC's efforts to codify "net neutrality" and establish common carriage protections for broadband service. Thus far, the US Congress has failed to reestablish the landmark Internet privacy protections and the net neutrality rules established during Obama's presidency that were summarily repealed under Trump.

It's not as if we haven't been repeatedly warned about the problems we would be facing throughout the span of cloud policy's history. H. D. Lloyd linked the "railroad problem" and the forces of the trusts to the future of American democracy in the nineteenth century, writing back in 1881 that "the forces of capital and industry have outgrown the forces of our government."[32] Douglas Parkhill presciently argued back in 1966 that among the many critical needs for the growing "computer utility" was the need to understand the related economic and social issues, particularly as it "broadens and merges with the myriad other challenges that are endemic to our modern society—the promise and threat of automation, the struggle for racial and social justice, and finally, the problem of survival in a divided and nuclear-armed world."[33] At the same time, Congress was holding hearings about the dangers of the government controlling citizen's data. Their 1968 report from the "Computer and Invasion of Privacy" proceedings cautioned against "a suffocating sense of surveillance" in which democratic principles could be sacrificed and freedom could not survive. At that point, Congress demanded that the federal government and the "computer community" guarantee all Americans that "the tonic of high speed information handling does not contain a toxic which will kill privacy."[34] And of course, Frank Church's 1975 prophecy of doom has now become our reality: technological advances have undeniably been "turned around on the American people" to facilitate government surveillance, and we are left without our privacy, just as he predicted.

As a result, in the policy space we are now faced with the consequences of Jill Lepore's question: "What if the future forgets its past?"[35]

All historians struggle with the implicit job requirement of remaining hopeful despite knowing too much, as "history does not offer a happier lesson very often."[36] That much is certain—the warnings of Chris Pyle, Neil Gallagher, Frank Church, and Frances Haugen are among the many testimonials affirming this point. They all told us so, as did Eugene V. Debs, Michael J. Copps, and the community of public interest advocates who have been working for better cloud policy since the analog era. And yet, as Edward Snowden reminds us, "awareness alone is not enough."[37] These twenty-first-century cloud infrastructures are shackled to policy that has become ruinous for society, and there is much work to be done. So we look to the boldness of rancher Thomas Carter and privacy activist Max Schrems, underdogs in the arc of cloud policy who took on Goliaths and prevailed. We take inspiration from Ida Tarbell, who successfully went after the most powerful trust in the world in an era when women did not even have the right to vote. And we try to live up to the legacies of Newton Minow and Nicholas Johnson, both tireless champions of the public interest who continued their respective crusades throughout their lifetimes. Although the villains usually get all the press, the history of cloud policy is replete with unsung heroes. May they be the ones to guide our fight going forward.

Appendix: Notable Investigations and Actions against Big Tech, 2017–June 2023

2017

- June—European Commission fines Google €2.7 billion euros for anticompetitive practices on their Google Shopping platform.

2018

- July—European Commission fines Google €4.34 billion for anticompetitive use of Android mobile O/S and installing apps on cell phones without permission.

2019

- January—French data protection authorities fine Google €50 million for failure to disclose data collection practices.
- March—European Commission fines Google €1.49 billion for stifling competition in the online advertising market.
- June—US House Judiciary Committee announces a bipartisan investigation into competition in digital markets. The investigation focuses on Amazon, Apple, Facebook, and Google and includes seven hearings, "Online Platforms and Market Power" in 2019–2020, producing a 450-page report in October 2020.
- July—The Securities and Exchange Commission (SEC) fines Facebook $100 million for "making misleading disclosures to investors about the risks of misuse of user data." These charges stemmed from the fact that the company knew about the misuse of user data by Cambridge Analytica since

2015 but did nothing about it for more than two years, and characterized the threat of improper use of data as a mere hypothetical for their investors and the news media.
- July—The Federal Trade Commission (FTC) fines Facebook $5 billion and orders it to create new layers of oversight after the Cambridge Analytica scandal revealed that the company had deceived its users about the privacy of their personal data.
- July—The FTC opens an antitrust investigation against Facebook.
- September—YouTube (owned by Google parent company Alphabet) is fined $170 million by the FTC for a COPPA violation of illegally collecting children's personal information.
- September—A group of attorneys general from forty-eight US states plus Puerto Rico and the District of Columbia begin investigating Google's dominance of the ad market and use of consumer data.
- December—The FTC begins investigating Amazon's retail business and cloud business (AWS).

2020

- October—The Department of Justice, along with eleven state attorneys general, opens *U.S. v Google*, the first antitrust case brought against Big Tech since *U.S. v. Microsoft* in 1998. Google is charged with unlawfully maintaining monopolies through anticompetitive and exclusionary practices in the search and search advertising markets.
- November—EU regulators file charges against Amazon regarding its anticompetitive use of data from third-party sellers to develop its own products.
- December—The FTC and a coalition of forty-six state attorneys general, the District of Columbia, and Guam, sue Facebook for its "systematic strategy" to eliminate threats to its monopoly with the "anticompetitive acquisitions" of Instagram in April 2012 and WhatsApp in February 2014. The case is dismissed a month later.
- December—Thirty-eight state attorneys general, the District of Columbia, and the territories of Guam and Puerto Rico sue Google over anticompetitive practices related to Google Search.

Appendix

- December—Ten state attorneys general sue Google over anticompetitive practices in their advertising business and collusion with Facebook; four more states and Puerto Rico join in 2021.
- December—The French Data Protection Agency fines Google €100 million and Amazon €35 million for illegal uses of advertising trackers (cookies) without user consent.

2021

- February—Facebook (Meta) settles a privacy class-action suit for $650 million after violating Illinois' biometric laws with its facial recognition technology that created scans without user consent.
- March—The UK's Competition and Markets Authority (CMA) opens an investigation of Apple regarding anticompetitive terms used in its App Store.
- May—The District of Columbia sues Amazon for abusing its monopoly power in the online retail market.
- June—Google agrees to pay French antitrust regulators $270 million in fines and change its practices to settle a case regarding their dominance of the online advertising market.
- June—The European Commission and British authorities began antitrust investigation of Facebook over its Marketplace classifieds service.
- July—Thirty-six state attorneys general and the District of Columbia sue Google over anticompetitive practices in its app store, Google Play.
- July—Luxembourg's National Commission for Data Protection fines Amazon €750 million for advertising violations.
- September—Ireland's Data Protection Commission fines Facebook (Meta) €225 million for transparency violations regarding WhatsApp data collections.

2022

- January—Three states (Texas, Indiana, Washington) and the District of Columbia sue Google over location tracking practices that invade consumer privacy.

- January—The FTC proceeds with a second attempt of its 2020 suit against Facebook for abusing their monopoly power when purchasing Instagram and WhatsApp.
- January—France fines Google €150 million and Facebook €60 million for making it more difficult to refuse cookies (which track browsing habits) than to accept them.
- May—The European Commission brings antitrust charges against Apple for restricting access to competitors in its Apple Pay digital wallet system.
- July—The FTC sues to block Facebook parent company Meta's acquisition of the VR company Within.
- July—The UK's Competition and Markets Authority (CMA) opens investigation of Amazon's digital marketplace for anticompetitive behavior and monopoly abuses.
- September—Google faces a €25 billion lawsuit in the UK and EU for anticompetitive conduct in the digital advertising market.
- September—The Irish Data Protection Commission fines Meta €405 million euros for breaking EU privacy laws with its handling of children's data on Instagram.
- October—Google settles with the state of Arizona for $85 million for deceptive and unfair practices related to user location data.
- October—British authorities force Meta to sell Giphy (purchased in 2020 for $315 million).
- November—Google agrees to pay $391.5 million to forty states to settle an investigation into privacy violations related to its location tracking features, the largest privacy settlement in US history. This is unrelated to the January 2022 lawsuit over location tracking practices.
- November—Facebook (Meta) reaches a $90 million privacy settlement in a US class action case regarding allegations of tracking users' online activity after they were logged out of the site.
- November—Irish Data Protection Commission fines Facebook (Meta) €265 million for major data breach.

2023

- January—The Department of Justice launches its second lawsuit against Google, this time for monopolizing digital advertising technologies in

violation of antitrust laws. The DOJ is joined by attorneys general from California, Colorado, Connecticut, New Jersey, New York, Rhode Island, Tennessee, and Virginia.
- January—Irish Data Protection Commission fines Meta €390 million in two cases regarding privacy violations, one involving Facebook and another involving Instagram.
- May—Meta is fined €1.2 billion for violating EU data protection rules and ordered to stop transferring data collected from Facebook users in Europe to the US.

Notes

Introduction

1. David Garland, "What Is a 'History of the Present'? On Foucault's Genealogies and Their Critical Preconditions," *Punishment & Society* 16, no. 4 (2014): 373, https://doi.org/10.1177/1462474514541711.

2. See, for example, Wendy Brown, *In the Ruins of Neoliberalism: The Rise of Antidemocratic Politics in the West* (New York: Columbia University Press, 2019).

3. Shoshana Zuboff, *The Age of Surveillance Capitalism* (New York: PublicAffairs, 2018), 53–54.

4. Also see Des Freedman, *The Politics of Media Policy* (Malden, MA: Polity Press, 2008), 97–100.

5. Harvey J. Levin, "Television's Second Chance: A Retrospective Look at the Sloan Cable Commission," *Bell Journal of Economics and Management Science* 4, no. 1 (Spring 1973): 343.

6. Levin, "Television's Second Chance," p. 362.

7. The Cabinet Committee on Cable Communications, Office of Telecommunications Policy, *Cable: Report to the President* (Washington, DC: US Government Printing Office, 1974), 11.

8. Tom Wheeler, interview by Jennifer Holt, February 28, 2023.

9. Sandra Braman, *Change of State: Information Policy and Power* (Cambridge, MA: MIT Press, 2007).

10. Julie Cohen, *Between Truth and Power: The Legal Constructions of Informational Capitalism* (New York: Oxford University Press, 2019).

11. Ithiel de Sola Pool, *Technologies of Freedom* (Cambridge, MA: Harvard University Press, 1983), 7.

12. Tung-Hui Hu, *The Prehistory of the Cloud* (Cambridge, MA: MIT Press, 2015), xxii.

13. Hu, *Prehistory of the Cloud*, xxv.

14. See Vincent Mosco, *To the Cloud: Big Data in a Turbulent World* (Boulder, CO: Paradigm Publishers, 2014), 16.

15. Compaq Computer Corporation, "Internet Solutions Division Strategy for Cloud Computing," Internal Document, November 14, 1996, https://s3.amazonaws.com/files.technologyreview.com/p/pub/legacy/compaq_cst_1996_0.pdf.

16. Quoted in Simon Garfinkel, "The Cloud Imperative," *MIT Technology Review*, October 3, 2011, https://www.technologyreview.com/2011/10/03/190237/the-cloud-imperative/.

17. Kevin Werbach, "The Network Utility," *Duke Law Journal* 60, no. 8 (May 2011): 1794.

18. Douglas F. Parkhill, *The Challenge of the Computer Utility* (Palo Alto, CA: Addison-Wesley Publishing, 1966), 52.

19. Parkhill, *Challenge of the Computer Utility*, 145–148.

20. Paul N. Edwards, "Some Say the Internet Should Never Have Happened," in *Media, Technology, and Society: Theories of Media Evolution*, ed. W. Russell Neuman (Ann Arbor: University of Michigan Press, 2010), 149.

21. See Janet Abbate, *Inventing the Internet* (Cambridge, MA: MIT Press, 1999), 65; Alexander A. McKenzie, "Oral History Interview with Alexander A. McKenzie," interview by Judy O'Neill, March 13, 1990, Cambridge, MA, https://conservancy.umn.edu/handle/11299/107489.

22. Jean-Christophe Plantin et al., "Infrastructure Studies Meet Platform Studies in the Age of Google and Facebook," *New Media & Society* 20, no. 1 (2018): 300, https://doi.org/10.1177/1461444816661553.

23. Lawrence Lessig, *Code and Other Laws of Cyberspace, Version 2.0* (New York: Basic Books, 2006), 350.

24. Benjamin Peters, *How Not to Network a Nation* (Cambridge, MA: MIT Press, 2016), 104.

25. Peters, *How Not to Network a Nation*, 104–105.

26. See for example J. C. R. Licklider, "Memorandum for Members and Affiliates of the Intergalactic Computer Network" (official memorandum, Washington, DC: Advanced Research Projects Agency, April 23, 1963), https://www.kurzweilai.net/memorandum-for-members-and-affiliates-of-the-intergalactic-computer-network.

27. Abbate, *Inventing the Internet*, 135.

Notes to Introduction

28. Stewart Brand, "Founding Father," *Wired*, March 1, 2001, https://www.wired.com/2001/03/baran/.

29. Katie Hafner and Matthew Lyon, *Where Wizards Stay Up Late: The Origins of the Internet* (New York: Touchstone, 1996), 62–64.

30. Interestingly, NEC was in part founded by AT&T's Western Electric subsidiary in 1896.

31. Kōji Kobayashi, *Computers and Communications: A Vision of C&C* (Cambridge, MA: MIT Press, 1986), 186.

32. Gartner, "Gartner Forecasts Worldwide Public Cloud End-User Spending to Reach Nearly $600 Billion in 2023," press release, October 31, 2022, https://www.gartner.com/en/newsroom/press-releases/2022-10-31-gartner-forecasts-worldwide-public-cloud-end-user-spending-to-reach-nearly-600-billion-in-2023; "Total Size of the Public Cloud Computing Market from 2008 to 2020," Statista, June 26, 2019, https://www.statista.com/statistics/510350/worldwide-public-cloud-computing/.

33. "Size of the Cloud Storage Market Worldwide from 2021 to 2029," Statista, September 20, 2022, https://www.statista.com/statistics/1322710/global-cloud-storage-market-size/.

34. See Molly Wood, "We Need to Talk about 'Cloud Neutrality,'" *Wired*, February 10, 2020, https://www.wired.com/story/we-need-to-talk-about-cloud-neutrality/; Aran Ali, "AWS: Powering the Internet and Amazon's Profits," *Visual Capitalist*, July 10, 2022, https://www.visualcapitalist.com/aws-powering-the-internet-and-amazons-profits/.

35. Sarah Perez, "Pandemic Accelerated Cord Cutting, Making 2020 the Worst-Ever Year for Pay TV," *TechCrunch*, September 21, 2020, https://techcrunch.com/2020/09/21/pandemic-accelerated-cord-cutting-making-2020-the-worst-ever-year-for-pay-tv/. This is quite a steep fall from just ten years earlier, when 105 million US TV households were pay-TV subscribers, a penetration of over 90 percent of TV homes. See Brad Adgate, "The Rise and Fall of Cable Television," *Forbes*, November 2, 2020, https://www.forbes.com/sites/bradadgate/2020/11/02/the-rise-and-fall-of-cable-television/?sh=39d574796b31.

36. See Cynthia Littleton, "How Hollywood Is Racing to Catch Up with Netflix," *Variety*, August 21, 2018, https://variety.com/2018/digital/features/media-streaming-services-netflix-disney-comcast-att-1202910463/.

37. One survey in 2012 revealed that 95 percent of those who thought they were not using the cloud, actually were—whether in the act of shopping, banking, or gaming online, using social networks, streaming media, or storing music/photos/videos. At that time, more than half believed that stormy weather would interfere with their cloud usage, and one in five of those who answered confessed to being "cloud imposters," pretending to know what the cloud is or how it works when they actually had

no idea. See Wakefield Research, "Citrix Cloud Survey Guide," August 2012, https://contentpit.files.wordpress.com/2012/09/citrix-cloud-survey-guide.pdf.

38. Hu, *Prehistory of the Cloud*, ix.

39. Lisa Parks, "Around the Antenna Tree: The Politics of Infrastructural Visibility," *Flow*, March 6, 2009, https://www.flowjournal.org/2010/03/flow-favorites-around-the-antenna-tree-the-politics-of-infrastructural-visibilitylisa-parks-uc-santa-barbara/.

40. John Durham Peters, *The Marvelous Clouds: Toward a Philosophy of Elemental Media* (Chicago: University of Chicago Press, 2015), 136.

41. Hu, *Prehistory of the Cloud*, x.

42. Hu, *Prehistory of the Cloud*, 66.

43. Asta Vonderau, "Technologies of Imagination: Locating the Cloud in Sweden's North," *Imaginations Journal of Cross-Cultural Image Studies* 8, no. 2 (2017): 11, https://doi.org/10.17742/IMAGE.LD.8.2.2.

44. Ghislain Thibault, "Bolts and Waves: Representing Radio Signals," *Early Popular Visual Culture* 16, no. 1 (2018): 39–56, https://doi.org/10.1080/17460654.2018.1472621.

45. Thomas Streeter, *Selling the Air: A Critique of the Policy of Commercial Broadcasting in the United States* (Chicago: University of Chicago Press, 1996), 223.

46. For the definitive analysis of corporate liberalism as a set of values governing broadcast policy in the US, see Streeter, *Selling the Air*, especially chapter 2.

47. Susan J. Douglas, *Inventing American Broadcasting, 1899–1922* (Baltimore, MD: Johns Hopkins University Press, 1987), 218.

48. Douglas, *Inventing American Broadcasting*, 218.

49. Ed Koops, former network engineer for Nextlink, Netstream, and Sprint. Quoted in Joanna Glasner, "High Bandwith Bureaucracy," *Wired*, August 24, 1999, https://ecfsapi.fcc.gov/file/6009553069.pdf.

50. Wolfgang Schivelbusch, *The Railway Journey: The Industrialization of Time and Space in the 19th Century* (Berkeley: University of California Press, 1986), 29.

51. Brian Larkin, *Signal and Noise: Media, Infrastructure, and Urban Culture in Nigeria* (Durham, NC: Duke University Press, 2008), 6, 252. Here, Larkin was drawing on Henri Lefebvre's ideas in *The Production of Space* (Cambridge, MA: Blackwell, 1991).

52. Shannon Mattern, "Deep Time of Media Infrastructure," in *Signal Traffic: Critical Studies of Media Infrastructures*, ed. Lisa Parks and Nicole Starosielski (Chicago: University of Illinois Press, 2015), 105–106.

53. Carolyn Marvin, *When Old Technologies Were New: Thinking about Electric Communication in the Late Nineteenth Century* (New York: Oxford University Press, 1990);

Lisa Gitelman, *Always Already New: Media, History and the Data of Culture* (Cambridge, MA: MIT Press, 2006); David A. Banks, "Lines of Power: Availability to Networks as a Social Phenomenon," *First Monday* 20, no. 11 (2015), https://doi.org/10.5210/fm.v20i11.6283.

54. Shannon Christine Mattern, *Code and Clay, Data and Dirt: Five Thousand Years of Urban Media* (Minneapolis: University of Minnesota Press, 2017), xi–xii.

55. Mattern, *Code and Clay*, xxviii.

56. Carrier hotels are large, dense hubs of interconnection in downtown urban areas, housing cloud infrastructure for multiple providers.

57. Jill Schachner Chanen, "In Chicago, from Printing Plant to Technology Hub," *New York Times*, June 11, 2000, https://www.nytimes.com/2000/06/11/realestate/in-chicago-from-printing-plant-to-technology-hub.html.

58. Mattern, *Code and Clay*, xxviii.

59. See Mattern, *Code and Clay*, xxxii, for example.

60. See Lisa Gitelman, *Scripts, Grooves, and Writing Machines* (Stanford, CA: Stanford University Press, 1999), 2, 13.

61. Christian Sandvig, "The Internet as Infrastructure," in *The Oxford Handbook of Internet Studies*, ed. William H. Dutton (Oxford: Oxford University Press, 2013), 93.

62. Mél Hogan and Tamara Shepherd, "Information Ownership and Materiality in an Age of Big Data Surveillance," *Journal of Information Policy* 5 (2015): 8, https://doi.org/10.5325/jinfopoli.5.2015.0006.

63. Lisa Parks, "Stuff You Can Kick: Toward a Theory of Media Infrastructures," in *Between Humanities and the Digital*, ed. Patrik Svensson and David Theo Goldberg (Cambridge, MA: MIT Press, 2015), 355.

64. Mara Einstein, *Media Diversity: Economics, Ownership, and the FCC* (Mahwah, NJ: Lawrence Erlbaum Associates, 2004), 10.

65. Thomas Winslow Hazlett, *The Political Spectrum: The Tumultuous Liberation of Wireless Technology, from Herbert Hoover to the Smartphone* (New Haven, CT: Yale University Press, 2017), 6, 8.

66. Hazlett, *The Political Spectrum*, 21.

67. Blair Levin, interview by Jennifer Holt, May 22, 2012, Washington, DC.

68. Newton N. Minow, Nell Minow, and Martha Minow, "Social Media, Distrust, and Regulation," in *Social Media, Freedom of Speech, and the Future of Our Democracy*, ed. Lee C. Bollinger and Geoffrey R. Stone (New York: Oxford University Press, 2022), 287.

69. Robert Britt Horwitz, *The Irony of Regulatory Reform: The Deregulation of American Telecommunications* (New York: Oxford University Press, 1989), 29.

70. The ICC was also a powerful vector of racism in American society. See Katie McCabe, "Making History in a Segregated Washington," *Journal of the Bar Association of the District of Columbia* 42, no. 1 (May 2011): 67–97.

71. Reed Hundt "Speech By Reed Hundt, Chairman, Federal Communications Commission, Center for National Policy" (speech, Washington, DC, May 6, 1996), https://transition.fcc.gov/Speeches/Hundt/spreh624.txt. The "revolving door" between policy offices and the private sector is another related contributor to this corrosive dynamic. Former FCC Chairman Kevin Martin, for example, is currently lobbying his former colleagues as Facebook's head of public policy and former FCC commissioner Meredith Attwell Baker is leading the wireless industry's major lobby group, the CTIA. Before taking office in 2001, FCC Chairman Michael Powell served as a lobbyist for the telecommunications industry, the very one that he was in charge of regulating; he is now the president and CEO of the Internet & Television Association, the cable industry's largest lobbying organization. Trump's FCC chairman, Ajit Pai, who came to the office from his post as a lawyer for Verizon, aligned himself with his former industry from day one, leaving the public behind in actions ranging from destroying net neutrality to abandoning almost all government-enforced privacy protections for US broadband users.

72. Jean-François Blanchette, "Introduction: Computing's Infrastructural Moment," in *Regulating the Cloud: Policy for Computing Infrastructure*, ed. Christopher S. Yoo and Jean-François Blanchette (Cambridge, MA: MIT Press, 2015), 2–3.

73. Bruce Schneier, *Data and Goliath: The Hidden Battles to Collect Your Data and Control Your World* (New York: W. W. Norton & Company, 2015), 78.

74. "The Many Lives of Herbert O. Yardley," *NSA Cryptologic Spectrum* 11, no. 4 (Fall 1981), https://www.nsa.gov/portals/75/documents/news-features/declassified-documents/cryptologic-spectrum/many_lives.pdf. *Cryptologic Spectrum* was an internal journal published by the NSA established in 1969. A selection of declassified articles from issues published between 1969 and 1981 can be found indexed at https://www.nsa.gov/news-features/declassified-documents/cryptologic-spectrum/.

75. Alex Urbelis, "After a Century of Mass Government Surveillance, It's Time for New Limits," *The Intercept*, September 22, 2015, https://theintercept.com/2015/09/22/history-of-us-surveillance-shows-need-for-new-limits/.

76. See the Church Committee report: Select Committee to Study Governmental Operations, *Supplementary Detailed Staff Reports on Intelligence Activities and the Rights of Americans: Book III*, S. Rep. 69–684, 94th Cong., 2d sess. (April 23, 1976), 765–776, https://aarclibrary.org/publib/church/reports/book3/html/ChurchB3_0386a.htm.

77. Colin Agur, "Negotiated Order: The Fourth Amendment, Telephone Surveillance, and Social Interactions, 1878–1968," *Information & Culture* 48, no. 4 (2013): 419, 424.

78. Mickie Edwardson, "James Lawrence Fly, the FBI, and Wiretapping," *The Historian* 61, no. 2 (Winter 1999): 364, https://www.jstor.org/stable/24449708.

79. Olmstead v. United States 277 U.S. 438 (1928); Katz v. United States 389 U.S. 347 (1967).

80. See Olmstead v. United States, 277 U.S. 438 (1928).

81. Zuboff, *Age of Surveillance Capitalism*, 11.

82. UNICEF, "Two Thirds of the World's School-Age Children Have No Internet Access at Home, New UNICEF-ITU Report Says," November 30, 2020, https://www.unicef.org/press-releases/two-thirds-worlds-school-age-children-have-no-internet-access-home-new-unicef-itu.

83. Robin Lake and Alvin Makori, "The Digital Divide among Students During COVID-19: Who Has Access? Who Doesn't?" Center on Reinventing Public Education, June 16, 2020, https://crpe.org/the-digital-divide-among-students-during-covid-19-who-has-access-who-doesnt/.

84. See Tim Wu, *The Curse of Bigness: Antitrust in the New Gilded Age* (New York: Columbia Global Reports, 2018); Zephyr Teachout, *Break 'Em Up: Recovering Our Freedom from Big Ag, Big Tech, and Big Money* (New York: All Points Books, 2020).

85. Wu, *Curse of Bigness*, 17.

86. Lina M. Khan, "Amazon's Antitrust Paradox," *Yale Law Journal* 126, no. 3 (January 2017): 710–805, https://www.yalelawjournal.org/note/amazons-antitrust-paradox.

87. Khan, "Amazon's Antitrust Paradox."

88. See David Streitfeld, "Amazon's Antitrust Antagonist Has a Breakthrough Idea," *New York Times*, September 7, 2018, https://www.nytimes.com/2018/09/07/technology/monopoly-antitrust-lina-khan-amazon.html.

89. Hearing before US Senate Committee on the Judiciary and Committee on Commerce, Science and Transportation, *Facebook, Social Media Privacy, and the Use and Abuse of Data*, 115th Cong., 2d sess. (April 10, 2018), https://www.washingtonpost.com/news/the-switch/wp/2018/04/10/transcript-of-mark-zuckerbergs-senate-hearing/.

90. See Jane Mayer, "Dianne Feinstein's Missteps Raise a Painful Age Question among Senate Democrats," *New Yorker*, December 10, 2020, https://www.newyorker.com/news/news-desk/dianne-feinsteins-missteps-raise-a-painful-age-question-among-senate-democrats.

91. Teachout, *Break 'Em Up*, 10.

92. Paul Starr, "How Neoliberal Policy Shaped the Internet—and What to Do About It Now," *American Prospect*, October 2, 2019, https://prospect.org/power/how-neoliberal-policy-shaped-internet-surveillance-monopoly/.

93. Richard R. John, "The Founders Never Intended the U.S. Postal Service to Be Managed like a Business," *Washington Post*, April 27, 2020, https://www.washingtonpost.com/outlook/2020/04/27/founders-never-intended-postal-service-be-managed-like-business/#comments-wrapper.

Chapter 1

1. This includes DSL, fiber optic, coaxial cable, satellite, and wireless networks.

2. Graham and Marvin, *Splintering Urbanism*, 73.

3. Graham and Marvin, *Splintering Urbanism*, 74, 52.

4. While the "last mile" in the US is owned by a tiny cabal of corporations, a handful of Big Tech giants like Google, Amazon, Facebook, and Microsoft now control a growing percentage of undersea cables. These cables were largely laid by telecommunications companies until Silicon Valley got involved in the 2010s. Within a decade, Silicon Valley outfits owned or leased more than half of the undersea bandwidth. See Adam Stariano, "How the Internet Travels across Oceans," *New York Times*, March 10, 2019, https://www.nytimes.com/interactive/2019/03/10/technology/internet-cables-oceans.html. For the most comprehensive analysis of global undersea cables and their environmental, political, policy, and cultural implications, see Nicole Starosielski, *The Undersea Network* (Durham, NC: Duke University Press, 2015).

5. See Susan Crawford, *Fiber: The Coming Tech Revolution—and Why America Might Miss It* (New Haven, CT: Yale University Press, 2018); Christopher Ali, *Farm Fresh Broadband: The Politics of Rural Connectivity* (Cambridge, MA: MIT Press, 2021).

6. Charles Franklin Phillips, *The Regulation of Public Utilities: Theory and Practice* (Arlington, VA: Public Utilities Reports, 1988), 83.

7. Munn v. Illinois 94 U.S. 113 (1877).

8. Philip M. Nichols, "Redefining 'Common Carrier': The FCC's Attempt at Deregulation by Redefinition," *Duke Law Journal* 1987: 509, https://doi.org/10.2307/1372565.

9. Susan Crawford, *Captive Audience: The Telecom Industry and Monopoly Power in the New Gilded Age* (New Haven, CT: Yale University Press, 2013), 32. De Sola Pool has also argued that similar benefits led to the telegraph having common carrier obligations. He noted that the Post Roads Act of 1866 allowed telegraph companies the privilege "to run their lines freely along post roads and across public lands. It also permitted them to fell trees for poles on public lands" in order to encourage the expansion of the system. In exchange, they had to provide service like a common carrier; Ithiel de Sola Pool, *Technologies of Freedom* (Cambridge, MA: Harvard University Press, 1983), 95.

10. Western Union Telegraph Co. v. Call Publishing Co., 181 U.S. 92, 98 (1901).

11. Christopher S. Yoo, "Common Carriage's Domain," *Yale Journal on Regulation* 35 (2018): 994–997, https://scholarship.law.upenn.edu/faculty_scholarship/2016.

12. See Communications Act of 1934, Pub. L. No. 416, 73d Cong., Section 3 (h) (1934).

13. National Association of Regulatory Utility Commissioners v. Federal Communications Commission, 525 F.2d 630 (D.C. Cir. 1976).

14. Robert Britt Horwitz, *The Irony of Regulatory Reform: The Deregulation of American Telecommunications* (New York: Oxford University Press, 1989), 13–14.

15. Jack M. Balkin, "To Reform Social Media, Reform International Capitalism," in *Social Media, Freedom of Speech, and the Future of Democracy*, ed. Lee C. Bollinger and Geoffrey R. Stone, (Cambridge, UK: Oxford University Press, 2022), 233–254 (236).

16. Harold Feld, "My Insanely Long Field Guide to Common Carriage, Public Utility, Public Forum—And Why the Differences Matter," *Wetmachine* (blog), September 5, 2017, https://wetmachine.com/tales-of-the-sausage-factory/my-insanely-long-field-guide-to-common-carriage-public-utility-public-forum-and-why-the-differences-matter/.

17. Crawford, *Captive Audience*, 33–34.

18. See "Telecom Services & Equipment: Lobbying," OpenSecrets, https://www.opensecrets.org/industries/lobbying.php?cycle=2020&ind=B09.

19. Werner Troesken, "Regime Change and Corruption: A History of Public Utility Regulation," in *Corruption and Reform: Lessons from America's Economic History*, ed. Edward L. Glaeser and Claudia Goldin (Chicago: University of Chicago Press, 2006), 259–281. Also see Matt Stoller, *Goliath: The 100-Year War between Monopoly Power and Democracy* (New York: Simon and Schuster, 2019).

20. Horace M. Gray, "The Passing of the Public Utility Concept," *Journal of Land & Public Utility Economics* 16, no. 1 (1940): 8–9, https://doi.org/10.2307/3158751.

21. Gray, "Passing of the Public Utility Concept," 15.

22. Texas has deregulated its electricity market, which has led to $28 billion more in direct costs to consumers in recent years and an abysmal failure of the power grid that left millions without heat or electricity for days during a storm with freezing temperatures in February 2021. See Tom McGinty and Scott Patterson, "Texas Electric Bills Were $28 Billion Higher under Deregulation," *Wall Street Journal*, February 24, 2021, https://www.wsj.com/articles/texas-electric-bills-were-28-billion-higher-under-deregulation-11614162780; Katherine Blunt and Russell Gold, "The Texas Freeze: Why the Power Grid Failed," *Wall Street Journal*, February 19, 2021, https://www.wsj.com/articles/texas-freeze-power-grid-failure-electricity-market-incentives-11613777856?mod=article_inline.

23. Harold Feld, "Broadband Access as Public Utility" (speech, Personal Democracy Forum, June 4, 2015), transcript, https://wetmachine.com/tales-of-the-sausage-factory/broadband-access-as-public-utility-my-speech-at-personal-democracy-forum/.

24. Richard R. John, *Network Nation: Inventing American Telecommunications* (Cambridge, MA: Harvard University Press, 2010), 32.

25. John, *Network Nation*, 60–61.

26. Dan Schiller, "Reconstructing Public Utility Networks: A Program for Action," *International Journal of Communication* 14 (2020): 4991, https://ijoc.org/index.php/ijoc/article/view/16242. Also see Richard R. John, "Recasting the Information Infrastructure for the Industrial Age," in *A Nation Transformed by Information: How Information Has Shaped the United States from Colonial Times to the Present*, ed. Alfred D. Chandler and James W. Cortada (New York: Oxford University Press, 2000), 78–79, 98–99.

27. US Post Office Department Postmaster General, *Government Ownership of Electrical Means of Communication: Letter from the Postmaster General, Transmitting, in Response to a Senate Resolution of January 12, 1914, a Report Entitled "Government Ownership of Electrical Means of Communication,"* S. Doc. 399, 63d Cong., 2d sess. (January 31, 1914), 5, 13.

28. See Michael A. Janson and Christopher S. Yoo, "The Wires Go to War: The U.S. Experiment with Government Ownership of the Telephone System During World War I," *Texas Law Review* 91 (2013): 983–1050, https://doi.org/10.2139/ssrn.2033124.

29. Eugene V. Debs, "'Better to Buy Books Than Beer'" (speech, Music Hall, Buffalo, NY, January 15, 1896), https://www.marxists.org/archive/debs/works/1896/960115-debs-speechatbuffalo.pdf.

30. See, for example, Franklin D. Roosevelt, "Power: Protection of the Public Interest" (speech, Portland, OR, September 21, 1932), FDR Library, http://www.fdrlibrary.marist.edu/_resources/images/msf/msf00530.

31. Roosevelt, "Power."

32. Franklin D. Roosevelt, *Message from the President of the United States Recommending That Congress Create a New Agency to Be Known as the Federal Communications Commission*, S. Doc. No. 144, 73d Cong, 2d sess. (February 26, 1934).

33. See Horwitz, *Irony of Regulatory Reform*.

34. "42 Million Americans Don't Have High-Speed Internet. Local Providers May Be the Key," *All Things Considered*, hosted by David Condos and produced by NPR, May 11, 2022, https://www.npr.org/2022/05/11/1098368187/42-million-americans-dont-have-high-speed-internet-local-providers-may-be-the-ke.

35. Federal Communication Commission, "Universal Service," Federal Communication Commission Telecommunications Access Policy Division, nd, https://www.fcc.gov/general/universal-service.

36. The subsidized program was known as "lifeline" telephone service. Funding came from cross-subsidies and essentially passing the charges on to long-distance and business customers.

37. US Post Office Department Postmaster General, *Government Ownership of Electrical Means*, 19–24.

38. US Post Office Department Postmaster General, *Government Ownership of Electrical Means*, 10.

39. Roland Marchand, *Creating the Corporate Soul: The Rise of Public Relations and Corporate Imagery in American Big Business* (Los Angeles: University of California Press, 1998), 50. See also John, *Network Nation*; Matthew Lasar, "How AT&T Conquered the 20th Century," *Wired*, September 3, 2011, https://www.wired.com/2011/09/att-conquered-20th-century/; Mueller, *Universal Service*.

40. AT&T Inc., *1910 Annual Report* (1911), 23.

41. John, "Recasting the Information Infrastructure," 96.

42. Marchand, *Creating the Corporate Soul*, 48.

43. For outstanding examples of this campaign, see chapter 2 of Marchand, *Creating the Corporate Soul*, 48–87.

44. Mueller, *Universal Service*, 5.

45. Adam D. Thierer, "Unnatural Monopoly: Critical Moments in the Development of the Bell System Monopoly," *Cato Journal* 14, no. 2 (1994): 278, https://www.cato.org/sites/cato.org/files/serials/files/cato-journal/1994/11/cj14n2-6.pdf.

46. AT&T Inc., *1907 Annual Report* (1908), 18.

47. AT&T, *1910 Annual Report*, 39.

48. AT&T, *1910 Annual Report*, 42–54.

49. Fred W. Henck and Bernard Strassburg, *A Slippery Slope: The Long Road to the Breakup of AT&T* (Westport, CT: Greenwood Press, 1988), xi.

50. Robert W. Crandall, *After the Breakup: U.S. Telecommunications in a More Competitive Era* (Washington, DC: Brookings Institution Press, 1991), 41.

51. Adam D. Thierer, "Unnatural Monopoly: Critical Moments in the Development of the Bell System Monopoly," *Cato Journal* 14, no. 2 (1994): 268–269, https://www.cato.org/sites/cato.org/files/serials/files/cato-journal/1994/11/cj14n2-6.pdf.

52. Robert MacDougall, "Long Lines: AT&T's Long-Distance Network as an Organizational and Political Strategy," *Business History Review* 80, no. 2 (July 2006): 318, https://doi.org/10.1017/S0007680500035509.

53. *Saturday Night Live*, season 2, episode 1, "Lily Tomlin / James Taylor," aired September 18, 1976, on NBC.

54. David Hochfelder, "Constructing an Industrial Divide: Western Union, AT&T, and the Federal Government, 1876–1971," *Business History Review* 76, no. 4 (Winter 2002): 721, https://doi.org/10.2307/4127707.

55. See Dan Schiller, "The Hidden History of US Public Service Telecommunications, 1919–1956," *Info* 9, no. 2/3 (2007): 17–28, https://doi.org/10.1108/14636690710734625.

56. See MacDougall, "Long Lines"; John, *Network Nation*.

57. John Brooks, *Telephone: The First Hundred Years* (New York: Harper & Row, 1975), 11.

58. Mueller, *Universal Service*, 133.

59. Brian Fung, "This 100-Year-Old Deal Birthed the Modern Phone System. And It's All About to End," *Washington Post*, December 19, 2013, https://www.washingtonpost.com/news/the-switch/wp/2013/12/19/this-100-year-old-deal-birthed-the-modern-phone-system-and-its-all-about-to-end/.

60. Schiller, "Hidden History," 18.

61. FCC, *Investigation of the Telephone Industry in the United States. Letter from the Chairman of the Federal Communications Commission*, H. Doc. 340, 76th Cong., 1st sess. (June 14, 1939), xvii, https://archive.org/details/InvestigationOfTheTelephoneIndustry/page/n25/mode/2up.

62. FCC, *Investigation of the Telephone Industry*, xvii.

63. FCC, *Investigation of the Telephone Industry*, 578.

64. FCC, *Investigation of the Telephone Industry*, 597.

65. "A.T.& T. Assails Walker Report," *New York Times*, December 6, 1938, 35.

66. Dan Schiller, *Crossed Wires* (New York: Oxford University Press, 2023), 215.

67. Jon Gertner, *The Idea Factory: Bell Labs and the Great Age of American Innovation* (New York: Penguin Press, 2012), 159.

68. Harry S. Truman, Letter to Leroy A. Wilson, President of AT&T, May 13, 1949. Necah S. Furman, *Contracting in the National Interest: Establishing the Legal Framework for the Interaction of Science, Government, and Industry at a Nuclear Weapons Laboratory*, Rep. No. SAND87–1651 UC–13 (Albuquerque, NM: Sandia National Laboratories for the United States Department of Energy, April 1988), 4. https://www.sandia.gov/about/history/_assets/documents/FurmanContractingInTheNationalInterest871651.pdf.

69. See Necah S. Furman, *Contracting in the National Interest*, 7. It is later noted on page 17 that this provision was deleted in 1983, largely the result of the impending breakup of the company.

70. Brooks, *Telephone*, 252–253, 273–278.

71. Fung, "This 100-Year-Old Deal."

72. Gertner, *Idea Factory*, 157.

73. Gertner, *Idea Factory*, 158.

74. Report of U.S. Congress, House of Representatives, Committee on the Judiciary Antitrust Subcommittee, *Consent Decree Program of the Department of Justice*, H.R. Rep. 33261, 86th Cong., 1st sess. (January 30, 1959), 90.

75. US Congress, *Consent Decree Program of the Department of Justice*, Appendix VII, 339.

76. US Congress, *Consent Decree Program*, Appendix VII, 339.

77. US Congress, *Consent Decree Program*, 341.

78. US Congress, *Consent Decree Program*, 57.

79. US Congress, *Consent Decree Program*, 51, 292.

80. US Congress, *Consent Decree Program*, 57.

81. There were also several other provisions related to licensing patents and furnishing technical information for its competitors. See United States v. Western Elec. Co., 1956 Trade Cas. (CCH) P68, 246 (D.N.J. January 24, 1956).

82. Amy Klobuchar, *Antitrust: Taking on Monopoly Power from the Gilded Age to the Digital Age* (New York: Alfred A. Knopf, 2021), 139. Also see Anthony Lewis, "Brownell Linked to A.T.&T. Decree," *New York Times*, March 27, 1958, 19. In this story, Brownell is characterized as working with AT&T on ways to settle the case by consent decree.

83. US Congress, *Consent Decree Program*, 290.

84. See Brooks, *Telephone*, 251–256; US Congress, *Consent Decree Program*, 42–73.

85. Quoted in Stewart Brand, "Founding Father," *Wired*, March 1, 2001, https://www.wired.com/2001/03/baran/.

86. Robert Cannon, "The Legacy of the Federal Communications Commission's Computer Inquiries," *Federal Communications Law Journal* 55, no. 2 (2003): 169, https://www.repository.law.indiana.edu/fclj/vol55/iss2/2.

87. FCC, *Regulatory and Policy Problems Presented by the Interdependence of Computer and Communication Services & Facilities*, Notice of Inquiry, 7 FCC 2d 11 (1966), 2. Also known as the *First Computer Inquiry*.

88. For an early discussion of this in relation to the Internet, see Robert Reilly, "Mapping Legal Metaphors in Cyberspace: Evolving the Underlying Paradigm," *Journal of Information Technology & Privacy Law* 16, no. 3 (Spring 1998): 579–596, https://repository.law.uic.edu/jitpl/vol16/iss3/3.

89. FCC, *Regulatory and Policy Problems Presented by the Interdependence of Computer and Communication Services & Facilities,* Notice of Inquiry, 13.

90. FCC, *Regulatory and Policy Problems*, Notice of Inquiry, 1.

91. Bernard Strassburg, "Competition and Monopoly in the Computer and Data Transmission Industries," *Antitrust Bulletin* 13 (1968): 991.

92. For analyses of the inquiries, see John Blevins, "The FCC and the 'Pre-Internet,'" *Indiana Law Journal* 91, no. 4 (2016): 1309–1362, https://www.repository.law.indiana.edu/ilj/vol91/iss4/6; Cannon, "Legacy."

93. Kevin Werbach, "The Federal Computer Commission," *North Carolina Law Review* 84, no. 1 (2005): 16, https://scholarship.law.unc.edu/nclr/vol84/iss1/3.

94. Bernard Strassburg, interview by James Pelkey, Computer History Museum, May 3, 1998, Washington, DC, 15, https://archive.computerhistory.org/resources/access/text/2015/11/102738016-05-01-acc.pdf.

95. The FCC also created a "hybrid" category for services that included elements of both communications and data services, which was dealt with on a case-by-case basis.

96. Cannon, "Legacy," 174.

97. Blevins, "FCC and the 'Pre Internet,'" 1317.

98. De Sola Pool, *Technologies of Freedom*, 23.

99. FCC, *Regulatory and Policy Problems*, Final Decision and Order, 28 FCC 2d 267, 12 (1971).

100. FCC, *Regulatory and Policy Problems*, Final Decision and Order, 10.

101. FCC, *Regulatory and Policy Problems*, Final Decision and Order, 4, 11, 30.

102. Dan Schiller, "Reconstructing Public Utility Networks: A Program for Action," *International Journal of Communication* 14 (2020): 4993, https://ijoc.org/index.php/ijoc/article/view/16242.

103. FCC, *Regulatory and Policy Problems*, Notice of Inquiry, 22.

104. FCC, *Regulatory and Policy Problems*, Notice of Inquiry, 23, 24.

105. See Nicholas Johnson, "Carterfone: My Story," *Santa Clara High Technology Law Journal* 25, no. 3 (2008): 683, https://digitalcommons.law.scu.edu/chtlj/vol25/iss3/5.

Notes to Chapter 1

106. Andrew Pollack, "The Man Who Beat A.T.&T.," *New York Times*, July 14, 1982, D1, D5.

107. See "In the Matter of Use of the Carterfone Device in Message Toll Telephone Service," 13 FCC 2d 420 (1968). Tom Carter was also the plaintiff in the Hush-A-Phone case that began in 1948, see Hush-A-Phone Corp v. United Sates, 238 F. 2d 266 (D.C. Cir. 1956). In this case, Carter filed a complaint with the FCC because AT&T claimed that the rubber cup he developed to connect to the phone's mouthpiece to act as a silencer was a "foreign attachment," which was prohibited by the company. The FCC agreed, arguing that the Hush-a-Phone was "deleterious to the telephone system and injures the service rendered by it." Kevin Werbach has called Hush-a-Phone "the high water mark of the FCC's willingness to defend the AT&T monopoly" (see Werbach, "Federal Computer Commission," 18). Carter was undaunted and challenged the decision in court, which ruled in his favor, noting "AT&T and the FCC have no business protecting callers from themselves" (Hush-a-Phone 238 F. 2d 269).

108. GTE was acquired by Bell Atlantic in 2000 and the new company was renamed Verizon Communications.

109. Matthew Lasar, "Any Lawful Device," *Ars Technica*, December 13, 2017, https://arstechnica.com/tech-policy/2017/12/carterfone-40-years/.

110. Lasar, "Any Lawful Device."

111. Andrew Pollack, "The Man Who Beat A.T.& T," D1, https://www.nytimes.com/1982/07/14/business/the-man-who-beat-at-t.html.

112. Sandra Braman, *Change of State: Information, Policy, and Power* (Cambridge, MA: MIT Press, 2007), 198.

113. "In the Matter of Use of the Carterfone Device in Message Toll Telephone Service," 13 FCC 2d 420 (1968).

114. This was further solidified in the 1980 Computer II Final Order, which deregulated all customer equipment.

115. Randal Picker, "The Arc of Monopoly: A Case Study in Computing," *University of Chicago Law Review* 87, no. 2 (March 2020): 535, https://chicagounbound.uchicago.edu/uclrev/vol87/iss2/9.

116. "Fighting Bell," *Wall Street Journal*, November 21, 1974, 1.

117. Peter Temin, *The Fall of the Bell System* (New York: Cambridge University Press, 1987), 224.

118. See James B. Stewart, "Whales and Sharks," *New Yorker*, February 15, 1993, 37–43, 38.

119. Robert Pear, "New Antitrust Leader Vows to Break Up AT&T," *New York Times*, April 9, 1981, A1.

120. See Louis M. Kohlmeier, "Testimony in Deepening ITT Antitrust Case Links Controversy Directly with Nixon," *Wall Street Journal*, March 10, 1974, 4. This is often referred to as the "Dita Beard affair," after the lobbyist who wrote the memo outlining the connection between the ITT funds donated to the convention and the settling of the government's antitrust suits against the company.

121. Temin, *Fall of the Bell System*, 223.

122. Tim Wu, *The Curse of Bigness* (New York: Columbia Global Reports, 2018), 83.

123. Richard Hofstadter, "What Happened to the Antitrust Movement," reprinted in *Richard Hofstadter: Anti-Intellectualism in American Life, The Paranoid Style in American Politics, Uncollected Essays 1956–1965* (New York: Library of America, 2020), 659.

124. Stoller, *Goliath*, 238.

125. FCC, *Report and Order and Notice of Proposed Rulemaking*, Rep. No. FCC 05-150 (August 5, 2005), 15.

126. FCC, *Computer II Final Decision*, Final Decision and Order, 77 FCC 2d 384, 5 (1980).

127. FCC, *Computer II Final Decision*, 84.

128. See FCC, *Computer II Final Decision*, 13. Also see Charles Ferris's comments regarding the FCC's "desire to allow AT&T to participate in the evolving communications/ data processing markets in spite of the 1956 Consent Decree." Separate Statement of Chairman Charles D. Ferris, FCC, *Computer II Final Decision*, 500.

129. FCC, *Computer II Final Decision*, 12.

130. Werbach, "Federal Computer Commission," 24.

131. Concurring Statement of FCC Commissioner James H. Quello, FCC, *Computer II Final Decision*, 503.

132. Separate Statement of FCC Commissioner Charles Ferris, FCC, *Computer II Final Decision*, 503.

133. Blevins, "FCC and the 'Pre Internet,'" 1345.

134. FCC, *Computer II Final Decision*, 107.

135. The original Baby Bell companies were US West, Pacific Telesis, Bell Atlantic, NYNEX, Bell South, Southwestern Bell, and Ameritech.

136. Merrill Brown and Caroline E. Mayer, "U.S. Ends Antitrust Suits against AT&T and IBM," *Washington Post*, January 9, 1982.

137. AT&T voluntarily divested Western Electric (along with holdings in Bell Labs) to create Lucent Technologies in 1995.

138. Barry G. Cole, ed., *After the Breakup: Assessing the New Post-AT&T Divestiture Era* (New York: Columbia University Press, 1991), 2.

139. Stewart, "Whales and Sharks," 38.

140. Steve Lohr, "Antitrust: Big Business Breathes Easier," *New York Times*, February 15, 1981, 1, Section 3, https://www.nytimes.com/1981/02/15/business/antitrust-big-business-breathes-easier.html.

141. Steve Lohr, "Antitrust," 1.

142. Paul Taylor, "Law Firm Waged 13-Year War for IBM," *Washington Post*, January 24, 1982.

143. "U.S. vs. I.B.M.," *New York Times*, February 15, 1981, 22, Section 3, https://www.nytimes.com/1981/02/15/business/us-vsibm.html.

144. Edward T. Pound, "Why Baxter Dropped the I.B.M. Suit," *New York Times*, January 9, 1982, 37, Section 2, https://www.nytimes.com/1982/01/09/business/why-baxter-dropped-the-ibm-suit.html.

145. Computer III also moved toward replacing the structural separation requirement with other safeguards, and allowed the newly formed "Baby Bells" or newly spun off Bell Operating Companies (BOCs) back into the "enhanced services" markets. However, those efforts wound up mired in the courts, muting most of the impact such proposed remedies would have. See Third Computer Inquiry (Computer III): FCC, *Amendment of Section 64.702 of the Commission's Rules and Regulations (Third Computer Inquiry)*, Report and Order, 104 FCC 2d 958 (1986). Also see Cannon, "Legacy," 199–204; Russell A. Newman, *The Paradoxes of Network Neutralities* (Cambridge, MA: MIT Press, 2019), 54. Further dimensions of Computer III extend beyond the purposes of this discussion.

146. Abbate, *Inventing the Internet* (Cambridge, MA: MIT Press, 1999), 111.

147. See Shane Greenstein, *How the Internet Became Commercial* (Princeton, NJ: Princeton University Press, 2015).

148. Katie Hafner, "The Internet's Invisible Hand; At a Public Utility Serving the World, No One's Really in Charge. Does It Matter?," *New York Times*, January 10, 2002, Section G1, https://www.nytimes.com/2002/01/10/technology/internet-s-invisible-hand-public-utility-serving-world-no-one-s-really-charge.html.

149. Quoted in Craig Timberg, "A Flaw in the Design," *Washington Post*, May 30, 2015, https://www.washingtonpost.com/sf/business/2015/05/30/net-of-insecurity-part-1/.

150. AT&T, *1910 Annual Report*, 23.

151. Bill Clinton, "Technology, The Engine of Economic Growth," Clinton-Gore National Campaign Headquarters, September 21, 1992, https://www.ibiblio.org/nii/tech-posit.html.

152. Matthew Crain, *Profit over Privacy: How Surveillance Advertising Conquered the Internet* (Minneapolis: University of Minnesota Press, 2021), 23.

153. Clinton, "Technology." Also see William J. Broad, "Clinton to Promote High Technology, with Gore in Charge," *New York Times*, November 10, 1992, https://www.nytimes.com/1992/11/10/science/clinton-to-promote-high-technology-with-gore-in-charge.html.

154. MOSAIC was developed in part because of funds from the High Performance Computing and Communication Act of 1991, also known as "the Gore Bill."

155. John Naughton, "The Evolution of the Internet: From Military Experiment to General Purpose Technology," *Journal of Cyber Policy* 1, no. 1 (2016): 14, https://doi.org/10.1080/23738871.2016.1157619.

156. Peter H. Lewis, "Attention Shoppers: The Internet Is Open," *New York Times*, August 12, 1994, D1, https://www.nytimes.com/1994/08/12/business/attention-shoppers-internet-is-open.html.

157. Communications Assistance for Law Enforcement Act, Pub. L. No. 103–414, 108 Stat. 4279 (1994). Katz v. United States (1967) had already made warrants once again required for wiretapping and extended Fourth Amendment rights and protections to individuals, not simply property. For more on the history of Katz and the history of wiretapping in general, see Brian Hochman, *The Listeners: A History of Wiretapping in the United States* (Cambridge, MA: Harvard University Press, 2022).

158. For excellent detail on this process, see Crain, *Profit over Privacy*, chapter 2; Leslie David Simon, *NetPolicy.Com: Public Agenda for a Digital World* (Washington, DC: Woodrow Wilson Center Press, 2000), chapter 13.

159. See Richard Posner, "The Decline and Fall of AT&T: A Personal Recollection," *Federal Communications Bar Journal* 61 (2008): 12–13, https://chicagounbound.uchicago.edu/journal_articles/6780. Posner served on Johnson's task force before he was a federal judge.

160. William Domnarski, *Richard Posner* (New York: Oxford University Press, 2016), 51–52. Also see Eugene V. Rostow, *President's Task Force on Communications Policy Final Report* (Washington, DC: President's Task Force on Communications Policy, December 7, 1968), https://files.eric.ed.gov/fulltext/ED034417.pdf.

161. See Rostow, *President's Task Force*, chapter 9, 26.

162. The OTP was later absorbed by the Commerce Department's National Telecommunications and Information Administration (NTIA) under President Carter.

163. Hearings before the US Congress, Senate Subcommittee on Antitrust and Monopoly of the Committee on the Judiciary, *The Industrial Reorganization Act. Part 6: The Communications Industry*, 93d Cong., 2d sess. (July 9, July 30, and July 31, 1974), 3840.

The chair of Johnson's task force, former Under Secretary of State Eugene Rostow, appeared as a witness in support of AT&T at these hearings about the concentration of power in industries such as communications, chemicals, and automobiles.

164. See William H. Jones, "Mass Media Laws Changes Proposed," *Washington Post*, March 30, 1979, D1.

165. Merrill Brown, "Communications Act Being Revamped," *Washington Post*, January 23, 1980, B1.

166. The full name of the bill was the Consumer Communications Reform Act (CCRA) of 1976. See Peter Temin, "Fateful Choices: AT&T in the 1970s," *Business and Economic History* 27, no. 1 (Fall 1998): 61–77, https://doi.org/https://www.jstor.org/stable/23703063. Also see John Eger, "The Future of Communications," *New York Times*, October 4, 1976, https://www.nytimes.com/1976/10/04/archives/the-future-of-communications.html.

167. Newton N. Minow and Craig L. Lamay, *Abandoned in the Wasteland* (New York: Hill and Wang, 1995), 5.

168. James Glassman, "Costly Bill-Busting by the Seven Baby Bells," *Washington Post*, September 28, 1994, F1, F10.

169. In 1991, the Baby Bells created from the 1984 breakup of AT&T were allowed to provide cable service or "video dial tone," as it was then called, *outside* their local area.

170. Patricia Aufderheide, *Communications Policy and the Public Interest: The Telecommunications Act of 1996* (New York: The Guilford Press, 1999), 27.

171. For the foremost work on the politics and history of industrial deregulation, see Robert Horwitz, *Irony of Regulatory Reform*.

172. Aufderheide, *Communications Policy*, 41.

173. Esther Dyson et al., "Cyberspace and the American Dream: A Magna Carta for the Knowledge Age," The Progress and Freedom Foundation, August 22, 1994, http://www.pff.org/issues-pubs/futureinsights/fi1.2magnacarta.html.

174. Fred Turner, *From Counterculture to Cyberculture* (Chicago: University of Chicago Press, 2006), 222. For an excellent scholarly analysis of the manifesto, see Richard K. Moore, "Cyberspace Inc. and the Robber Baron Age: An Analysis of PFF's 'Magna Carta.'" *Information Society* 12, no. 3 (1996): 315–323, https://doi.org/10.1080/019722496129503.

175. Mitchell Kapor and Jerry Berman, "A Superhighway through the Wasteland?," *New York Times*, November 24, 1993, A25.

176. John Markoff, "Building the Electronic Superhighway," *New York Times*, January 24, 1993, https://www.nytimes.com/1993/01/24/business/building-the-electronic-superhighway.html.

177. Al Gore, "Innovation Delayed Is Innovation Denied," *Computer* 27, no. 12 (December 1994): 47, https://doi.org/10.1109/2.335728.

178. John Heilemann, "The Making of the President 2000," *Wired*, December 1, 1995, https://www.wired.com/1995/12/gorenewt/.

179. Rep. John Conyers, Conference Report on *Telecommunications Act of 1996*, S. 652, 104th Cong., 2d sess., *Congressional Record* 142, No. 14 (February 1, 1996), https://www.congress.gov/congressional-record/1996/02/01/house-section/article/H1145–6.

180. Telecommunications Act of 1996, Pub L. No. 104–104, 110 Stat. 56 (1996).

181. John Perry Barlow, "A Declaration of the Independence of Cyberspace," Electronic Frontier Foundation, February 8, 1996, https://www.eff.org/cyberspace-independence.

182. Hannah Bloch-Wehba, "Global Platform Governance: Private Power in the Shadow of the State," *SMU Law Review* 72, no. 1 (2019): 34, https://scholar.smu.edu/smulr/vol72/iss1/9; see 34–39 for other critiques of Barlow.

183. Barlow, "Declaration," 1996.

184. Jill Lepore, "Edward Snowden and the Rise of Whistle-Blower Culture," *New Yorker*, September 16, 2019, https://www.newyorker.com/magazine/2019/09/23/edward-snowden-and-the-rise-of-whistle-blower-culture.

185. Gilder is also the author of *Wealth and Poverty*, a love letter to capitalism and supply-side economics which became "a sacred text" for members of the Reagan administration, and the winner of *Time* magazine and the National Organization for Women's "Male Chauvinist Pig of the Year" award in 1973—which he declared "a triumph I could not exceed." See Paul Gray, "Inside the Minds of Gingrich's Gurus," *Time*, January 23, 1995, http://content.time.com/time/subscriber/article/0,33009,982259-1,00.html; Katie Hafner, "The Revolution Is Coming, Eventually," *New York Times*, October 19, 2003, https://www.nytimes.com/2003/10/19/business/the-revolution-is-coming-eventually.html.

186. William J. Clinton and Albert Gore Jr., "A Framework for Global Electronic Commerce," July 1997, https://clintonwhitehouse4.archives.gov/WH/New/Commerce/read.html.

187. Based on statistics in "Number of Broadband Internet Subscribers in the US . . ." Statista, July 12, 2021, https://www.statista.com/statistics/217348/us-broadband-internet-susbcribers-by-cable-provider/.

188. S. Derek Turner, "Price Too High and Rising: The Facts about America's Broadband Affordability Gap," *Free Press*, May 20, 2021, 2, https://www.freepress.net/sites/default/files/202105/prices_too_high_and_rising_free_press_report.pdf.

189. Blake Morgan, "The Top 5 Industries Most Hated by Consumers," *Forbes*, October 16, 2018, https://www.forbes.com/sites/blakemorgan/2018/10/16/top-5-most-hated-industries-by-customers/?sh=7febd04c90b5.

Notes to Chapter 1

190. Information services are defined as "the offering of a capability for generating, acquiring, storing, transforming, processing, retrieving, utilizing, or making available information via telecommunications, and includes electronic publishing, but does not include any use of any such capability for the management, control, or operation of a telecommunications system or the management of a telecommunications service." Communications Act of 1934, 47 U.S.C. § 153(20) (1934).

191. Tim Wu, "How the FCC's Net Neutrality Plan Breaks with 50 Years of History," *Wired*, December 6, 2017, https://www.wired.com/story/how-the-fccs-net-neutrality-plan-breaks-with-50-years-of-history/.

192. Victor Pickard and David Elliot Berman, *After Net Neutrality: A New Deal for the Digital Age* (New Haven, CT: Yale University Press, 2019), 4.

193. Laura DeNardis, "Hidden Levers of Internet Control: An Infrastructure-Based Theory of Internet Governance," *Information, Communication & Society* 15, no. 5 (2012): 735, https://doi.org/10.1080/1369118X.2012.659199.

194. Quoted in Rebecca R. Ruiz and Steve Lohr, "F.C.C. Approves Net Neutrality Rules, Classifying Broadband Internet Service as a Utility," *New York Times*, February 26, 2015, https://www.nytimes.com/2015/02/27/technology/net-neutrality-fcc-vote-internet-utility.html.

195. For more detailed analyses of this history, see Pickard and Berman, *After Net Neutrality*; Newman, *Paradoxes of Network Neutralities*; Danny Kimball, *Net Neutrality and the Battle for the Open Internet* (Ann Arbor: University of Michigan Press, 2022); Harold Feld, "The History of Net Neutrality in 13 Years of Tales of the Sausage Factory (with a Few Additions). Part I," *Wetmachine* (blog) January 10, 2018, https://wetmachine.com/tales-of-the-sausage-factory/the-history-of-net-neutrality-in-13-years-of-tales-of-the-sausage-factory-with-a-few-additions-part-i/; Becky Lentz and Allison Perlman, eds., "Net Neutrality" special issue of *International Journal of Communication* 10 (2016).

196. National Cable and Telecommunications Association v. Brand X Internet Services, 545 U.S. 967 (2005).

197. FCC, *Policy Statement*, Rep. No. FCC FCC 05-151 (August 5, 2005), https://docs.fcc.gov/public/attachments/fcc-05-151a1.pdf.

198. Matthew Lasar, "Comcast 1, FCC 0: What to Look for in the Inevitable Rematch," *Ars Technica*, April 7, 2010, https://arstechnica.com/tech-policy/2010/04/comcast-1-fcc-0-what-to-look-for-in-the-inevitable-rematch/.

199. Lawrence Lessig, "A Deregulation Debacle for the Internet," *New York Times*, August 9, 2010, https://www.nytimes.com/roomfordebate/2010/08/09/who-gets-priority-on-the-web/a-deregulation-debacle-for-the-internet.

200. Michael J. Copps, "Statement of Commissioner Michael J. Copps on Chairman Genachowski's Announcement to Reclassify Broadband," FCC, May 6, 2010, https://docs.fcc.gov/public/attachments/doc-297946A1.pdf.

201. Jennifer Holt, "Net Neutrality and the Public Interest: An Interview with Gene Kimmelman, President and CEO of Public Knowledge," *International Journal of Communication* 10 (2016): 5799, https://ijoc.org/index.php/ijoc/article/view/5394.

202. Newman, *Paradoxes of Network Neutralities*, 4.

203. Tom Wheeler, *Statement of Chairman Tom Wheeler: FCC Releases Open Internet Order. GN Docket No. 14–28*, FCC (March 12, 2015), https://www.fcc.gov/document/fcc-releases-open-internet-order/wheeler-statement.

204. Ruiz and Lohr, "F.C.C. Approves Net Neutrality Rules."

205. Geoff West, "Money Flows into Net Neutrality Debate ahead of FCC Vote," Opensecrets.org, December 14, 2017, https://www.opensecrets.org/news/2017/12/money-flows-into-net-neutrality-debate-ahead-of-fcc-vote/.

206. Kaleigh Rogers, "99.7 Percent of Unique FCC Comments Favored Net Neutrality," *Vice*, October 15, 2018, https://www.vice.com/en/article/3kmedj/997-percent-of-unique-fcc-comments-favored-net-neutrality.

207. "FCC Internal Investigation Shows Ajit Pai Knew DDoS Attack Was Bogus Months Ago," Fight for the Future, August 6, 2018, https://tumblr.fightforthefuture.org/post/176708614273/breaking-fcc-finally-admits-that-alleged-ddos.

208. Those states are California, Colorado, Maine, New Jersey, Oregon, Vermont, and Washington.

209. Specifically Section 222 of the Communications Act as amended by the Telecommunications Act of 1996. For more, see Travis LeBlanc and Lindsay DeFrancesco, "The Federal Communications Commission as Privacy Regulator," in *The Cambridge Handbook of Surveillance Law*, ed. David Gray and Stephen E. Henderson (Cambridge, UK: Cambridge University Press, 2017), 727–756, 736–737.

210. Quoted in Kimberly Kindy, "How Congress Dismantled Federal Internet Privacy Rules," *Washington Post*, May 20, 2017, https://www.washingtonpost.com/politics/how-congress-dismantled-federal-internet-privacy-rules/2017/05/29/7ad06e14-2f5b-11e7-8674-437ddb6e813e_story.html.

211. Avi-Asher Schapiro, "Coronavirus Crisis Threatens Internet Opportunity for Native Americans," Reuters, July 27, 2020, https://www.reuters.com/article/us-health-coronavirus-usa-rights-trfn/coronavirus-crisis-threatens-internet-opportunity-for-native-americans-idUSKCN24T06B.

212. Emily A. Vogels, "Some Digital Divides Persist between Rural, Urban and Suburban America," Pew Research Center, August 19, 2021, https://www.pewresearch.org/fact-tank/2021/08/19/some-digital-divides-persist-between-rural-urban-and-suburban-america/.

213. Pope Francis address to the World Meeting of Popular Movements, October 16, 2021.

214. See David D. Clark, *Designing an Internet* (Cambridge, MA: MIT Press, 2018); Janet Abbate, *Inventing the Internet*.

215. See Barbara van Schewick, *Internet Architecture and Innovation* (Cambridge, MA: MIT Press, 2010) for the definitive discussion on this topic. Also see Janet Abbate, *Inventing the Internet*; David D. Clark, *Designing an Internet*.

216. Hafner, "Internet's Invisible Hand."

217. Christian Sandvig, "Network Neutrality Is the New Common Carriage," *Info* 9, no. 2/3 (2007): 137, https://doi.org/10.1108/14636690710734751.

218. See, for example, Holt, "Regulating Connected Viewing," in *Connected Viewing: Selling, Streaming, & Sharing Media in the Digital Age*, ed. Jennifer Holt and Kevin Sanson (New York: Routledge, 2014), 19–39 (25, 27); Ramon Lobato, *Netflix Nations* (New York: New York University Press, 2019), 92–100.

219. Julie Cohen, *Between Truth and Power: The Legal Constructions of Informational Capitalism* (New York: Oxford University Press, 2019), 176.

220. Pickard and Berman, *After Net Neutrality*, 104.

221. Quoted in John, *Network Nation*, 19–20.

222. See Newman, *Paradoxes of Network Neutralities*.

223. Johnson, "Carterfone: My Story," 699.

224. Quoted in Mark Dent, "AT&T's 'Cowboy Swagger' Led to Its Hollywood Misadventure," *Texas Monthly*, May 18, 2021, https://www.texasmonthly.com/news-politics/att-warnermedia-hbo-mistake-discovery-merger.

225. Crawford, *Fiber*, 44–45.

226. MacDougall, "Long Lines," 318.

227. Less than 20 million out of roughly 128 million US households subscribed as of 2019. See Jon Brodkin, "50% of US Homes Still Won't Have Fiber Broadband by 2025, Study Says," *Ars Technica*, September 18, 2019, https://arstechnica.com/tech-policy/2019/09/50-of-us-homes-still-wont-have-fiber-broadband-by-2025-study-says/.

228. Tyler Cooper, "FCC Report Concludes US Internet Speeds Are 'Among Worst in the Developed World,'" *Broadband Now*, May 18, 2021, https://broadbandnow.com/report/2018-fcc-international-data-insights/.

Chapter 2

1. US House of Representatives, Committee on the Judiciary, Subcommittee on Antitrust, Commercial and Administrative Law, *Investigation of Competition in*

Digital Markets: Majority Staff Report and Recommendations: Part I, H.R. Rep. CP 117–8, 117th Cong., 2d sess. (July 2020), 10.

2. US House of Representatives, *Investigation of Competition in Digital Markets*, 6–7. Despite the gravity of their findings, these hearings have yet to produce any meaningful congressional action or legislation.

3. US House of Representatives, *Investigation of Competition*, 11.

4. US House of Representatives, *Investigation of Competition*, 261, 254–262.

5. Jonathan Vanian, "Amazon's Advertising Business," CNBC.com, February 2, 2023, https://www.cnbc.com/2023/02/02/amazons-advertising-business-grew-19percent-unlike-google-meta.html; Afef Abrougui et al., "Key Findings from the 2022 Ranking Digital Rights Big Tech Scorecard," Ranking Digital Rights, https://rankingdigitalrights.org/mini-report/key-findings-2022/. In 2021, Amazon earned $31.2 billion from digital advertising. While this sector is only 7 percent of their overall revenue of roughly $470 billion, it is the company's third largest source of income after e-commerce and cloud computing, and steadily rising.

6. All figures taken from 2021 corporate annual reports.

7. See Terry Flew, *Regulating Platforms* (Medford, MA: Polity, 2021), 26.

8. Emma Roth, "Facebook's Plan to Offer Free Internet in Developing Countries Ended Up Costing Users," *The Verge*, January 25, 2022, https://www.theverge.com/2022/1/25/22900924/facebooks-free-internet-less-developed-costing-users-wsj.

9. Elisa Shearer, "More Than Eight-in-Ten Americans Get News from Digital Devices," Pew Research Center, January 12, 2021, https://www.pewresearch.org/fact-tank/2021/01/12/more-than-eight-in-ten-americans-get-news-from-digital-devices/; Mason Walker and Katerina Eva Matsa, "News Consumption across Social Media in 2021," Pew Research Center, September 20, 2021, https://www.pewresearch.org/journalism/2021/09/20/news-consumption-across-social-media-in-2021/.

10. Matt Stoller, *Goliath: The 100-Year War between Monopoly Power and Democracy* (New York: Simon and Schuster, 2019), 449.

11. See Testimony of Tim Kent, Hearing of US House of Representatives, Committee on Energy and Commerce, Subcommittee on Consumer Protection and Commerce, *Mainstreaming Extremism: Social Media's Role in Radicalizing America*, 116th Cong., 2d sess. (September 24, 2020), https://energycommerce.house.gov/committee-activity/hearings/hearing-on-mainstreaming-extremism-social-media-s-role-in-radicalizing; Timothy B. Lee, "YouTube Should Stop Recommending Garbage Videos to Users," *Ars Technica*, August 12, 2019, https://arstechnica.com/tech-policy/2019/08/youtube-should-stop-recommending-garbage-videos-to-users/.

Notes to Chapter 2

12. Shoshana Zuboff, "The Coup We Are Not Talking About," *New York Times*, January 29, 2021, https://www.nytimes.com/2021/01/29/opinion/sunday/facebook-surveillance-society-technology.html.

13. Adam Satariano, "The World's First Ambassador to the Tech Industry," *New York Times*, September 3, 2019, https://www.nytimes.com/2019/09/03/technology/denmark-tech-ambassador.html. Also see "The Techplomacy Approach," Ministry of Foreign Affairs of Denmark, Office of Denmark's Tech Ambassador, https://techamb.um.dk/the-techplomacy-approach.

14. Open Secrets, "Annual Lobbing on Internet, 2021," https://www.opensecrets.org/federal-lobbying/industries/summary?cycle=2021&id=B13.

15. Pawel Popiel, "The Tech Lobby: Tracing the Contours of New Media Elite Lobbying Power," *Communication, Culture and Critique* 11, no. 4 (2018): 573, https://doi.org/10.1093/ccc/tcy027.

16. Derek Thompson, "Google's CEO: 'The Laws Are Written by Lobbyists,'" *The Atlantic*, October 1, 2010, https://www.theatlantic.com/technology/archive/2010/10/googles-ceo-the-laws-are-written-by-lobbyists/63908/#video.

17. Cory Doctorow, "Money Is Power," *Pluralistic*, February 3, 2022, https://pluralistic.net/2022/02/03/liquidation-preference/#sweet-sweet-corruption.

18. Laura DeNardis and A. M. Hackl, "Internet Governance by Social Media Platforms," *Telecommunications Policy* 39, no. 9 (October 2015): 769, https://doi.org/10.1016/j.telpol.2015.04.003.

19. Alan Z. Rozenshtein, "Surveillance Intermediaries," *Stanford Law Review* 70 (January 2018): 188.

20. Some key examples include Tim Wu, *The Attention Merchants* (New York: Alfred A. Knopf, 2016); Siva Vaidhyanathan, *Antisocial Media* (Oxford: Oxford University Press, 2018); Zuboff, *Age of Surveillance Capitalism*; Matthew Crain, *Profit over Privacy* (Minneapolis: University of Minnesota Press, 2021).

21. See Nick Couldry and Ulises A. Mejias, *The Costs of Connection: How Data Is Colonizing Human Life and Appropriating It for Capitalism* (Stanford, CA: Stanford University Press, 2019), chapter 1.

22. See Whitney Phillips, *This Is Why We Can't Have Nice Things* (Cambridge, MA: MIT Press, 2015); Safiya Noble, *Algorithms of Oppression* (New York: New York University Press, 2018); Danielle Keats Citron, *Hate Crimes in Cyberspace* (Cambridge, MA: Harvard University Press, 2014); Sarah Roberts, *Behind the Screen* (New Haven, CT: Yale University Press, 2019); Philip N. Howard, *Lie Machines* (New Haven, CT: Yale University Press, 2020).

23. David Cicilline, "Statement on Hearing 'Online Platforms and Market Power, Part 6: Examining the Dominance of Amazon, Apple, Facebook, and Google,'" press release, July 29, 2020.

24. See the European Commission, "e-Commerce Directive Policy," https://digital-strategy.ec.europa.eu/en/policies/e-commerce-directive.

25. Quoted in Article 19, *Internet Intermediaries: Dilemma of Liability*, 2013, 10, https://www.article19.org/data/files/Intermediaries_ENGLISH.pdf. Also see Frank LaRue et al., "International Mechanisms for Promoting Freedom of Expression: Joint Declaration on Freedom of Expression and the Internet," press release, June 1, 2011, p. 2, https://www.article19.org/data/files/pdfs/press/international-mechanisms-for-promoting-freedom-of-expression.pdf. The signatories are the United Nations (UN) Special Rapporteur on Freedom of Opinion and Expression; the Organization for Security and Co-operation in Europe (OSCE) Representative on Freedom of the Media; the Organization of American States (OAS) Special Rapporteur on Freedom of Expression; and the African Commission on Human and Peoples' Rights (ACHPR) Special Rapporteur on Freedom of Expression and Access to Information.

26. Testimony of Mark Zuckerberg, Hearing of US Senate Committee on the Judiciary and Committee on Commerce, Science and Transportation, *Facebook, Social Media Privacy, and the Use and Abuse of Data*, 115th Cong., 2d sess. (April 10, 2018), 111, https://www.congress.gov/115/chrg/CHRG-115shrg37801/CHRG-115shrg37801.pdf.

27. Taylor Hatmaker, "Facebook Will Lure Creators with $1 Billion in Payments," *TechCrunch*, July 14, 2021, https://techcrunch.com/2021/07/14/facebook-creator-bonuses/.

28. Brad McCarty, "Amazon's CTO: "Amazon Is a Technology Company. We Just Happen to Do Retail," *The Next Web*, October 5, 2011, https://thenextweb.com/news/amazons-cto-amazon-is-a-technology-company-we-just-happen-to-do-retail.

29. For an excellent analysis of the impact of Amazon Studios and the company's streaming platforms on the revenues of its other market sectors, see Tyler Klatt, "The Streaming Industry and the Great Disruption: How Winning a Golden Globe Helps Amazon Sell More Shoes," *Media, Culture & Society* 44, no. 8 (2022), 1541–1558.

30. See Philip M. Napoli, *Social Media and the Public Interest: Media Regulation in the Disinformation Age* (New York: Columbia University Press, 2019), 5–15; Philip Napoli and Robyn Caplan, "Why Media Companies Insist They're Not Media Companies, Why They're Wrong, and Why It Matters," *First Monday* 22, no. 5 (2017), https://doi.org/10.5210/fm.v22i5.7051; Terry Flew and Fiona R. Martin, "Introduction," in *Digital Platform Regulation: Global Perspectives on Internet Governance*, ed. Terry Flew and Fiona R. Martin (London: Palgrave Macmillan, 2022), 5–6.

31. Dwayne Winseck, "The Broken Internet and Platform Regulation: Promises and Perils," in *Digital Platform Regulation: Global Perspectives on Internet Governance*, ed. Terry Flew and Fiona R. Martin (London: Palgrave Macmillan, 2022), 238.

32. US House of Representatives, *Investigation of Competition*, 10.

33. US House of Representatives, *Investigation of Competition*, 6.

34. US House of Representatives, *Investigation of Competition*, 15.

35. US House of Representatives, *Investigation of Competition*, 322, 380.

36. For a useful survey of literature and research agendas in platform governance studies, see Robert Gorwa, "What Is Platform Governance?," *Information, Communication & Society* 22, no. 6 (2019): 854–871, https://doi.org/10.1080/1369118X.2019.1573914.

37. Flew and Martin, "Introduction," 7.

38. Napoli, *Social Media and the Public Interest*, 17.

39. Lawrence Lessig, *Code Version 2.0* (New York: Basic Books, 2006), 5–7.

40. All quotes taken directly from transcript of Hearing of US Senate Committee on the Judiciary and Committee on Commerce, Science and Transportation, *Facebook, Social Media Privacy, and the Use and Abuse of Data*, 115th Cong., 2d sess. (April 10, 2018), https://www.washingtonpost.com/news/the-switch/wp/2018/04/10/transcript-of-mark-zuckerbergs-senate-hearing/ and US House of Representatives Committee on Energy and Commerce. *Facebook: Transparency and Use of Consumer Data*. 115th Congress, 2d sess., April 11, 2018.

41. See Paul N. Edwards, *The Closed World* (Cambridge, MA: MIT Press, 1996); Benjamin Peters, *How Not to Network a Nation* (Cambridge, MA: MIT Press, 2016); Fred Turner, *From Counterculture to Cyberculture: Stewart Brand, the Whole Earth Network, and the Rise of Digital Utopianism* (Chicago: University of Chicago Press, 2006); Alice Marwick, "Silicon Valley and the Social Media Industry," in *The Sage Handbook of Social Media*, ed. Jean Burgess, Alice E. Marwick, and Thomas Poell (Thousand Oaks, CA: Sage Publications, 2018), 314–329; Margaret O'Mara, *The Code* (New York: Penguin Books, 2019); Esther Dyson et al., "Cyberspace and the American Dream: A Magna Carta for the Knowledge Age," The Progress and Freedom Foundation, August 22, 1994, http://www.pff.org/issues-pubs/futureinsights/fi1.2magnacarta.html; Barlow, "A Declaration of the Independence of Cyberspace," 1996; Thomas Streeter, *The Net Effect* (New York: New York University Press, 2011).

42. Jonathan Haidt, "Why the Past 10 Years of American Life Have Been Uniquely Stupid," *The Atlantic*, April 11, 2022, https://www.theatlantic.com/magazine/archive/2022/05/social-media-democracy-trust-babel/629369/.

43. Richard Barbrook and Andy Cameron, "The Californian Ideology," *Science as Culture* 6, no. 1 (1996): 44–72, https://doi.org/10.1080/09505439609526455. See, for example, Turner, *From Counterculture to Cyberculture*, and Terry Flew, *Regulating Platforms*.

44. Barbrook and Cameron, "The Californian Ideology" 49.

45. Barbrook and Cameron, "The Californian Ideology," 53, 55. In *Regulating Platforms*, Flew has noted that Barbrook and Cameron had a particularly strong influence on what he calls the "libertarian internet" of 1990–2005 that ultimately gave way to the "platformized internet" of today, which demands more in the way of governmental attention after decades of regulatory neglect.

46. Streeter, *Net Effect*, 70.

47. The company was hoping to raise $5 billion and wound up raising over $16 billion.

48. Mark Zuckerberg letter to investors, Reuters, February 1, 2012, https://www.reuters.com/article/us-facebook-letter/zuckerbergs-letter-to-investors-idUSTRE8102MT20120201. Also See Khadeeja Safdar, "Facebook, One Year Later," *The Atlantic*, May 20, 2013, https://www.theatlantic.com/business/archive/2013/05/facebook-one-year-later-what-really-happened-in-the-biggest-ipo-flop-ever/275987/ for a discussion of the corruption plaguing this IPO, which many called "the most anticipated initial public offering in history."

49. Fred Turner, "Machine Politics," *Harper's Magazine*, December 30, 2020, https://harpers.org/archive/2019/01/machine-politics-facebook-political-polarization/.

50. Craig Timberg, Elizabeth Dwoskin, and Reed Albergotti, "Inside Facebook, Jan. 6 Violence Fueled Anger, Regret over Missed Warning Signs," *Washington Post*, October 22, 2021, https://www.washingtonpost.com/technology/2021/10/22/jan-6-capitol-riot-facebook/.

51. Andrienne LaFrance, "The Facebook Papers," *The Atlantic*, October 25, 2021, https://www.theatlantic.com/ideas/archive/2021/10/facebook-papers-democracy-election-zuckerberg/620478/.

52. Sarah Igo, *The Known Citizen* (Cambridge, MA: Harvard University Press, 2019), 3.

53. Jeannie Suk Gersen, "Why the 'Privacy' Wars Rage On," *New Yorker*, June 20, 2022.

54. Many scholars have addressed their limited vision of this right, as it was clearly rooted in a white, patriarchal, bourgeois sensibility. See Igo, *The Known Citizen*, 39; Anita Allen and Erin Mack, "How Privacy Got Its Gender," *Northern Illinois University Law Review* 10 (1991): 441–478, https://scholarship.law.upenn.edu/faculty_scholarship/1309.

55. Samuel D. Warren and Louis D. Brandeis, "The Right to Privacy," *Harvard Law Review* 4, no. 5 (1890): 196.

56. Roe v. Wade (1973) was overturned by the Roberts court in Dobbs v. Jackson (2022), taking away a fundamental right for the first time in US Supreme Court history.

57. Examples include the Fair Credit Reporting Act (FCRA, 1970), the Privacy Act of 1974, the Family Educational Rights and Privacy Act (FERPA, 1974), the Right to Financial Privacy Act (RFPA, 1978), the Health Insurance Portability and Accountability Act

of 1996 (HIPAA), and the Electronic Communications Privacy Act (ECPA) of 1986, which is discussed at length in chapter 3.

58. The United States Department of Justice, "Overview of the Privacy Act: 2020 Edition," https://www.justice.gov/opcl/overview-privacy-act-1974-2020-edition/introduction#LegHistory. The Privacy Act established practices regarding the "collection, maintenance, use, and dissemination of information about individuals that is maintained in systems of records by federal agencies." However, its significant loopholes and exceptions left citizens much less protected than what the law initially promised. See Electronic Privacy Information Center, "The Privacy Act of 1974," https://epic.org/the-privacy-act-of-1974/.

59. O'Mara, *The Code*, 124–125.

60. Crain, *Profit over Privacy*, 22; also see 26–34.

61. Calhoun, et al. v. Google LLC, Case No. 4:20-cv-5146-YGR, unsealed documents filed September 19, 2022, 35. I am grateful to Mike Schmidt for sharing this material with me.

62. Frank Pasquale, *The Black Box Society: The Secret Algorithms That Control Money and Information* (Cambridge, MA: Harvard University Press, 2015), 143.

63. Dominic Rushe, "Google: Don't Expect Privacy When Sending to Gmail," *Guardian*, August 14, 2013, http://www.theguardian.com/technology/2013/aug/14/google-gmail-users-privacy-email-lawsuit; Ken Fisher, "AT&T Threatens to Disconnect Subscribers Who Criticize the Company," *Ars Technica*, September 30, 2007, https://arstechnica.com/tech-policy/2007/09/att-threatens-to-disconnect-subscribers-who-are-critical-of-the-company/; Geoffrey A. Fowler, "Alexa Has Been Eavesdropping on You This Whole Time," *Washington Post*, May 6, 2019, https://www.washingtonpost.com/technology/2019/05/06/alexa-has-been-eavesdropping-you-this-whole-time/; Vuctor Luckerson, "7 Surprising Things Lurking in Online 'Terms of Service' Agreements," *Time*, August 28, 2012, https://business.time.com/2012/08/28/7-surprising-things-lurking-in-online-terms-of-service-agreements/; Mitchell Clark, "Google, Like Amazon, Will Let Police See Your Video without a Warrant," *The Verge*, July 26, 2022, https://www.theverge.com/2022/7/26/23279562/arlo-apple-wyze-eufy-google-ring-security-camera-foortage-warrant; "Facebook Data Policy," https://www.facebook.com/about/privacy/; "Facebook Cookies & Other Storage Technologies Policy," https://www.facebook.com/policies/cookies/; Terms of Service: Didn't Read, https://tosdr.org/.

64. Jose van Dijck, *The Culture of Connectivity: A Critical History of Social Media* (Oxford: Oxford University Press, 2013), 171.

65. All of these companies have issued denials regarding their participation in and/or knowledge of the PRISM program, but their roles have been well documented. The earliest stories breaking this news were Glenn Greenwald and Ewan MacAskill, "NSA

Prism Program Taps in to User Data of Apple, Google, and Others," *The Guardian*, June 6, 2013, https://www.theguardian.com/world/2013/jun/06/us-tech-giants-nsa-data; Barton Gellman and Laura Poitras, "U.S., British Intelligence Mining Data from Nine U.S. Internet Companies in Broad Secret Program," *Washington Post*, June 7, 2013, https://www.washingtonpost.com/investigations/us-intelligence-mining-data-from-nine-us-internet-companies-in-broad-secret-program/2013/06/06/3a0c0da8-cebf-11e2-8845-d970ccb04497_story.html.

66. Barton Gellman, *Dark Mirror: Edward Snowden and the American Surveillance State* (New York: Penguin Press, 2020), 121.

67. Quoted in Rebecca MacKinnon, *Consent of the Networked: The Worldwide Struggle for Internet Freedom* (New York: Basic Books, 2012), 134.

68. Nicholas D. Kristof, "Chinas's Cyberdissidents and the Yahoos at Yahoo," *New York Times*, February 19, 2006, https://www.nytimes.com/2006/02/19/opinion/chinas-cyberdissidents-and-the-yahoos-at-yahoo.html.

69. See, for example, Ryan Gallagher, "How U.S. Tech Giants Are Helping to Build China's Surveillance State," *The Intercept*, July 11, 2019, https://theintercept.com/2019/07/11/china-surveillance-google-ibm-semptian/; Jack Nicas, Raymond Zhong, and Daisuke Wakabayashi, "Censorship, Surveillance and Profits: A Hard Bargain for Apple in China," *New York Times*, May 17, 2021, https://www.nytimes.com/2021/05/17/technology/apple-china-censorship-data.html; Paul Mozur, "Joining Apple, Amazon's China Cloud Service Bows to Censors," *New York Times*, August 1, 2017, https://www.nytimes.com/2017/08/01/business/amazon-china-internet-censors-apple.html.

70. Statement of Representative Christopher H. Smith (R-NJ), Hearing of US House of Representatives Committee on International Relations, Subcommittee on Africa, Global Human Rights and International Operations and the Subcommittee on Asia and the Pacific, *The Internet in China: A Tool for Freedom or Suppression?*, 109th Cong., 2d sess. (February 15, 2006), 2–3, https://chrissmith.house.gov/uploadedfiles/2006.02.15_the_internet_in_china_-_a_tool_for_freedom_or_suppression.pdf.

71. See Federal Trade Commission, *Privacy Online: Fair Information Practices in the Electronic Marketplace. A Report to Congress* (May 2000), https://www.ftc.gov/sites/default/files/documents/reports/privacy-online-fair-information-practices-electronic-marketplace-federal-trade-commission-report/privacy2000.pdf; Federal Trade Commission, *Protecting Consumer Privacy in an Era of Rapid Change. Recommendations for Businesses and Policymakers. FTC Report* (March 2012), https://www.ftc.gov/sites/default/files/documents/reports/federal-trade-commission-report-protecting-consumer-privacy-era-rapid-change-recommendations/120326privacyreport.pdf.

72. These breaches include the data of 3.5 billion records from Yahoo user accounts in 2013–2014; over 150 million customers of credit reporting company Equifax in 2017; 500 million records from customers of Marriott International in 2018; the personal

data of 533 million Facebook users in 2019; and information about 700 million users on LinkedIn in 2021. Sony Pictures (2011) and the US government (2006–2011) have also been the subjects of high profile hacks.

73. For a history of this failure, see Glenn Fleisman, "How the Tragic Death of Do Not Track Ruined the Web for Everyone," *Fast Company*, March 17, 2019, https://www.fastcompany.com/90308068/how-the-tragic-death-of-do-not-track-ruined-the-web-for-everyone.

74. Calhoun, et al. v. Google LLC, Case No. 4:20-cv-5146-YGR, unsealed documents filed September 19, 2022, 35.

75. The two parties in this case were Google Spain SL and Google Inc. v. Agencia Española de Protección de Datos (AEPD) and Mario Costeja González.

76. Alan Travis and Charles Arthur, "EU Court Backs 'Right to Be Forgotten,'" *The Guardian*, May 13, 2014, https://www.theguardian.com/technology/2014/may/13/right-to-be-forgotten-eu-court-google-search-results.

77. Article 17, General Data Protection Regulation, 2016/679 of the European Parliament and of the Council (2016), https://gdpr-info.eu/art-17-gdpr/.

78. Meg Leta Jones, *Ctrl+Z: The Right to Be Forgotten* (New York: New York University Press, 2016), 2.

79. Alan F. Westin, *Privacy and Freedom* (New York: Atheneum, 1967), 7.

80. See, for example, Linda Kinstler, "Into Oblivion: How News Outlets Are Handling the Right to Be Forgotten," *Columbia Journalism Review*, October 5, 2021, https://www.cjr.org/special_report/right-to-be-forgotten.php.

81. Jones, *Ctrl+Z*, 21.

82. Zuboff, "The Coup We Are Not Talking About."

83. Jack M. Balkin, "Old-School/New-School Speech Regulation," *Harvard Law Review* 127, no. 8 (June 2014): 2297, https://harvardlawreview.org/print/vol-127/old-schoolnew-school-speech-regulation/.

84. See Kate Klonick, "The New Governors: The People, Rules, and Processes Governing Online Speech," *Harvard Law Review* 131, no. 6 (April 2018): 1598–1670, https://harvardlawreview.org/print/vol-131/the-new-governors-the-people-rules-and-processes-governing-online-speech/; Shaun B. Spencer, "The First Amendment and the Regulation of Speech Intermediaries," *Marquette Law Review* 106, no. 1 (Fall 2022): 1–71, https://scholarship.law.marquette.edu/mulr/vol106/iss1/2.

85. See, for example, Philip M. Napoli, "What If More Speech Is No Longer the Solution? First Amendment Theory Meets Fake News and the Filter Bubble," *Federal Communications Law Journal* 70, no. 1 (2017): 55; Tim Wu, "Is the First Amendment

Obsolete?," *Michigan Law Review* 117, no. 3 (2018): 547, https://doi.org/https://doi.org/10.36644/mlr.117.3.first.

86. Communications Decency Act of 1996, 47 U.S.C. § 230(c) (1) (1996).

87. See Valerie C. Brannon and Eric N. Holmes, *Section 230: An Overview*, CRS Rep. No. R46751 (Washington, DC: Congressional Research Service, 2021), 3, 8, https://crsreports.congress.gov/product/pdf/R/R46751#, for a sampling of court decisions that have led to the interpretation that both Internet service providers (pipelines) and website operators (platforms) are included under the umbrella of "interactive computer services."

88. The exceptions include intellectual property rights, inducing or developing illegal content, and sex trafficking offenses (as of 2018, added by the FOSTA-SESTA legislation package).

89. Stratton Oakmont, Inc. v. Prodigy Services Co., 1995 WL 323710 (N.Y. 1995).

90. Stratton Oakmont, Inc. v. Prodigy Services Co., 1995 WL 323710 (N.Y. 1995).

91. Cubby, Inc. v. CompuServe Inc., 776 F. Supp. 135 (S.D.N.Y. 1991).

92. Stratton Oakmont, Inc. v. Prodigy Services Co., 1995 WL 323710 (N.Y. 1995).

93. Danielle Keats Citron, *The Fight for Privacy: Protecting Dignity, Identity, and Love in the Digital Age* (New York: W.W. Norton & Company, 2022), 86.

94. The ACLU immediately sued over the vague and overly broad definitions for criminalized speech, winning their case when the Supreme Court unanimously ruled that the censorship stipulations on indecency in the CDA were unconstitutional violations of free speech. The rest of the CDA was struck down, but Section 230 has endured. See Reno v. American Civil Liberties Union, 521 U.S. 844 (1997). For more on the history of legislating indecency in the CDA, see Robert Cannon, "The Legislative History of Senator Exon's Communications Decency Act: Regulating Barbarians on the Information Superhighway," *Federal Communications Law Journal* 49, no. 1 (1996): 51–94, https://www.repository.law.indiana.edu/fclj/vol49/iss1/3.

95. Communications Decency Act of 1996, 47 U.S.C. § 230(c) (2) (1996).

96. Tarleton Gillespie, *Custodians of the Internet: Platforms, Content Moderation, and the Hidden Decisions That Shape Social Media* (New Haven, CT: Yale University Press, 2018), 30–31.

97. See Rebecca Tushnet, "Power without Responsibility: Intermediaries and the First Amendment," *George Washington Law Review* 76, no. 4 (June 2008): 1002, https://www.gwlr.org/wp-content/uploads/2012/08/76-4-Tushnet.pdf.

98. Danielle Keats Citron and Mary Anne Franks, "The Internet as a Speech Machine and Other Myths Confounding Section 230 Reform," University of Chicago Legal

Forum 2020 (December 1, 2020): 47, https://chicagounbound.uchicago.edu/uclf/vol20 20/iss1/3.

99. US Senate Committee on Commerce, Science, and Transportation, *Telecommunications Competition and Deregulation Act of 1995*, S. Rep. 104–23, 104th Cong., 1st sess. (March 30, 1995), 59, https://www.congress.gov/104/crpt/srpt23/CRPT-104srpt23.pdf.

100. Kosseff, *The Twenty-Six Words That Created the Internet* (Ithaca, NY: Cornell University Press, 2019), 145.

101. See Rebecca MacKinnon et al., *Fostering Freedom Online: The Role of Internet Intermediaries*, Series on Internet Freedom (UNESCO, 2014), https://unesdoc.unesco.org/ark:/48223/pf0000231162?posInSet=1&queryId=N-EXPLORE-bfa151f5-d485-4dde-99b1-7f203576819a, on three global models of liability for intermediaries, ranging from strict (e.g., China) to conditional (e.g., the EU) to broad immunity (e.g., Section 230 in the US).

102. Julie Cohen, "Law for the Platform Economy," *UC Davis Law Review* 51 (2017): 164–165, https://scholarship.law.georgetown.edu/facpub/2015.

103. Calhoun, et al. v. Google LLC, Case No. 4:20-cv-5146-YGR, unsealed documents filed September 19, 2022, 35.

104. Roberts, *Behind the Screen*. Also see the haunting documentary *The Cleaners* (Block and Riesewieck, 2018)

105. Zuboff, "The Coup We Are Not Talking About."

106. Evan Osnos, "Can Mark Zuckerberg Fix Facebook before It Breaks Democracy?," *New Yorker*, September 10, 2018, https://www.newyorker.com/magazine/2018/09/17/can-mark-zuckerberg-fix-facebook-before-it-breaks-democracy. Osnos also reported that Facebook knew about the data breach for years but only acknowledged it once it was leaked to the press. Also see Christopher Wylie, *Mindf*ck: Inside Cambridge Analytica's Plot to Break the World* (London: Profile Books, 2019).

107. See, for example, Craig Silverman et al., "Facebook Hosted Surge of Misinformation and Insurrection Threats in Months Leading Up to Jan. 6 Attack, Records Show," *ProPublica*, January 4, 2022, https://www.propublica.org/article/facebook-hosted-surge-of-misinformation-and-insurrection-threats-in-months-leading-up-to-jan-6-attack-records-show; Donie O'Sullivan, Tara Subramaniam, and Clare Duffy, "Not Stopping 'Stop the Steal': Facebook Papers Paint Damning Picture of Company's Role in Insurrection," CNN.com, October 24, 2021, https://www.cnn.com/2021/10/22/business/january-6-insurrection-facebook-papers/index.html.

108. Howard, *Lie Machines*.

109. These taboos included "sex relationships between the white and black races," men and women in bed together, surgical operations, and ridicule of the clergy. This

developed into the self-censorship guidelines known as the Production Code, first adopted in 1930 and then enforced more carefully from 1934 to 1968, at which time the MPAA replaced the code with its rating system, which has endured in modified form to the present day.

110. Sheldon Whitehouse, "Section 230 Reforms," in *Social Media, Freedom of Speech, and the Future of Our Democracy*, ed. Lee C. Bollinger and Geoffrey R. Stone (Oxford: Oxford University Press, 2022), 104.

111. Nathalie Maréchal and Ellery Roberts Biddle, *It's Not Just the Content, It's the Business Model: Democracy's Online Speech Challenge*, Ranking Digital Rights Report (New America Open Technology Institute, March 17, 2020), 10, http://newamerica.org/oti/reports/its-not-just-content-its-business-model/.

112. US Senate Committee on the Judiciary, *Breaking the News: Censorship, Suppression, and the 2020 Election*, 116th Cong., 2d sess. (November 17, 2020), https://www.judiciary.senate.gov/meetings/breaking-the-news-censorship-suppression-and-the-2020-election.

113. See Twitter v. Taamneh, 598 U.S. 471 (2023) and Gonzalez v. Google, 598 U.S. 617 (2022).

114. European Commission, "Digital Services Act: Commission Welcomes Political Agreement on Rules Ensuring a Safe and Accountable Online Environment," press release, April 23, 2022, https://ec.europa.eu/commission/presscorner/detail/en/ip_22_2545.

115. Andrew Ross Sorkin et al., "A Big Swing at Big Tech," *New York Times*, March 25, 2022, https://www.nytimes.com/2022/03/25/business/dealbook/eu-tech-law.html.

116. European Commission, "Digital Services Act: Commission Welcomes Political Agreement on Rules Ensuring a Safe and Accountable Online Environment," press release, April 23, 2022, https://ec.europa.eu/commission/presscorner/detail/en/ip_22_2545.

117. Adam Satariano, "E.U. Takes Aim at Big Tech's Power with Landmark Digital Act," *New York Times*, March 24, 2022, https://www.nytimes.com/2022/03/24/technology/eu-regulation-apple-meta-google.html.

118. Jens Pohlmann, "Platform Regulation and the Digital Sphere: Comparing the Discourse in German and the United States" (lecture, MIT Graduate Program in Comparative Media Studies, March 31, 2022).

119. Sonja Solomun, Maryna Polataiko, and Helen A. Hayes, "Platform Responsibility and Regulation in Canada: Considerations on Transparency, Legislative Clarity, and Design," *Harvard Journal of Law & Technology Digest* 34 (Spring 2021): 1–18, https://jolt.law.harvard.edu/digest/platform-responsibility-and-regulation-in-canada-considerations-on-transparency-legislative-clarity-and-design.

120. Flew, *Regulating Platforms*, 146.

121. See Jennifer Holt and Lisa Parks, "The Labor of Digital Privacy Advocacy in an Era of Big Tech," *Media Industries* 8, no. 1 (2021): 1–25, https://doi.org/10.3998/mij.93.

122. "The Manila Principles on Intermediary Liability Background Paper," Electronic Frontier Foundation, 2015, https://www.manilaprinciples.org; "The Santa Clara Principles on Transparency and Accountability in Content Moderation," New America, 2018, https://newamericadotorg.s3.amazonaws.com/documents/Santa_Clara_Principles.pdf. Also see "Santa Clara Principles 2.0," https://santaclaraprinciples.org/, which were updated in 2021.

123. "Corporate Accountability Index," Ranking Digital Rights, 2020, https://rankingdigitalrights.org/index2020.

124. Amy Brouillette, "Key findings: Companies Are Improving in Principle, but Failing in Practice," Ranking Digital Rights, 2020 Corporate Accountability Index, https://rankingdigitalrights.org/index2020/key-findings#ttnt_ref1.

125. Alexandra Reeve Givens, interview by Jennifer Holt, October 11, 2022.

126. H. D. Lloyd, "The Story of a Great Monopoly," *Atlantic*, March 1881, https://www.theatlantic.com/magazine/archive/1881/03/the-story-of-a-great-monopoly/306019/. This work also contributed to a national debate about monopolies and antitrust legislation, which ultimately produced the Interstate Commerce Act of 1887 and the Sherman Antitrust Act of 1890.

127. See US House of Representatives, *Investigation of Competition in Digital Markets*, 15–16, 248, 268–330.

128. See Lina M. Khan, "The Separation of Platforms and Commerce," *Columbia Law Review* 119 (May 15, 2019): 1007–1013, 1061–1062, https://scholarship.law.columbia.edu/faculty_scholarship/2789/.

129. Franklin D. Roosevelt, "Acceptance Speech for the Renomination for the Presidency" (speech, Philadelphia, PA, June 27, 1936), archived at The American Presidency Project, UC Santa Barbara, https://www.presidency.ucsb.edu/documents/acceptance-speech-for-the-renomination-for-the-presidency-philadelphia-pa.

130. US Senate Committee on Commerce, *Appointments to the Regulatory Agencies: The Federal Communications Commission and the Federal Trade Commission (1949–1974)*, S. Rep. 62-119, 94th Cong., 2d sess. (April 1976), 205.

131. US Senate Committee on Small Business, Subcommittee on Retailing, Distribution, and Fair Trade Practices, *The Role of Private Antitrust Enforcement in Protecting Small Business*, 85th Cong., 2d sess. (March 3, 1958), 6.

132. Thomas Piraino, "Reconciling the Harvard and Chicago Schools: A New Antitrust Approach for the 21st Century," *Indiana Law Journal* 82, no. 2 (Spring 2007): 345–409, https://www.repository.law.indiana.edu/cgi/viewcontent.cgi?article=1354&context=ilj.

133. Zephyr Teachout, *Break 'Em Up: Recovering Our Freedom from Big Ag, Big Tech, and Big Money* (New York: All Points Books, 2020), 7.

134. "Facebook Presentation for Investors," cited in US House of Representatives, *Investigation of Competition in Digital Markets*, 138.

135. Robert H. Bork, "Legislative Intent and the Policy of the Sherman Act," *Journal of Law & Economics* 9 (October 1966): 7–48, https://www.jstor.org/stable/724991.

136. Tim Wu, *The Curse of Bigness: Antitrust in the New Gilded Age* (New York: Columbia Global Reports, 2018), 89.

137. Stoller, *Goliath*, 248.

138. Derek Thompson, "America's Monopoly Problem," *The Atlantic*, October 2016, https://www.theatlantic.com/magazine/archive/2016/10/americas-monopoly-problem/497549/.

139. Lina M. Khan, "Amazon's Antitrust Paradox," *Yale Law Journal* 126, no. 3 (January 2017): 710–805, https://www.yalelawjournal.org/note/amazons-antitrust-paradox.

140. Khan, "Amazon's Antitrust Paradox."

141. Stoller, *Goliath*, 444.

142. US House of Representatives, *Investigation of Competition in Digital Markets*, 6.

143. Cecilia Kang, Jack Nicas, and David McCabe, "Amazon, Apple, Facebook and Google Prepare for Their 'Big Tobacco Moment,'" *New York Times*, July 28, 2020, https://www.nytimes.com/2020/07/28/technology/amazon-apple-facebook-google-antitrust-hearing.html.

144. US House of Representatives, *Investigation of Competition*, 6, 14.

145. US House of Representatives, *Investigation of Competition*, 13–14, 151–156.

146. US House of Representatives, *Investigation of Competition*, 6.

147. Peter Thiel, "Competition Is for Losers," *Wall Street Journal*, September 12, 2014, https://www.wsj.com/articles/peter-thiel-competition-is-for-losers-1410535536.

148. US House of Representatives, *Investigation of Competition*, 406–450.

149. See Mark Glick, Catherine Ruetschlin, and Darren Bush, "Big Tech's Buying Spree and the Failed Ideology of Competition Law," *Hastings Law Journal* 72, no. 2 (2021): 467, https://repository.uclawsf.edu/hastings_law_journal/vol72/iss2/1.

150. Alex Sherman and Lauren Feiner, "Amazon, Microsoft and Alphabet Went on a Buying Spree in 2021 Despite D.C.'s Vow to Take on Big Tech," CNBC.com, January 22, 2022, https://www.cnbc.com/2022/01/22/amazon-microsoft-alphabet-set-more-deals-in-2021-than-last-10-years.html.

151. US House of Representatives, *Investigation of Competition*, 388.

152. US House of Representatives, *Investigation of Competition*, 377–405.

153. Microsoft was sued in 1998 by the DOJ and twenty state attorneys general for abusing its monopoly powers in the software and browsing markets. Its Windows operating system was on 90 percent of PCs, Internet Explorer was 98 percent of the browser market, and Microsoft Office had already become standard word processing and spreadsheet software. The district court found that Microsoft had violated parts of the Sherman Antitrust Act, stifling innovation and competition in their industry, and deterring investment in technology and businesses that threatened Microsoft. It was ordered to break up into two companies—one for Windows O/S and one for its other businesses in 1999/2000. Upon appeal, the decision was overturned in 2001, resulting in a settlement in 2002 with less severe penalties but core ruling regarding anticompetitive behavior remained. See United States v. Microsoft Corporation, 65 F. Supp. 2d 1 (D.D.C. 1999); Victor Luckerson, "'Crush Them': An Oral History of the Lawsuit That Upended Silicon Valley," *The Ringer*, May 18, 2018, https://www.theringer.com/tech/2018/5/18/17362452/microsoft-antitrust-lawsuit-netscape-internet-explorer-20-years.

154. These include the Senate's Platform Competition and Opportunity Act, the American Innovation and Choice Online Act, the Open App Markets Act, and the House of Representative's Ending Platform Monopolies Act.

155. See, for example, Cat Zakrzewski and Elizabeth Dwoskin, "Facebook Quietly Bankrolled Small, Grass-Roots Groups to Fight Its Battles in Washington," *Washington Post*, May 17, 2022, https://www.washingtonpost.com/technology/2022/05/17/american-edge-facebook-regulation/.

156. Tom Wheeler, "History Repeats Itself with Big Tech's Misleading Advertising," *Brookings Institute TechTank* (blog), June 15, 2022, https://www.brookings.edu/blog/techtank/2022/06/15/history-repeats-itself-with-big-techs-misleading-advertising/. Here, Wheeler was citing Doris Kearns Goodwin, *The Bully Pulpit: Theodore Roosevelt, William Howard Taft, and the Golden Age of Journalism* (New York: Simon & Schuster, 2013).

157. Elisa Jillson, "Aiming for Truth, Fairness, and Equity in Your Company's Use of AI," Federal Trade Commission Business Blog, April 19, 2021, https://www.ftc.gov/business-guidance/blog/2021/04/aiming-truth-fairness-equity-your-companys-use-ai. Also see Andrew D. Selbst and Solon Barocas, "Unfair Artificial Intelligence: How FTC Intervention Can Overcome the Limitations of Discrimination Law," *University of Pennsylvania Law Review* 171 (forthcoming), for a discussion of the FTC's potential role and interventions in this area.

158. Ashley Gold, "Former Google Exec: Antitrust Enforcement Is Key to Online Privacy," *Axios*, June 22, 2022, https://www.axios.com/2022/06/22/google-antitrust-privacy-sridhar-ramaswamy.

159. Nelson Lichtenstein, "America's 40-Year Experiment with Big Business Is Over," *New York Times*, July 13, 2021, https://www.nytimes.com/2021/07/13/opinion/biden-executive-order-antitrust.html.

160. Alphabet, Inc., *2021 Annual Report*, 2022, https://abc.xyz/investor/static/pdf/2021_alphabet_annual_report.pdf?cache=3a96f54.

161. See Courtney C. Radsch, *Making Big Tech Pay for the News They Use* (Center for International Media Assistance Report, July 7, 2022), https://www.cima.ned.org/publication/making-big-tech-pay-for-the-news-they-use/; Keach Hagey, "How Google Edged Out Rivals and Built the World's Dominant Ad Machine," *Wall Street Journal*, November 7, 2019, https://www.wsj.com/articles/how-google-edged-out-rivals-and-built-the-worlds-dominant-ad-machine-a-visual-guide-11573142071?mod=article_inline; Crain, *Profit over Privacy*.

162. Federal Trade Commission, "Federal Trade Commission Closes Google/DoubleClick Investigation," press release, December 20, 2007, https://www.ftc.gov/news-events/news/press-releases/2007/12/federal-trade-commission-closes-googledoubleclick-investigation.

163. See Dina Srinivasan, "Why Google Dominates Advertising Markets: Competition Policy Should Lean on the Principles of Financial Market Regulation," *Stanford Law School* 24, no. 1 (December 7, 2020): 62, 94–98, https://law.stanford.edu/publications/why-google-dominates-advertising-markets/.

164. See Robert W. McChesney, "Press-Radio Relations and the Emergence of Network, Commercial Broadcasting in the United States, 1930–1935," *Historical Journal of Film, Radio and Television* 11, no. 1 (1991): 41–57, https://doi.org/10.1080/01439689100260031. Also see Gwenyth Jackaway, "America's Press-Radio War of the 1930s: A Case Study in Battles between Old and New Media," *Historical Journal of Film, Radio and Television* 14, no. 3 (1994): 299–314, https://doi.org/10.1080/01439689400260211.

165. McChesney, "Press-Radio Relations," 53.

166. "Local News Deserts Are Expanding," *Washington Post*, November 30, 2021, https://www.washingtonpost.com/magazine/interactive/2021/local-news-deserts-expanding/.

167. Stoller, *Goliath*, 442.

168. "News Consumption across Social Media in 2021," Pew Research Center, September 20, 2021, https://www.pewresearch.org/journalism/2021/09/20/news-consumption-across-social-media-in-2021/.

169. Shoshana Zuboff, "You Are the Object of a Secret Extraction Operation," *New York Times*, November 12, 2021, https://www.nytimes.com/2021/11/12/opinion/facebook-privacy.html.

170. See, for example, Craig Silverman, "This Analysis Shows How Viral Fake Election News Stories Outperformed Real News On Facebook," *BuzzFeed News*, November 16, 2016, https://www.buzzfeednews.com/article/craigsilverman/viral-fake-election-news-outperformed-real-news-on-facebook; Tara McGowan, "Democracy Dies behind a Paywall," *Poynter*, July 15, 2022, https://www.poynter.org/commentary/2022/all-news-election-articles-should-be-free/.

171. Peter Dizikes, "Study: On Twitter, False News Travels Faster Than True Stories," *MIT News Office*, March 8, 2018, https://news.mit.edu/2018/study-twitter-false-news-travels-faster-true-stories-0308.

172. Keach Hagey and Jeff Horwitz, "Facebook Tried to Make Its Platform a Healthier Place. It Got Angrier Instead," *Wall Street Journal*, September 15, 2021, A1.

173. Hagey and Horwitz, "Facebook Tried," A1.

174. Statement of Frances Haugen before the United States Senate Committee on Commerce, Science and Transportation, Subcommittee on Consumer Protection, Product Safety, and Data Security, *Protecting Kids Online: Testimony from a Facebook Whistleblower*, 117th Cong., 1st sess. (October 4, 2021), https://www.commerce.senate.gov/services/files/FC8A558E-824E-4914-BEDB-3A7B1190BD49.

175. The myriad problems plaguing US news media are beyond the scope of this study but excellent sources include Robert W. McChesney, *Rich Media, Poor Democracy* (New York: The New Press, 2016); Martha Minow, *Saving the News* (New York: Oxford University Press, 2021); Robert W. McChesney and Victor Pickard, eds., *Will the Last Reporter Please Turn Out the Lights* (New York: The New Press, 2011).

176. McKay Coppins, "A Secretive Hedge Fund Is Gutting Newsrooms," *The Atlantic*, October 14, 2021, https://www.theatlantic.com/magazine/archive/2021/11/alden-global-capital-killing-americas-newspapers/620171/. When Alden took over the Tribune Publishing newspapers in 2021, they immediately laid off a quarter of the *Chicago Tribune* newsroom the purchase has been called "a disaster for Chicago, democracy, and society at large." See David Folkenflik, "'Vulture' Fund Alden Global, Known for Slashing Newsrooms, Buys Tribune Papers," NPR, May 21, 2021, https://www.npr.org/2021/05/21/998730863/vulture-fund-alden-global-known-for-slashing-newsrooms-buys-tribune-papers.

177. US House of Representatives, *Investigation of Competition*, 389–390.

178. *Senate Report No. 92*, 102d Congress, 1st Session (1991), 35.

179. Flew, *Regulating Platforms*, xiii.

180. Josh Taylor, "Google Threatens to Shut Down Search in Australia If Digital News Code Goes Ahead," *The Guardian*, January 21, 2021, https://www.theguardian.com/media/2021/jan/22/google-threatens-to-shut-down-search-in-australia-if-digital-news-code-goes-ahead.

181. Sara Fischer, "Meta Officially Cuts Funding for U.S. News Publishers," *Axios*, July 28, 2022, https://www.axios.com/2022/07/28/meta-publishers-news-funding-cut.

182. Clarksburg Publishing Company v. Google LLC, et al., C.A. No. 1:21-00051 (N.D.W. Va. 2021), 13–14.

183. Clarksburg Publishing Company v. Google LLC, et al., C.A. No. 1:21-00051 (N.D.W. Va. 2021), 15–16. The revelations about Jedi Blue first emerged in the antitrust lawsuit led by the state of Texas in December, 2020. They have since spawned new antitrust investigations in the EU and UK. See Natasha Lomas, "'Jedi Blue' Ad Deal between Google and Facebook Sparks New Antitrust Probes in EU and UK," *TechCrunch*, March 11, 2022, https://techcrunch.com/2022/03/11/google-meta-jedi-blue-eu-uk-antitrust-probes/.

184. Anne Applebaum and Peter Pomerantsev, "How to Put Out Democracy's Dumpster Fire," *The Atlantic*, April 2021, https://www.theatlantic.com/magazine/archive/2021/04/the-internet-doesnt-have-to-be-awful/618079/.

185. Nicholas Suzor, *Lawless* (New York: Cambridge University Press, 2019).

186. Digital Platform Commission Act of 2022, S. Res. 4201, 117th Cong. (2022), https://www.congress.gov/117/bills/s4201/BILLS-117s4201is.pdf. As of this writing, no further action has been taken on this bill.

187. Tom Wheeler, Phil Verveer, and Gene Kimmelman, *New Digital Realities; New Oversight Solutions in the U.S.* (Shorenstein Center on Media, Politics and Public Policy, August 2020), https://shorensteincenter.org/wp-content/uploads/2020/08/New-Digital-Realities_August-2020.pdf.

188. Tom Wheeler, interview by Jennifer Holt, February 28, 2023.

189. Victor Pickard, "A New Social Contract for Platforms," in *Regulating Big Tech: Policy Responses to Digital Dominance*, ed. Martin Moore and Damian Tambini (New York: Oxford University Press, 2022), 324, 325–326.

190. Philip M. Napoli, "Treating Dominant Digital Platforms as Public Trustees," in *Regulating Big Tech: Policy Responses to Digital Dominance*, ed. Martin Moore and Damian Tambini (New York: Oxford University Press, 2022), 153, 154–155, 163.

191. See Jack M. Balkin, "Information Fiduciaries and the First Amendment," *UC Davis Law Review* 49, no. 4 (2016): 1183–1234.

192. Applebaum and Pomerantsev, "How to Put Out Democracy's Dumpster Fire."

193. Applebaum and Pomerantsev, "How to Put Out Democracy's Dumpster Fire."

194. Lina Khan, "The Separation of Platforms and Commerce," 973.

195. See Spotify's summary of its claims at "Time to Play Fair: The Case," https://timetoplayfair.com/the-case/.

196. Lina Khan, "The Separation of Platforms and Commerce," 973.

197. US House of Representatives, *Investigation of Competition*, 380.

198. Timothy Karr and Craig Aaron, *Beyond Fixing Facebook*, Free Press Research Report (February 2019), 5, 7–9, https://www.freepress.net/policy-library/beyond-fixing-facebook.

199. Courtney C. Radsch, *Making Big Tech Pay for the News They Use*, Report for the Center for International Media Assistance (CIMA), July 7, 2022, https://www.cima.ned.org/publication/making-big-tech-pay-for-the-news-they-use/#cima_footnote_20.

200. See, for example, Testimony of Gene Kimmelman, President and CEO of Public Knowledge, before the U.S. House of Representatives Committee on the Judiciary, Subcommittee on Antitrust, Commercial, and Administrative Law, *Online Platforms and Market Power, Part 1: The Free and Diverse Press*, 116th Cong., 1st sess. (June 11, 2019), https://docs.house.gov/meetings/JU/JU05/20190611/109616/HHRG-116-JU05-Wstate-KimmelmanG-20190611.pdf.

201. Mariana Mazzucato, Josh Entsminger, and Rainer Kattel, "Reshaping Platform-Driven Digital Markets," in *Regulating Big Tech: Policy Responses to Digital Dominance*, ed. Martin Moore and Damian Tambini (New York: Oxford University Press, 2021), 17.

Chapter 3

1. See "Senate Select Committee to Study Governmental Operations with Respect to Intelligence Activities," United States Senate, nd, https://www.senate.gov/artandhistory/history/common/investigations/ChurchCommittee.htm. The Senate page goes on to note that, "in the course of their work, investigators identified programs that had never before been known to the American public, including NSA's Projects SHAMROCK and MINARET, programs which monitored wire communications to and from the United States and shared some of that data with other intelligence agencies." This surveillance took place without any warrants. For more on Project SHAMROCK, including a link to the full report, see Nate Anderson, "How a 30-Year-Old Lawyer Exposed NSA Mass Surveillance of Americans—in 1975," *Ars Technica*, June 30, 2013, https://arstechnica.com/tech-policy/2013/06/how-a-30-year-old-lawyer-exposed-nsa-mass-surveillance-of-americans-in-1975/. For more on SHAMROCK's sister program, Project MINARET (which actually targeted Senator Church himself), see Matthew M. Aid and William Burr, eds., "'Disreputable If Not Outright Illegal': The National Security Agency versus Martin Luther King, Muhammad Ali, Art Buchwald, Frank Church, et al.," National Security Archive Electronic Briefing Book no. 441, September 25, 2013, https://nsarchive2.gwu.edu/NSAEBB/NSAEBB441/; see specifically "The Watch List and MINARET" and "The Targets" sections. These reports were also partially inspired by bombshell reporting of Carl Bernstein and Bob Woodward, "Five Held in Plot to Bug Democratic Office," *Washington Post*, June 18, 1972, http://www.washingtonpost.com

/wp-srv/politics/special/watergate/part1.html, on the Watergate break-in and corruption in the executive branch. Also see exposés such as the front-page article written by Seymour M. Hersh, "Huge C.I.A. Operation Reported in U.S. against Antiwar Forces, Other Dissidents in Nixon Years," *New York Times*, December 22, 1974, which contributed to the cultural awareness about the systemic abuse of power and intelligence services that had been taking place for decades.

2. Christopher Pyle, "CONUS Intelligence: The Army Watches Civilian Politics," *Washington Monthly*, January 1970, 7.

3. James Reston, "Government by Outrage," *New York Times*, September 2, 1973, E13.

4. The Church Committee was formally known as the Senate Select Committee to Study Governmental Operations with Respect to Intelligence Activities. The Pike Committee refers to the House Select Committee on Intelligence and their 1975–1976 investigation of the US intelligence community. See David Medine and Esteban Morin, "Privacy and Civil Liberties Oversight Board," in *The Cambridge Handbook of Surveillance Law*, ed. David Gray and Stephen E. Henderson (New York: Cambridge University Press, October 12, 2017), 687.

5. See the main contents of the report here: Commission on CIA Activities within the United States (Rockefeller Commission), *Report to the President by the Commission on CIA Activities within the United States* (June 6, 1975), https://history-matters.com/archive/contents/church/contents_church_reports_rockcomm.htm. Also see the eighty-six-page removed section on CIA assassination plots and other edits to the report by then-deputy White House Chief of Staff Dick Cheney here: https://nsarchive.gwu.edu/briefing-book/intelligence/2016-02-29/gerald-ford-white-house-altered-rockefeller-commission-report.

6. Select Committee to Study Governmental Operations (Church Committee), *Intelligence Activities and the Rights of Americans: Book II*, S. Rep. 94–755, 94th Cong., 2d sess. (April 26, 1976), http://www.aarclibrary.org/publib/contents/church/contents_church_reports_book2.htm. Documentation of the targeting of Dr. King and those in the "Women's Liberation Movement" begins on page 7.

7. Stuart Taylor Jr., "The Big Snoop: Life, Liberty, and the Pursuit of Terrorists," *The Brookings Essay*, April 29, 2014, http://www.brookings.edu/research/essays/2014/the-big-snoop.

8. Select Committee to Study Governmental Operations, *Intelligence Activities and the Rights of Americans: Book II*, 3.

9. Select Committee to Study Governmental Operations, *Intelligence Activities and the Rights of Americans: Book II*, 4.

10. Select Committee to Study Governmental Operations, *Intelligence Activities and the Rights of Americans: Book II*, 5.

Notes to Chapter 3

11. David Rohde, *In Deep: The FBI, the CIA, and the Truth about America's "Deep State"* (New York: W. W. Norton & Company, 2020), 10. Church's comments on the August 17, 1975, episode were also replayed on a roundtable about surveillance in the wake of the Snowden revelations on *Meet the Press*, aired August 4, 2013, on NBC. The Church quotation also appears in the front matter (and title) of Glenn Greenwald's book about Edward Snowden, *No Place to Hide: Edward Snowden, the NSA, and the U.S. Surveillance State* (New York: Metropolitan Books, 2014).

12. Also quoted in Rohde, *In Deep*, 10.

13. United Kingdom National Infrastructure Commission, *Data for the Public Good* (December 2017), 38, https://nic.org.uk/app/uploads/Data-for-the-Public-Good-NIC-Report.pdf.

14. Tom Vanderbilt, "Data Center Overload," *New York Times*, June 8, 2009, https://www.nytimes.com/2009/06/14/magazine/14search-t.html.

15. See the Google gallery here: https://www.google.com/about/datacenters/gallery/#/. See Holt and Vonderau on the related "hyperpolitics of visibility" in, Jennifer Holt and Patrick Vonderau, "Where the Internet Lives: Data Centers as Digital Media Infrastructure," in *Signal Traffic: Critical Studies of Media Infrastructures*, ed. Lisa Parks and Nicole Starosielski (Urbana: University of Illinois Press, 2015), 75–81.

16. Parks, "Around the Antenna Tree," 2009.

17. Alix Johnson and Mél Hogan, "Introducing Location and Dislocation: Global Geographies of Digital Data," *Imaginations Journal* 8, no. 2 (September 5, 2017): 4–7.

18. Lewis Mumford, *The City in History: Its Origins, Its Transformations, and Its Prospects* (Orlando, FL: Houghton Mifflin Harcourt, 1961), 563. Quoted in Nicole Starosielski, "'Warning: Do Not Dig': Negotiating the Visibility of Critical Infrastructures," *Journal of Visual Culture* 11, no. 1 (2012): 39, https://doi.org/10.1177/1470412911430465.

19. Starosielski, "'Warning: Do Not Dig,'" 39–40.

20. For an overview, see Jeffrey Ritter and Anna Mayer, "Regulating Data as Property: A New Construct for Moving Forward," *Duke Law & Technology Review* 16, no. 1 (2018): Part III, https://scholarship.law.duke.edu/dltr/vol16/iss1/7. Also see Aaron Perzanowski and Jason Schultz, *The End of Ownership: Personal Property in the Digital Economy* (Cambridge, MA: MIT Press, 2016); Lothar Determann, "No One Owns Data," *UC Law Journal* 70, no. 1 (2019): 1–44, https://repository.uclawsf.edu/hastings_law_journal/vol70/iss1/1.

21. Katharina Pistor, "Rule by Data: The End of Markets?" *Law and Contemporary Problems* 83, no. 2 (2020): 107, https://scholarship.law.duke.edu/lcp/vol83/iss2/6.

22. Mike McKenzie, Deputy Bureau Chief and Senior Advisor for New Technology, interview by Jennifer Holt, May 21, 2012, Washington, DC.

23. See, for example, Andrew Keane Woods, "Against Data Exceptionalism," *Stanford Law Review* 68, no. 4 (April 2016): 729–789, https://www.stanfordlawreview.org/print/article/against-data-exceptionalism/.

24. David R. Johnson and David Post, "Law and Borders: The Rise of Law in Cyberspace," *Stanford Law Review* 48 (1996): 1370, 1367–1402.

25. John Perry Barlow, "A Declaration of the Independence of Cyberspace," Electronic Frontier Foundation, February 8, 1996, https://www.eff.org/cyberspace-independence.

26. Reno v. American Civil Liberties Union, 521 U.S. 844, 851 (1997).

27. See Jennifer Daskal, "The Un-Territoriality of Data," *Yale Law Journal* 125, no. 2 (November 2015): 397, https://www.yalelawjournal.org/article/the-un-territoriality-of-data.

28. Zachary D. Clopton, "Territoriality, Technology, and National Security," *University of Chicago Law Review* 83, no. 1 (2015): 49, https://chicagounbound.uchicago.edu/uclrev/vol83/iss1/3.

29. Paul M. Schwartz, "Legal Access to the Global Cloud," *Columbia Law Review* 118, no. 6 (October 2018): 1703, https://columbialawreview.org/content/legal-access-to-the-global-cloud/.

30. Vivek Kundra, *Federal Cloud Computing Strategy*, White House Report by US Chief Information Officer (February 8, 2011), 30, https://obamawhitehouse.archives.gov/sites/default/files/omb/assets/egov_docs/federal-cloud-computing-strategy.pdf.

31. Woods, "Against Data Exceptionalism," 734–735.

32. Sasha Segall, "Jurisdictional Challenges in the United States Government's Move to Cloud Computing Technology," *Fordham Intellectual Property, Media and Entertainment Law Journal* 23, no. 3 (Spring 2013): 1105–1153, https://ir.lawnet.fordham.edu/iplj/vol23/iss3/7.

33. Secil Bilgic, "Something Old, Something New, and Something Moot: The Privacy Crisis under the Cloud Act," *Harvard Journal of Law & Technology* 32, no. 1 (Fall 2018): 330, https://jolt.law.harvard.edu/assets/articlePDFs/v32/32HarvJLTech321.pdf.

34. Quoted in Tung-Hui Hu, *A Prehistory of the Cloud* (Cambridge, MA: MIT Press, 2016), xiii.

35. "Project Natick," Project Natick webpage, accessed May 4, 2020, https://natick.research.microsoft.com/.

36. See, for example, Nathaniel, "Underwater Data Centers Are Not the Answer," *Medium* (blog), July 2, 2019, https://medium.com/discourse/underwater-data-centers-are-not-the-answer-a789d072f614.

Notes to Chapter 3

37. Julia Velkova, "Data Centers as Impermanent Infrastructures," *Culture Machine* 18 (2019), https://culturemachine.net/vol-18-the-nature-of-data-centers/data-centers-as-impermanent/.

38. Henry McDonald, "Ireland Is Cool for Google as Its Data Servers Like the Weather," *The Guardian*, December 22, 2012, https://www.theguardian.com/technology/2012/dec/23/ireland-cool-google-data-servers-weather?CMP=share_btn_link.

39. Marcus Law, "Energy Efficiency Predictions for Data Centres in 2023," *DataCentre*, December 30, 2022, https://datacentremagazine.com/articles/efficiency-to-loom-large-for-data-centre-industry-in-2023; Timothy Rooks, "Data Centers Keep Energy Use Steady Despite Big Growth" DW.com, January 24, 2022, https://www.dw.com/en/data-centers-energy-consumption-steady-despite-big-growth-because-of-increasing-efficiency/a-60444548.

40. Richard Orange, "Facebook to Build Server Farm on Edge of Arctic Circle," *The Telegraph*, October 26, 2011, http://www.telegraph.co.uk/technology/facebook/8850575/Facebook-to-build-server-farm-on-edge-of-Arctic-Circle.html.

41. See Julia Velkova, "Data That Warms: Waste Heat, Infrastructural Convergence and the Computation Traffic Commodity," *Big Data & Society* 3, no. 2 (December 2016): 1–10, https://doi.org/10.1177/2053951716684144.

42. Justin Vela, "Helsinki Data Centre to Heat Homes," *The Guardian*, July 20, 2010, https://www.theguardian.com/environment/2010/jul/20/helsinki-data-centre-heat-homes.

43. See He Zike, "Life amid the Guizhou Clouds," *Sixth Tone*, March 17, 2022, https://www.sixthtone.com/news/1009897/life-amid-the-guizhou-clouds; Xu Yanqui and Zhang Song, "Big Data: Guizhou's New Calling Card," *CGTN*, January 9, 2020, https://news.cgtn.com/news/2020-01-09/Big-data-Guizhou-s-new-calling-card-N7a4p7zgUU/index.html.

44. Adrian Shahbaz, Allie Funk, and Kian Vesteinsson, *Countering an Authoritarian Overhaul of the Internet*, Report (Freedom on the Net, 2022), https://freedomhouse.org/report/freedom-net/2022/countering-authoritarian-overhaul-internet.

45. See Sarah Igo, *The Known Citizen* (Cambridge, MA: Harvard University Press, 2019), chapter 2.

46. Chief Justice Warren, Concurring Opinion, Lopez v. United States, 373 U.S. 427, 441 (1963).

47. Arthur R. Miller, *The Assault on Privacy: Computers, Data Banks, and Dossiers* (Ann Arbor: University of Michigan Press, 1971), 2.

48. Alan F. Westin, *Privacy and Freedom* (New York: Atheneum, 1967), 4.

49. Westin, *Privacy and Freedom*, 7.

50. US House of Representatives, Committee on Government Operations, Special Subcommittee on Invasion of Privacy, *The Computer and Invasion of Privacy*, 89th Cong, 2d sess. (July 26, 1966), 3.

51. US House of Representatives, *Computer and Invasion of Privacy*, 5–6.

52. US House of Representatives, *Computer and Invasion of Privacy*, 125.

53. Testimony of Vance Packard before the US House of Representatives, *Computer and Invasion of Privacy*, 11–12.

54. See Ron Felber, *The Privacy War* (Montvale, NY: Croce Publishing, 2003).

55. Quoted in Felber, *Privacy War*, 176. Also see the Statement of Professor Arthur R. Miller, before the US Senate Committee on the Judiciary, Subcommittee on Administrative Practice and Procedure, *Computer Privacy*, 90th Cong., 1st sess. (March 14, 1967), 72–73.

56. US House of Representatives Committee on Government Operations, *Privacy and the National Data Bank Concept*, 9th Cong., 2d sess. (July 1968), x.

57. US House of Representatives, *Privacy and the National Data Bank Concept*, 6.

58. US House of Representatives, *Privacy and the National Data Bank Concept*, 5.

59. Jack Star, "The Computer Data Bank: Will It Kill Your Freedom?" *Look*, June 1968, 27.

60. Margaret O'Mara, "The End of Privacy Began in the 1960s," *New York Times*, December 5, 2018, https://www.nytimes.com/2018/12/05/opinion/google-facebook-privacy.html.

61. US Senate Committee on the Judiciary, Subcommittee on Administrative Practice and Procedure, *Computer Privacy*, 90th Cong., 1st sess. (March 14, 1967), 1.

62. US Department of Health, Education, and Welfare, Secretary's Advisory Committee on Automated Personal Data Systems, *Records, Computers, and the Rights of Citizens*, Rep. No. (OS) 73–94 (Cambridge, MA: DHEW Publication, July 1973), xx.

63. O'Mara, "End of Privacy Began."

64. Igo, *Known Citizen*, 257.

65. US Senate, Joint Hearing of the Subcommittee on Constitutional Rights of Committee on the Judiciary and the Special Subcommittee on Science, Technology, and Commerce of the Committee on Commerce, *Surveillance Technology*, 94th Cong., 1st sess. (June 23, September 9, and September 10, 1975), 3.

66. US Senate, *Surveillance Technology*, 1.

67. US Senate, *Surveillance Technology*, 104.

68. US Senate, *Surveillance Technology*, 107.

69. US Senate, *Surveillance Technology*, 2.

70. Uniting and Strengthening America by Providing Appropriate Tools Required to Intercept and Obstruct Terrorism (USA PATRIOT) Act of 2001, Pub. L. No. 107–56, 115 Stat. 272 (2001).

71. US Senate, *Surveillance Technology*, 2. Section 215 expired in 2020 and has not yet been reauthorized as of this writing. However, the law remains in effect for existing investigations and for new investigations into events that occurred before the expiration date.

72. Andrea Renda, "Cloud Privacy Law in the United States and the European Union," in *Regulating the Cloud: Policy for Computing Infrastructure*, ed. Christopher S. Yoo and Jean-François Blanchette (Cambridge, MA: MIT Press, 2015), 135–164 (141).

73. Segall, "Jurisdictional Challenges," 1134–1136.

74. Rohde, *In Deep*, 11.

75. Foreign Intelligence Surveillance Act of 1978, Pub. L. No. 95–511, 92 Stat. 1783 (1978), 20–22, https://fas.org/irp/agency/doj/fisa/hspci1978.pdf.

76. For example, no longer was it possible to determine that a US citizen is an agent of a foreign power "solely upon the basis of activities protected by the first amendment to the Constitution of the United States." Section 402(a)(1) of Foreign Intelligence Surveillance Act of 1978 (FISA), 50 U.S.C. § 1842(a)(1); Section 501(a)(1) and (a)(2)(B) of FISA, 50 U.S.C. § 1861(a)(1) and (a)(2)(B). This was precisely the rationalization for targeting Martin Luther King Jr. and women's rights protesters, among many others.

77. Quoted in Renda, "Cloud Privacy Law," 141.

78. Executive Order 12333 was also amended three times under George W. Bush to expand the surveillance operations and abilities of the NSA. For a discussion of whistleblower John Tye's arguments about Executive Order 12333, see Cyrus Farivar, "Meet John Tye: The Kinder, Gentler, and By-the-Book Whistleblower," *Ars Technica*, August 20, 2014, https://arstechnica.com/tech-policy/2014/08/meet-john-tye-the-kinder-gentler-and-by-the-book-whistleblower/.

79. FISA was amended in 2001 by the PATRIOT Act; in 2007 by the Protect America Act; in 2008 by the FISA Amendments Act; and in 2017 by the FISA Amendments Reauthorization Act.

80. Foreign Intelligence Surveillance Act of 1978 Amendments Act of 2008, Pub. L. No. 110–261, 122 Stat. 2436 (2008). Also see United States Office of the Director of National Intelligence, "Section 702 Overview," nd, 1, https://www.dni.gov/files/icotr/Section702-Basics-Infographic.pdf .

81. "'Incidental,' Not Accidental, Collection," Electronic Frontier Foundation, nd, https://www.eff.org/pages/Incidental-collection.

82. "Foreign Intelligence Surveillance Court, Foreign Intelligence Surveillance Court of Review, Current and Past Members," Foreign Intelligence Surveillance Court, May 2020, https://www.fisc.uscourts.gov/sites/default/files/FISC%20FISCR%20Judges%20Revised%20May%2029%202020%20200608.pdf.

83. Transcript available here: Dina Temple-Raston, "FISA Court Appears to Be Rubber Stamp for Government Requests," NPR, June 13, 2013, https://www.npr.org/2013/06/13/191226106/fisa-court-appears-to-be-rubberstamp-for-government-requests.

84. "Former Judge Admits Flaws with Secret FISA Court," *CBS News*, July 9, 2013, https://www.cbsnews.com/news/former-judge-admits-flaws-with-secret-fisa-court/.

85. Ezra Klein, "Did You Know John Roberts Is Also Chief Justice of the NSA's Surveillance State?" *Washington Post*, July 5, 2013, https://www.washingtonpost.com/news/wonk/wp/2013/07/05/did-you-know-john-roberts-is-also-chief-justice-of-the-nsas-surveillance-state/.

86. "Foreign Intelligence Surveillance Act Court Orders 1979–2017," Electronic Privacy Information Center, https://epic.org/privacy/surveillance/fisa/stats/default.html#foot2text.

87. Specifically through Section 505 of the PATRIOT Act, which lowered the threshold for situations in which NSLs may be issued and expanded personnel who could provide approval authority well beyond FBI headquarters, among other provisions. This led to much greater and easier use of this intrusive surveillance tool. See "National Security Letters," Electronic Privacy Information Center, https://epic.org/privacy/nsl/.

88. Andrew Nieland, "National Security Letters and the Amended Patriot Act," *Cornell Law Review* 92, no. 6 (September 2007): 1203, https://scholarship.law.cornell.edu/clr/vol92/iss6/4; US Department of Justice Office of the Inspector General, *A Review of the Federal Bureau of Investigation's Use of National Security Letters* (March 2007), 36, https://oig.justice.gov/special/s0703b/final.pdf. The true number was estimated to be even higher as the report estimated "approximately 8,850 NSL requests, or 6 percent of NSL requests issued by the FBI during this period, were missing from the database" (see US Department of Justice Office of the Inspector General, *A Review of the Federal Bureau of Investigation's Use*, 34).

89. See "Patriot Act Gives Foreigners Good Reason to Avoid US Clouds," *Wired*, December, 2011, https://www.wired.com/insights/2011/12/us-cloud/; ACLU, "Internal Report Finds Flagrant National Security Letter Abuse by FBI," press release, January 20, 2010, https://www.aclu.org/press-releases/internal-report-finds-flagrant-national-security-letter-abuse-fbi.

90. Electronic Communications Privacy Act of 1986, Pub. L. No. 99–508, 100 Stat. 1848 (1986).

Notes to Chapter 3

91. US Senate Committee on the Judiciary, *Electronic Communications Privacy Act of 1986 Report*, S. Rep. 99–541, 99th Cong., 2d sess. (October 17, 1986), 2, https://www.justice.gov/sites/default/files/jmd/legacy/2014/08/10/senaterept-99-541-1986.pdf.

92. US Senate Committee on the Judiciary, *Electronic Communications Privacy Act of 1986 Report*, 2.

93. US Senate Committee on the Judiciary, *Electronic Communications Privacy Act of 1986 Report*, 8.

94. Statement of Senator Patrick Leahy, before the US Senate Committee on the Judiciary, *The Electronic Communications Privacy Act: Promoting Security and Protecting Privacy in the Digital Age*, 111th Cong., 2d sess. (September 22, 2010), https://www.judiciary.senate.gov/imo/media/doc/leahy_statement_09_22_10.pdf.

95. Statement of Senator Patrick Leahy, before the US Senate Committee on the Judiciary, *Electronic Communications Privacy Act*.

96. United States v. Miller, 425 US 435 (1976); Smith v. Maryland, 442 US 735 (1979).

97. The ECPA has three titles. The Stored Communications Act (SCA) is Title II of the ECPA. The amendments to the Wiretap Act served as Title I. Title III addressed "pen register" and "trap and trace devices" that allowed for the surveillance of a subject's outgoing and incoming telephone communications. See Richard M. Thompson II and Jared P. Cole, *Stored Communications Act: Reform of the Electronic Communications Privacy Act (ECPA)*, CRS Rep No. R44036 (Washington, DC: Congressional Research Service, May 19, 2015), https://www.everycrsreport.com/reports/R44036.html. Everycrsreport.com is an outstanding resource for accessing the non-partisan Congressional Research Service (CRS) reports that are available to legislative staff and other government officials and journalists. The website is dedicated to promoting open legislative information and making reports free and available online for all.

98. Stored Communications Act, 18 U.S.C. §§ 2701–12 (1986).

99. H. K. Ramapriyan, "Evolution of Archival Storage (from Tape to Memory)," NASA Technical Reports Server, nd, https://ntrs.nasa.gov/api/citations/20150006826/downloads/20150006826.pdf.

100. Rainey Reitman, "Deep Dive: Updating the Electronic Communications Privacy Act," Electronic Frontier Foundation, December 6, 2012, https://www.eff.org/deeplinks/2012/12/deep-dive-updating-electronic-communications-privacy-act.

101. United States v. Reicherter, 647 F.2d 397 (1981).

102. See "Who We Are," Digital Due Process, https://digitaldueprocess.org/who-we-are/; "Our Principles," Digital Due Process, https://digitaldueprocess.org/our-principles/.

103. "About the Issue," Digital Due Process, https://digitaldueprocess.org/.

104. Statement of Brad Smith, General Counsel, Microsoft Corporation, "The Need for ECPA Reform and Advancing Cloud Computing," before US Senate Committee on the Judiciary, *Electronic Communications Privacy Act.*

105. Stuart Lauchlan, "Non-US Citizens to Get US Privacy Rights in the Cloud under Obama Big Data Overhaul?" *Diginomica*, May 6, 2014, http://diginomica.com/2014/05/06/non-us-citizens-privacy-rights/.

106. White House Press Office, "Fact Sheet: Big Data and Privacy Working Group Review," May 1, 2014, http://www.whitehouse.gov/the-press-office/2014/05/01/fact-sheet-big-data-and-privacy-working-group-review.

107. Orin Kerr, "What Legal Protections Apply to E-mail Stored Outside the U.S.?" *Washington Post*, July 7, 2014, https://www.washingtonpost.com/news/volokh-conspiracy/wp/2014/07/07/what-legal-protections-apply-to-e-mail-stored-outside-the-u-s/.

108. United States v. Microsoft Corp., 584 U.S. ___ (2018).

109. For an excellent analysis of the legal issues involved with U.S. v. Microsoft Corp and the role of the ECPA, see Andrew Keane Woods, "Litigating Data Sovereignty," *Yale Law Journal* 128, no. 2 (November 2018): 328–406, https://www.yalelawjournal.org/article/litigating-data-sovereignty.

110. US House of Representatives Committee on the Judiciary, *International Conflicts of Law and Their Implications for Cross Border Data Requests by Law Enforcement*, 114th Cong., 2d sess. (February 25, 2016), https://www.govinfo.gov/content/pkg/CHRG-114hhrg98827/html/CHRG-114hhrg98827.htm; US House of Representatives Committee on the Judiciary, *Data Stored Abroad: Ensuring Lawful Access and Privacy Protection in the Digital Era*, 115th Cong., 1st sess. (June 15, 2017), https://www.govinfo.gov/content/pkg/CHRG-115hhrg31564/pdf/CHRG-115hhrg31564.pdf; US Senate Committee on the Judiciary, Subcommittee on Crime and Terrorism, *Law Enforcement Access to Data Stored across Borders: Facilitating Cooperation and Protecting Rights*, 115th Cong., 1st sess. (May 24, 2017), https://www.judiciary.senate.gov/meetings/law-enforcement-access-to-data-stored-across-borders-facilitating-cooperation-and-protecting-rights.

111. There was no debate in Congress about the CLOUD Act per se, it was passed without a dedicated hearing as part of an omnibus spending bill.

112. Clarifying Lawful Overseas Use of Data (CLOUD) Act, Pub. L. No. 115–141, Div. V, 132 Stat. 1213 (2018).

113. Bilgic, "Something Old, Something New," 347.

114. Glenn Greenwald and Ewan MacAskill, "NSA Prism Program Taps in to User Data of Apple, Google, and Others," *The Guardian*, June 6, 2013, https://www.theguardian.com/world/2013/jun/06/us-tech-giants-nsa-data; Barton Gellman and Laura Poitras, "U.S., British Intelligence Mining Data from Nine U.S. Internet Companies in Broad

Secret Program," *Washington Post*, June 7, 2013, https://www.washingtonpost.com/investigations/us-intelligence-mining-data-from-nine-us-internet-companies-in-broad-secret-program/2013/06/06/3a0c0da8-cebf-11e2-8845-d970ccb04497_story.html.

115. Gellman and Poitras, "U.S., British Intelligence Mining Data," 2013.

116. Edward Snowden, *Permanent Record* (New York: Metropolitan Books, 2019), 1.

117. The company viewed the government's request for records as unconstitutional, but they ultimately lost their legal battle and became part of the PRISM machinery. See Craig Timberg, "U.S. Threatened Massive Fine to Force Yahoo to Release Data," *Washington Post*, September 11, 2014, https://www.washingtonpost.com/business/technology/us-threatened-massive-fine-to-force-yahoo-to-release-data/2014/09/11/38a7f69e-39e8-11e4-9c9f-ebb47272e40e_story.html.

118. Glenn Greenwald, "NSA Collecting Phone Records of Millions of Verizon Customers Daily," *The Guardian*, June 6, 2013, https://www.theguardian.com/world/2013/jun/06/nsa-phone-records-verizon-court-order.

119. Julia Angwin et al., "AT&T Helped U.S. Spy on Internet on a Vast Scale," *New York Times*, August 15, 2015, https://www.nytimes.com/2015/08/16/us/politics/att-helped-nsa-spy-on-an-array-of-internet-traffic.html; Leslie Cauley and John Diamond, "Telecoms Let NSA Spy on Calls," *USA Today*, February 5, 2006, http://usatoday30.usatoday.com/news/washington/2006-02-05-nsa-telecoms_x.htm.

120. "The Many Lives of Herbert O. Yardley," *NSA Cryptologic Spectrum* 11, no. 4 (Fall 1981), https://www.nsa.gov/portals/75/documents/news-features/declassified-documents/cryptologic-spectrum/many_lives.pdf. The *Cryptologic Spectrum* was an internal journal published by the NSA established in 1969. A selection of declassified articles from issues published between 1969–1981 can be found indexed at https://www.nsa.gov/news-features/declassified-documents/cryptologic-spectrum/.

121. James Risen and Eric Lichtblau, "Bush Lets U.S. Spy on Callers without Courts," *New York Times*, December 16, 2005, https://www.nytimes.com/2005/12/16/politics/bush-lets-us-spy-on-callers-without-courts.html; David E. Sanger, "Bush Says He Ordered Domestic Spying," *New York Times*, December 18, 2005, https://www.nytimes.com/2005/12/18/politics/bush-says-he-ordered-domestic-spying.html; Eric Lichtblau and James Risen, "Spy Agency Mined Vast Data Trove, Officials Report," *New York Times*, December 24, 2005, https://www.nytimes.com/2005/12/24/politics/spy-agency-mined-vast-data-trove-officials-report.html.

122. "Whistle-Blower Outs NSA Spy Room," *Wired*, April 7, 2006, https://www.wired.com/2006/04/whistle-blower-outs-nsa-spy-room-2/.

123. See Hepting v. AT&T Corp., 439 F. Supp. 2d 974 (N.D. Cal. 2006).

124. See the Electronic Frontier Foundation's discussion of Hepting v. AT&T, along with related documents, https://www.eff.org/cases/hepting.

125. Cauley and Diamond, "Telecoms Let NSA Spy"; Leslie Cauley, "NSA Has Massive Database of Americans' Calls," *USA Today*, May 11, 2006.

126. John Stokes, "Sprint Fed Customer GPS Data to Cops over 8 Million Times," *Ars Technica*, December 1, 2009, https://arstechnica.com/tech-policy/2009/12/sprint-fed-customer-gps-data-to-leos-over-8-million-times/.

127. See original complaint, ACLU, et al. v. National Security Agency, https://www.aclu.org/files/pdfs/safefree/nsacomplaint.011706.pdf. The rest of the materials, including amicus briefs, district court memorandums and rulings, and all documents related to the unsuccessful Sixth Circuit Court of Appeals case can be found at https://www.aclu.org/other/legal-documents-challenge-illegal-nsa-spying.

128. See original complaint, Jewel v. National Security Agency, https://www.eff.org/files/filenode/jewel/jewel.complaint.pdf.

129. Cindy Cohn, "EFF's Flagship Jewel v. NSA Dragnet Spying Case Rejected by the Supreme Court," June 13, 2022, https://www.eff.org/deeplinks/2022/06/effs-flagship-jewel-v-nsa-dragnet-spying-case-rejected-supreme-court.

130. Offices of Inspectors General, *Unclassified Report on the President's Surveillance Program*, Rep. No. 2009–0013–AS (July 10, 2009), https://fas.org/irp/eprint/psp.pdf.

131. Cohn, "EFF's Flagship Jewel v. NSA Dragnet Spying Case Rejected."

132. See Charlie Savage, "File Says N.S.A. Found Way to Replace Email Program," *New York Times*, November 19, 2015, https://www.nytimes.com/2015/11/20/us/politics/records-show-email-analysis-continued-after-nsa-program-ended.html. The telephone records collection stopped in 2015 as part of the USA Freedom Act, but a revamped version continued for three more years. The mass Internet metadata program was shut down "for operational and resource reasons" in 2011.

133. Schneier, *Data and Goliath*, 78.

134. Chris Pyle, interview by Juan González, *Democracy Now*, June 13, 2013, transcript, https://www.democracynow.org/2013/6/13/chris_pyle_whistleblower_on_cia_domestic; "U.S. Intelligence Community Budget," Office of the Director of National Intelligence, https://www.dni.gov/index.php/what-we-do/ic-budget.

135. Tim Shorrock, "The Corporate Takeover of U.S. Intelligence," *Salon*, June 1, 2007, https://www.salon.com/2007/06/01/intel_contractors/#:~:text=On%20May%2014%2C%20at%20an,contracts%3A%20a%20whopping%2070%20percent.

136. See, for example, Mark Hertsgaard, "How the Pentagon Punished NSA Whistleblowers," *The Guardian*, May 22, 2016, https://www.theguardian.com/us-news/2016/may/22/how-pentagon-punished-nsa-whistleblowers.

137. Jill Lepore, "Edward Snowden and the Rise of Whistle-Blower Culture," *New Yorker*, September 16, 2019, https://www.newyorker.com/magazine/2019/09/23/edward-snowden-and-the-rise-of-whistle-blower-culture.

138. Schwartz, "Legal Access to the Global Cloud," 1691–1692.

139. For example, the EU originated a data protection directive (also known as Directive 95/46/EC) back in 1995 that established seven principles aimed at protecting the privacy of EU data, and required any data exported out of the European Union to be adequately protected by the receiving country. For a more granular look at these policies, see Jennifer Holt and Steven Malčić, "The Privacy Ecosystem: Regulating Digital Identity in the United States and European Union," *Journal of Information Policy* 5 (2015): 155–178, https://doi.org/10.5325/jinfopoli.5.2015.0155.

140. Kashmir Hill, "Law Student of the Day: Max Schrems," *Above the Law*, February 8, 2012, https://abovethelaw.com/2012/02/law-student-of-the-day-max-schrems/?rf=1.

141. Hill, "Law Student of the Day."

142. Maximillian Schrems v. Data Protection Commissioner and Digital Rights Ireland, Ltd., C-362/14, ECLI:EU:C:2015:650 (C.J.E.U. 2015).

143. Paul M. Schwartz, "The EU-U.S. Privacy Collision: A Turn to Institutions and Procedures," *Harvard Law Review* 126, no. 7 (May 2013): 1966–2009, https://www.jstor.org/stable/23415063. See also Rolf H. Weber, "Transborder Data Transfers: Concepts, Regulatory Approaches and New Legislative Initiatives," *International Data Privacy Law* 3, no. 2 (May 2013): 117–130, https://doi.org/10.1093/idpl/ipt001.

144. Danny O'Brien, "EU Court Again Rules That NSA Spying Makes U.S. Companies Inadequate for Privacy," Electronic Frontier Foundation, July 16, 2020, https://www.eff.org/deeplinks/2020/07/eu-court-again-rules-nsa-spying-makes-us-companies-inadequate-privacy.

145. Data Protection Commissioner v. Facebook Ireland Ltd., ECLI:EU:C:2020:559 (C.J.E.U. 2020).

146. Natasha Lomas, "Europe's Top Court Strikes Down Flagship EU-US Data Transfer Mechanism," *TechCrunch*, July 16, 2020, https://techcrunch.com/2020/07/16/europes-top-court-strikes-down-flagship-eu-us-data-transfer-mechanism/. See also O'Brien, "EU Court Again Rules."

147. Felix Richter, "Amazon Leads $100 Billion Cloud Market," Statista, February 11, 2020, https://www.statista.com/chart/18819/worldwide-market-share-of-leading-cloud-infrastructure-service-providers/. Also see Jay Chapel, "AWS vs Azure vs Google Cloud Market Share 2019: What the Latest Data Shows," *Medium* (blog), July 12, 2019, https://medium.com/@jaychapel/aws-vs-azure-vs-google-cloud-market-share-2019-what-the-latest-data-shows-dc21f137ff1c.

148. AWS lost their $10 billion contract with the Pentagon to Microsoft in October 2019, following intervention from the Trump White House. Amazon then filed a suit with the US Court of Federal Claims arguing the decision was politically motivated because of Trump's disdain for Amazon CEO Jeff Bezos who also owns the *Washington*

Post. The judge in the case ordered the Pentagon to halt work on the contract with Microsoft in March 2020. In December 2022, the Pentagon decided to split its cloud-computing contracts between the four top cloud providers—Amazon, Google, Microsoft, and Oracle. These deals run through 2028 and are known as the Joint Warfighting Cloud Capability (JWCC) initiative. See Maureen Farrell, "Pentagon Divides Big Cloud-Computing Deal among 4 Firms," *New York Times*, December 7, 2022, https://www.nytimes.com/2022/12/07/business/pentagon-cloud-contracts-jwcc.html; Ryan Browne, "Pentagon Watchdog Finds Defense Department Behaved Appropriately but Doesn't Rule on White House Influence over Controversial Cloud Contract," CNN, April 15, 2020, https://www.cnn.com/2020/04/15/politics/pentagon-inspector-general-cloud-report/index.html; Aaron Gregg, "Judge Says Amazon Is 'Likely to Succeed' on Key Argument in Pentagon Cloud Lawsuit," *Washington Post*, March 6, 2020, https://www.washingtonpost.com/business/2020/03/06/judge-says-amazon-likely-succeed-key-argument-pentagon-cloud-lawsuit/.

149. Nitasha Tiku, "Three Years of Misery inside Google, the Happiest Company in Tech," *Wired*, August 13, 2019, https://www.wired.com/story/inside-google-three-years-misery-happiest-company-tech/.

150. Lionel Sujay Vailshery, "Annual Revenue of Amazon Web Services (AWS) from 2013 to 2021," Statista, March 22, 2022, https://www.statista.com/statistics/233725/development-of-amazon-web-services-revenue/.

151. Aaron Perzanowski and Jason Schultz, *End of Ownership*, 58–59.

152. Perzanowski and Schultz, *End of Ownership*, 64.

153. Perzanowski and Schultz, *End of Ownership*, 123.

154. Kevin Litman-Navarro, "We Read 150 Privacy Policies. They Were an Incomprehensible Disaster," *New York Times*, June 12, 2019, https://www.nytimes.com/interactive/2019/06/12/opinion/facebook-google-privacy-policies.html.

155. Perzanowski and Schultz, *End of Ownership*, 59.

156. Litman-Navarro, "We Read 150 Privacy Policies."

157. Lina Khan, "Thrown Out of Court: How Corporations Became People You Can't Sue," *Washington Monthly*, June 6, 2014, https://washingtonmonthly.com/magazine/junejulyaug-2014/thrown-out-of-court/.

158. "CloudFront Key Features," Amazon Web Services, https://aws.amazon.com/cloudfront/features/?p=ugi&l=na.

159. "AWS Customer Agreement," Amazon Web Services, https://aws.amazon.com/agreement/.

160. Tim Cook, "A Message to Our Customers," Apple, February 16, 2016, https://www.apple.com/customer-letter/.

Notes to Chapter 3

161. Leander Kahney, "The FBI Wanted a Back Door to the iPhone. Tim Cook Said No," *Wired*, April 16, 2019, https://www.wired.com/story/the-time-tim-cook-stood-his-ground-against-fbi/.

162. Kahney, "The FBI Wanted a Back Door."

163. Kahney, "The FBI Wanted a Back Door."

164. Susan Landau, *Listening In: Cybersecurity in an Insecure Age* (New Haven, CT: Yale University Press, 2017), xi.

165. Landau, *Listening In*, xiii.

166. Eliza Sweren-Becker, "This Map Shows How the Apple-FBI Fight Was about Much More Than One Phone," *ACLU* (blog), March 30, 2016, https://www.aclu.org/blog/privacy-technology/internet-privacy/map-shows-how-apple-fbi-fight-was-about-much-more-one-phone. Also see map of "All Writs Act Orders for Assistance from Tech Companies," ACLU, nd, https://www.aclu.org/issues/privacy-technology/internet-privacy/all-writs-act-orders-assistance-tech-companies.

167. Danny Lewis, "What the All Writs Act of 1789 Has to Do with the iPhone," *Smithsonian Magazine*, February 24, 2106, https://www.smithsonianmag.com/smart-news/what-all-writs-act-1789-has-do-iphone-180958188/.

168. Susan Crawford, "The Law Is Clear: The FBI Cannot Make Apple Rewrite Its OS," *Wired*, March 16, 2016, https://www.wired.com/2016/03/the-law-is-clear-the-fbi-cannot-make-apple-rewrite-its-os-2/.

169. FCC, *First Report and Order and Further Notice of Proposed Rulemaking. Communications Assistance for Law Enforcement Act (CALEA) and Broadband Access and Services*, 20 FCC Rcd 14989 (September 23, 2005), https://www.fcc.gov/document/communications-assistance-law-enforcement-act-calea-and-broadband.

170. Steven Levy, "Battle of the Clipper Chip," *New York Times*, June 12, 1994, https://www.nytimes.com/1994/06/12/magazine/battle-of-the-clipper-chip.html.

171. Quoted in Matthias Schulze, "Clipper Meets Apple vs. FBI—A Comparison of the Cryptography Discourses from 1993 and 2016," *Media and Communication* 5, no. 1 (March 22, 2017): 55, https://doi.org/10.17645/mac.v5i1.805.

172. Rory Carroll, "Microsoft and Google to Sue over US Surveillance Requests," *The Guardian*, August 30, 2013, https://www.theguardian.com/law/2013/aug/31/microsoft-google-sue-us-fisa.

173. "Tech Companies Give First Look at Secret Gov't Data Requests," *CBS News*, February 3, 2014, https://www.cbsnews.com/news/google-microsoft-yahoo-facebook-linkedin-secret-government-nsa-data-requests/; "Twitter v. Holder," Reporters Committee for Freedom of the Press, nd, https://www.rcfp.org/briefs-comments/twitter-v-holder/.

174. Microsoft Corporation v. United States Department of Justice, Complaint for Declaratory Judgment (filed April 14, 2016), 1–2.

175. "Apple CEO Time Cook Lobs Daggers at Google, Facebook," *Ad Age*, October 24, 2018, https://adage.com/article/digital/apple-ceo-tim-cook-lobs-daggers-google-facebook/315377.

176. Jack Nicas, Raymond Zhong, and Daisuke Wakabayashi, "Censorship, Surveillance and Profits: A Hard Bargain for Apple in China," *New York Times*, May 17, 2021, https://www.nytimes.com/2021/05/17/technology/apple-china-censorship-data.html.

177. Alan Z. Rozenshtein, "Surveillance Intermediaries," *Stanford Law Review* 70 (January 2018): 187.

178. Pistor, "Rule by Data," 101.

179. Kristina Irion, "Government Cloud Computing and National Data Sovereignty: Government Cloud Computing and National Data Sovereignty," *Policy & Internet* 4, no. 3/4 (2012): 50, https://doi.org/10.1002/poi3.10.

180. Ryan Gallagher and Mark Bergen, "Google Scrapped Cloud Initiative in China, Other Markets," *Bloomberg*, July 8, 2020, https://www.bloomberg.com/news/articles/2020-07-08/google-scrapped-cloud-initiative-in-china-sensitive-markets.

181. For example, see "A Virtual Counter-Revolution," *The Economist*, September 2, 2010, https://www.economist.com/briefing/2010/09/02/a-virtual-counter-revolution; Keith Wright, "The Splinternet Is Already Here," *TechCrunch*, March 13, 2019, https://techcrunch.com/2019/03/13/the-splinternet-is-already-here/.

182. See John Selby, "Data Localization Laws: Trade Barriers or Legitimate Responses to Cybersecurity Risks, or Both?," *International Journal of Law and Information Technology* 25, no. 3 (September 2017): 213, http://resolver.scholarsportal.info/resolve/09670769/v25i0003/213_dlltbortcrob.xml.

183. Marc Santora, "Turkey Passes Law Extending Sweeping Powers over Social Media," *New York Times*, July 29, 2020, https://www.nytimes.com/2020/07/29/world/europe/turkey-social-media-control.html.

184. Schneier, *Data and Goliath*, 188.

185. Jennifer Daskal, "Law Enforcement Access to Data across Borders," *Journal of National Security Law & Policy* 8 (2016): 474–475, https://jnslp.com/2016/09/06/law-enforcement-access-data-across-borders-evolving-security-rights-issues/.

186. Schwartz, "Legal Access to the Global Cloud," 1684–1685.

187. Brad Smith, "The Collapse of the US-EU Safe Harbor: Solving the New Privacy Rubik's Cube," Microsoft (blog), October 20, 2015, https://blogs.microsoft.com/on

-the-issues/2015/10/20/the-collapse-of-the-us-eu-safe-harbor-solving-the-new-privacy-rubiks-cube/.

188. "National Cloud Deployments," Microsoft, accessed May 5, 2020, https://learn.microsoft.com/en-us/graph/deployments.

189. "AWS Global Infrastructure," Amazon Web Services, https://aws.amazon.com/about-aws/global-infrastructure/.

190. For a detailed discussion of the many issues related to data localization, particularly in relation to human rights, see Adrian Shahbaz, Allie Funk, and Andrea Hackl, "User Privacy or Cyber Sovereignty?" Freedom House, 2020, https://freedomhouse.org/report/special-report/2020/user-privacy-or-cyber-sovereignty#footnote16_f9rdeq3.

191. European Parliament. Policy Department C: Citizens' Rights and Constitutional Affairs. Directorate-General for Internal Policies, *Fighting Cyber Crime and Protecting Privacy in the Cloud*, Rep. No. PE 462.509 (October 2012), 30, https://www.europarl.europa.eu/RegData/etudes/etudes/join/2012/462509/IPOL-LIBE_ET%282012%29462509_EN.pdf.

192. Lionel Sujay Vailshery, "Cloud Infrastructure Services Vendor Market Share Worldwide," Statista, August 25, 2022, https://www.statista.com/statistics/967365/worldwide-cloud-infrastructure-services-market-share-vendor/.

193. General Data Protection Regulation, 2016/679 of the European Parliament and of the Council (2016), chapter 3, Articles 12–23, "Rights of the Data Subject," https://gdpr.eu/tag/chapter-3/.

194. Jeremy Kahn, "Amazon's Pitch to Europe: Your Data Is Safe from American Spies," *Bloomberg News*, January 7, 2016, https://www.bloomberg.com/news/articles/2016-01-07/amazon-s-pitch-to-europe-your-data-is-safe-from-american-spies.

195. "What's New: Amazon Web Services and Ningxia Western Cloud Data Technology Co. Ltd (NWCD) Announce a Second Amazon Web Services Region in China, Now Available to Customers," *AWS News*, December 12, 2017, https://www.amazonaws.cn/en/new/2017/whats-new-announcing-partnership-between-aws-and-nwcd-availability-of-ningxia-region/.

196. See Bilgic, "Something Old, Something New," 346; Shannon Liao, "Apple Officially Moves Its Chinese iCloud," *The Verge*, February 28, 2018, https://www.theverge.com/2018/2/28/17055088/apple-chinese-icloud-accounts-government-privacy-speed.

197. "Putin Signs Internet Isolation Bill into Law," *Moscow Times*, May 1, 2019, https://www.themoscowtimes.com/2019/05/01/putin-signs-internet-isolation-bill-into-law-a65461.

198. See, for example, Anthony Cuthbertson, "Russia Protests: Thousands March against Plans to Cut Off Internet from Rest of World," *The Independent*, March 11,

2019, https://www.independent.co.uk/life-style/gadgets-and-tech/news/russia-internet-protest-putin-online-censorship-privacy-a8817361.html.

199. Andrei Soldatov and Irina Borogan, "The New Iron Curtain Part 4: Russia's Sovereign Internet Takes Route," Center for European Policy Analysis (CEPA), April 5, 2022, https://cepa.org/article/the-new-iron-curtain-part-4-russias-sovereign-internet-takes-root/.

200. Anton Troianovski and Ivan Nechepurenko, "Could Navalny's 'Smart Voting' Strategy Shake up Russia's Election?" *New York Times*, September 15, 2021, https://www.nytimes.com/2021/09/15/world/europe/navalny-smart-voting-russia-election.html.

201. William Partlett, "Russia Is Building Its Own Kind of Sovereign Internet—with Help from Apple and Google," *The Conversation*, October 4, 2021, https://theconversation.com/russia-is-building-its-own-kind-of-sovereign-internet-with-help-from-apple-and-google-169115.

202. Edward Snowden, "Introduction," in *Little Brother & Homeland*, by Cory Doctorow (New York: Macmillan Publishing Group, 2020), 7.

Epilogue

1. See Nilay Patel, "The Mystery of Biden's Deadlocked FCC," *The Verge*, November 3, 2022, https://www.theverge.com/23437518/biden-fcc-gigi-sohn-fox-news-comcast-senate-democrats-midterms-election. Gigi Sohn withdrew her nomination in March 2023, stating, "It is a sad day for our country and our democracy when dominant industries, with assistance from unlimited dark money, get to choose their regulators."

2. Newton N. Minow, "Commemorative Messages," in *A Legislative History of the Communications Act of 1934*, ed. Max Paglin (New York: Oxford University Press, 1989), xv.

3. Des Freedman, "Media Policy Silences: The Hidden Face of Communications Decision Making," *International Journal of Press/Politics* 15, no. 3 (2010): 345, https://doi.org/10.1177/1940161210368292.

4. Tom Wheeler, interview by Jennifer Holt, February 28, 2023.

5. Adi Robertson, "Lots of Politicians Hate Section 230—But They Can't Agree on Why," *The Verge*, June 24, 2020, https://www.theverge.com/21294198/section-230-tech-congress-justice-department-white-house-trump-biden.

6. Veszna Wessenauer and Ellery Roberts Biddle, "The Shift to First-Party Tracking Is a Power Play by Apple and Google. What Will It Mean for Users' Rights?," *Ranking Digital Rights*, November 17, 2021, https://rankingdigitalrights.org/staging/5316/2021/11/17/first-party-tracking-power-play-apple-google-human-rights/.

7. Martin Moore and Damian Tambini, "Conclusion: Without a Holistic Vision, Democratic Media Reforms May Fail," in *Regulating Big Tech: Policy Responses to Digital Dominance*, ed. Martin Moore and Damian Tambini (New York: Oxford University Press, 2021), 339.

8. Christian Fuchs and Klaus Unterberger, eds., *The Public Service Media and Public Service Internet Manifesto* (London: University of Westminster Press, 2021), p. 10, https://www.uwestminsterpress.co.uk/site/books/e/10.16997/book60/.

9. See Holt and Parks, "Labor of Digital Privacy Advocacy."

10. See, for example, Colin John Bennett, "Chapter 1: Framing the Problem," in *The Privacy Advocates: Resisting the Spread of Surveillance* (Cambridge, MA: MIT Press, 2008), 1–24.

11. Alexandra Reeve Givens, president and CEO of the Center for Democracy and Technology, interview by Jennifer Holt, October 11, 2022.

12. Mary L. Gray and Siddharth Suri, *Ghost Work: How to Stop Silicon Valley from Building a New Global Underclass* (Boston: Houghton Mifflin Harcourt, 2019), 3.

13. See, for example, Nitasha Tiku et al., "From Amazon to Apple, Tech Giants Turn to Old-School Union-Busting," *Washington Post*, April 24, 2022, https://www.washingtonpost.com/technology/2022/04/24/amazon-apple-google-union-busting/.

14. Victor Pickard, *Democracy without Journalism?: Confronting the Misinformation Society* (New York: Oxford University Press, 2019), 167.

15. See Steve Weinberg, *Taking on the Trust: How Ida Tarbell Brought Down John D. Rockefeller and Standard Oil* (New York: W. W. Norton & Company, 2008).

16. Jane Mayer's *Dark Money: The Hidden History of the Billionaires behind the Rise of the Radical Right* (New York: Random House, 2016) is essential reading on this topic. Currently, the best place for tracking the money in US politics and its effects on elections and public policy is opensecrets.org created by the Center for Responsive Politics. It is notable that Frank Church (D-Idaho) was one of the founders of the center in 1983. FollowTheMoney.org is an equally outstanding resource for tracking private contributions to politics across all sectors of the economy.

17. See Dina Srinivasan, "Why Google Dominates Advertising Markets: Competition Policy Should Lean on the Principles of Financial Market Regulation," *Stanford Law School* 24, no. 1 (December 7, 2020): 65, https://law.stanford.edu/publications/why-google-dominates-advertising-markets/.

18. Alan F. Westin, *Privacy and Freedom* (New York: Atheneum, 1967), 399.

19. Quoted in Nick Romeo, "What Can America Learn from Europe about Regulating Big Tech?" *New Yorker*, August 18, 2020, https://www.newyorker.com/tech/annals-of-technology/what-can-america-learn-from-europe-about-regulating-big-tech.

20. Nicholas Johnson, "Carterfone: My Story," *Santa Clara High Technology Law Journal* 25, no. 3 (2008): 692, https://digitalcommons.law.scu.edu/chtlj/vol25/iss3/5.

21. See Barbara van Schewick, *Internet Architecture and Innovation* (Cambridge, MA: MIT Press, 2010); Janet Abbate, *Inventing the Internet* (Cambridge, MA: MIT Press, 1999).

22. Such bills in the US include the Algorithmic Justice and Online Transparency Act (2021), the Platform Accountability and Transparency Act (2021), and the Social Media Disclosure and Transparency of Advertisements Act of 2021 (also known as the Social Media DATA Act).

23. Shoshana Zuboff, "Salon Series—Shoshana Zuboff," Beyond the Web Salon Series, Ostrom Workshop at Indiana University, April 11, 2022.

24. Fred Turner, "Machine Politics," *Harper's Magazine*, December 30, 2020, https://harpers.org/archive/2019/01/machine-politics-facebook-political-polarization/.

25. Harold Feld, "Broadband Access as Public Utility" (speech, Personal Democracy Forum, June 4, 2015), transcript, https://wetmachine.com/tales-of-the-sausage-factory/broadband-access-as-public-utility-my-speech-at-personal-democracy-forum/.

26. Dan Schiller, "Reconstructing Public Utility Networks: A Program for Action," *International Journal of Communication* 14 (2020): 4494, https://ijoc.org/index.php/ijoc/article/view/16242.

27. Jathan Sadowski, "The Internet of Landlords Makes Renters of Us All," *The Reboot*, March 8, 2021.

28. Quoted in Edgar Llivisupa, "The Accidental Internet Scholar," *New York Review of Books*, June 11, 2022, https://www.nybooks.com/daily/2022/06/11/the-accidental-internet-scholar-ethan-zuckerman/.

29. Luzhou Li, "How to Think about Media Policy Silence," *Media, Culture & Society* 43, no. 2 (2021): 360, https://doi.org/10.1177/0163443720948004.

30. Freedman, "Media Policy Silences," 355.

31. Hannah Appel, Nikhil Anand, and Akhil Gupta, "Introduction: Temporality, Politics, and the Promise of Infrastructure," in *The Promise of Infrastructure*, ed. Nikhil Anand, Akhil Gupta, and Hannah Appel (Durham, NC: Duke University Press, 2018), 19.

32. H. D. Lloyd, "The Story of a Great Monopoly," *Atlantic*, March 1881, https://www.theatlantic.com/magazine/archive/1881/03/the-story-of-a-great-monopoly/306019/.

33. Parkhill, *Challenge of the Computer Utility*, 182.

34. US House of Representatives Committee on Government Operations, *Privacy and the National Data Bank Concept*, 9th Cong., 2d sess. (July 1968), x, 5.

35. Jill Lepore, *If Then: How the Simulmatics Corporation Invented the Future* (New York: Liveright Publishing, 2020), 327.

36. Tim Tyson, "Can Honest History Allow for Hope?" *The Atlantic*, December 18, 2015, https://www.theatlantic.com/politics/archive/2015/12/can-hope-and-history-coexist/420651/.

37. Edward Snowden, *Permanent Record* (New York: Metropolitan Books, 2019), 326.

Bibliography

Abbate, Janet. *Inventing the Internet*. Cambridge, MA: MIT Press, 1999.

Agur, Colin. "Negotiated Order: The Fourth Amendment, Telephone Surveillance, and Social Interactions, 1878–1968." *Information & Culture* 48, no. 4 (2013): 419–447.

Ali, Christopher. *Farm Fresh Broadband: The Politics of Rural Connectivity*. Cambridge, MA: MIT Press, 2021.

Allen, Anita, and Erin Mack. "How Privacy Got Its Gender." *Northern Illinois University Law Review* 10 (1991): 441–478. https://scholarship.law.upenn.edu/faculty_scholarship/1309.

Appel, Hannah, Nikhil Anand, and Akhil Gupta. "Introduction: Temporality, Politics, and the Promise of Infrastructure." In *The Promise of Infrastructure*, edited by Nikhil Anand, Akhil Gupta, and Hannah Appel, 1–38. Durham, NC: Duke University Press, 2018.

Aufderheide, Patricia. *Communications Policy and the Public Interest: The Telecommunications Act of 1996*. New York: Guilford Press, 1999.

Balkin, Jack M. "Information Fiduciaries and the First Amendment." *UC Davis Law Review* 49, no. 4 (2016): 1183–1234.

Balkin, Jack M. "Old-School/New-School Speech Regulation." *Harvard Law Review* 127, no. 8 (June 2014): 2296–2342. https://harvardlawreview.org/print/vol-127/old-schoolnew-school-speech-regulation/.

Banks, David A. "Lines of Power: Availability to Networks as a Social Phenomenon." *First Monday* 20, no. 11 (2015). https://doi.org/10.5210/fm.v20i11.6283.

Barbrook, Richard, and Andy Cameron. "The Californian Ideology." *Science as Culture* 6, no. 1 (1996): 44–72. https://doi.org/10.1080/09505439609526455.

Bennett, Colin John. *The Privacy Advocates: Resisting the Spread of Surveillance*. Cambridge, MA: MIT Press, 2008.

Bilgic, Secil. "Notes: Something Old, Something New, and Something Moot: The Privacy Crisis under the Cloud Act." *Harvard Journal of Law & Technology* 32, no. 1 (Fall 2018): 321–355. https://jolt.law.harvard.edu/assets/articlePDFs/v32/32HarvJLTech321.pdf.

Blanchette, Jean-François. "Introduction: Computing's Infrastructural Moment." In *Regulating the Cloud: Policy for Computing Infrastructure*, edited by Christopher S. Yoo and Jean-François Blanchette, 1–20. Cambridge, MA: MIT Press, 2015.

Blevins, John. "The FCC and the 'Pre-Internet.'" *Indiana Law Journal* 91, no. 4 (2016): 1309–1362. https://www.repository.law.indiana.edu/ilj/vol91/iss4/6.

Bloch-Wehba, Hannah. "Global Platform Governance: Private Power in the Shadow of the State." *SMU Law Review* 72, no. 1 (2019): 27–80. https://scholar.smu.edu/smulr/vol72/iss1/9.

Bork, Robert H. *The Antitrust Paradox*. New York: Basic Books, 1978.

Bork, Robert H. "Legislative Intent and the Policy of the Sherman Act." *Journal of Law & Economics* 9 (October 1966): 7–48. https://www.jstor.org/stable/724991.

Braman, Sandra. *Change of State: Information, Policy, and Power*. Cambridge, MA: MIT Press, 2007.

Brooks, John. *Telephone: The First Hundred Years*. New York: Harper & Row, 1975.

Brown, Wendy. *In the Ruins of Neoliberalism: The Rise of Antidemocratic Politics in the West*. New York: Columbia University Press, 2019.

Cannon, Robert. "The Legacy of the Federal Communications Commission's Computer Inquiries." *Federal Communications Law Journal* 55, no. 2 (2003): 167–205. https://www.repository.law.indiana.edu/fclj/vol55/iss2/2.

Citron, Danielle Keats. *Hate Crimes in Cyberspace*. Cambridge, MA: Harvard University Press, 2014.

Citron, Danielle Keats. *The Fight for Privacy: Protecting Dignity, Identity, and Love in the Digital Age*. New York: W.W. Norton & Company, 2022.

Citron, Danielle Keats, and Mary Anne Franks. "The Internet as a Speech Machine and Other Myths Confounding Section 230 Reform." *University of Chicago Legal Forum* 2020 (December 1): 45–75. https://chicagounbound.uchicago.edu/uclf/vol2020/iss1/3.

Clark, David D. *Designing an Internet*. Cambridge, MA: MIT Press, 2018.

Clopton, Zachary D. "Territoriality, Technology, and National Security." *University of Chicago Law Review* 83, no. 1 (2015): 45–63. https://chicagounbound.uchicago.edu/uclrev/vol83/iss1/3.

Cohen, Julie. "Law for the Platform Economy." *UC Davis Law Review* 51 (2017): 133–204. https://scholarship.law.georgetown.edu/facpub/2015.

Bibliography

Cole, Barry G., ed. *After the Breakup: Assessing the New Post-AT&T Divestiture Era.* New York: Columbia University Press, 1991.

Couldry, Nick, and Ulises A. Mejias. *The Costs of Connection: How Data Is Colonizing Human Life and Appropriating It for Capitalism.* Stanford, CA: Stanford University Press, 2019.

Crain, Matthew. *Profit over Privacy: How Surveillance Advertising Conquered the Internet.* Minneapolis: University of Minnesota Press, 2021.

Crandall, Robert W. *After the Breakup: U.S. Telecommunications in a More Competitive Era.* Washington, DC.: Brookings Institution Press, 1991.

Crawford, Susan. *Captive Audience: The Telecom Industry and Monopoly Power in the New Gilded Age.* New Haven, CT: Yale University Press, 2013.

Crawford, Susan. *Fiber: The Coming Tech Revolution—and Why America Might Miss It.* New Haven, CT: Yale University Press, 2018.

Daskal, Jennifer. "Law Enforcement Access to Data across Borders." *Journal of National Security Law & Policy* 8 (2016): 473–501. https://jnslp.com/2016/09/06/law-enforcement-access-data-across-borders-evolving-security-rights-issues/.

Daskal, Jennifer. "The Un-Territoriality of Data." *Yale Law Journal* 125, no. 2 (November 2015): 326–398. https://www.yalelawjournal.org/article/the-un-territoriality-of-data.

Debs, Eugene V. "Better to Buy Books Than Beer." *Speech*, Buffalo, NY, January 15, 1896.

DeNardis, Laura. "Hidden Levers of Internet Control: An Infrastructure-Based Theory of Internet Governance." *Information, Communication & Society* 15, no. 5 (2012): 720–738. https://doi.org/10.1080/1369118X.2012.659199.

DeNardis, Laura, and A. M. Hackl. "Internet Governance by Social Media Platforms." *Telecommunications Policy* 39, no. 9 (October 2015): 761–770. https://doi.org/10.1016/j.telpol.2015.04.003.

De Sola Pool, Ithiel. *Technologies of Freedom.* Cambridge, MA: Harvard University Press, 1983.

Determann, Lothar. "No One Owns Data." *UC Law Journal* 70, no. 1 (2019): 1–44. https://repository.uclawsf.edu/hastings_law_journal/vol70/iss1/1.

Dijck, Jose van. *The Culture of Connectivity: A Critical History of Social Media.* Oxford: Oxford University Press, 2013.

Domnarski, William. *Richard Posner.* New York: Oxford University Press, 2016.

Douglas, Susan J. *Inventing American Broadcasting, 1899–1922.* Baltimore, MD: Johns Hopkins University Press, 1987.

Edwards, Paul N. "Some Say the Internet Should Never Have Happened." In *Media, Technology, and Society: Theories of Media Evolution*, edited by W. Russell Neuman, 141–160. Ann Arbor: University of Michigan Press, 2010.

Edwards, Paul N. *The Closed World*. Cambridge, MA: MIT Press, 1996.

Edwardson, Mickie. "James Lawrence Fly, the FBI, and Wiretapping." *The Historian* 61, no. 2 (Winter 1999): 361–381. https://www.jstor.org/stable/24449708.

Einstein, Mara. *Media Diversity: Economics, Ownership, and the FCC*. Mahwah, NJ: Lawrence Erlbaum Associates, 2004.

Felber, Ron. *The Privacy War: One Congressman, J. Edgar Hoover, and the Fight for the Fourth Amendment*. Montvale, NY: Croce Publishing Group, 2003.

Flew, Terry. *Regulating Platforms*. Medford, MA: Polity, 2021.

Flew, Terry, and Fiona R. Martin. "Introduction." In *Digital Platform Regulation: Global Perspectives on Internet Governance*, edited by Terry Flew and Fiona R. Martin, 1–22. London: Palgrave Macmillan, 2022.

Freedman, Des. "Media Policy Silences: The Hidden Face of Communications Decision Making." *International Journal of Press/Politics* 15, no. 3 (July 2010): 344–361. https://doi.org/10.1177/1940161210368292.

Freedman, Des. *The Politics of Media Policy*. Malden, MA: Polity Press, 2008.

Fuchs, Christian, and Klaus Unterberger, eds. *The Public Service Media and Public Service Internet Manifesto*. London: University of Westminster Press, 2021. https://www.uwestminsterpress.co.uk/site/books/e/10.16997/book60/.

Garland, David. "What Is a 'History of the Present'? On Foucault's Genealogies and Their Critical Preconditions." *Punishment & Society* 16, no. 4 (October 2014): 365–384. https://doi.org/10.1177/1462474514541711.

Gellman, Barton. *Dark Mirror: Edward Snowden and the American Surveillance State*. New York: Penguin Press, 2020.

Gertner, Jon. *The Idea Factory: Bell Labs and the Great Age of American Innovation*. New York: Penguin Press, 2012.

Gillespie, Tarleton. *Custodians of the Internet: Platforms, Content Moderation, and the Hidden Decisions That Shape Social Media*. New Haven, CT: Yale University Press, 2018.

Gitelman, Lisa. *Always Already New: Media, History and the Data of Culture*. Cambridge, MA: MIT Press, 2006.

Gitelman, Lisa. *Scripts, Grooves, and Writing Machines: Representing Technology in the Edison Era*. Stanford, CA: Stanford University Press, 1999.

Bibliography

Glick, Mark, Catherine Ruetschlin, and Darren Bush. "Big Tech's Buying Spree and the Failed Ideology of Competition Law." *Hastings Law Journal* 72, no. 2 (2021): 465–516. https://repository.uclawsf.edu/hastings_law_journal/vol72/iss2/1.

Goodwin, Doris Kearns. *The Bully Pulpit: Theodore Roosevelt, William Howard Taft, and the Golden Age of Journalism.* New York: Simon and Schuster, 2013.

Gore, Al. "Innovation Delayed Is Innovation Denied." *Computer* 27, no. 12 (December 1994): 45–47. https://doi.org/10.1109/2.335728.

Gorwa, Robert. "What Is Platform Governance?" *Information, Communication & Society* 22, no. 6 (2019): 854–871. https://doi.org/10.1080/1369118X.2019.1573914.

Graham, Stephen, and Simon Marvin. *Splintering Urbanism: Networked Infrastructures, Technological Mobilities and the Urban Condition.* London: Routledge, 2001.

Gray, Horace M. "The Passing of the Public Utility Concept." *Journal of Land & Public Utility Economics* 16, no. 1 (1940): 8–20. https://doi.org/10.2307/3158751.

Gray, Mary L., and Siddharth Suri. *Ghost Work: How to Stop Silicon Valley from Building a New Global Underclass.* Boston: Houghton Mifflin Harcourt, 2019.

Greenstein, Shane. *How the Internet Became Commercial: Innovation, Privatization, and the Birth of a New Network.* Princeton, NJ: Princeton University Press, 2015.

Greenwald, Glenn. *No Place to Hide: Edward Snowden, the NSA, and the U.S. Surveillance State.* New York: Metropolitan Books, 2014.

Hazlett, Thomas Winslow. *The Political Spectrum: The Tumultuous Liberation of Wireless Technology, from Herbert Hoover to the Smartphone.* New Haven, CT: Yale University Press, 2017.

Henck, Fred W., and Bernard Strassburg. *A Slippery Slope: The Long Road to the Breakup of AT&T.* Westport, CT: Greenwood Press, 1988.

Hochfelder, David. "Constructing an Industrial Divide: Western Union, AT&T, and the Federal Government, 1876–1971." *Business History Review* 76, no. 4 (Winter 2002): 705–732. https://doi.org/10.2307/4127707.

Hochman, Brian. *The Listeners: A History of Wiretapping in the United States.* Cambridge, MA: Harvard University Press, 2022.

Hofstadter, Richard. "What Happened to the Antitrust Movement." In *Richard Hofstadter: Anti-Intellectualism in American Life, the Paranoid Style in American Politics, Uncollected Essays 1956–1965.* New York: Library of America, 2020.

Hogan, Mél, and Tamara Shepherd. "Information Ownership and Materiality in an Age of Big Data Surveillance." *Journal of Information Policy* 5 (2015): 6–31. https://doi.org/10.5325/jinfopoli.5.2015.0006.

Holt, Jennifer. "Net Neutrality and the Public Interest: An Interview with Gene Kimmelman, President and CEO of Public Knowledge." *International Journal of Communication* 10 (2016): 5795–5810. https://ijoc.org/index.php/ijoc/article/view/5394.

Holt, Jennifer. "Regulating Connected Viewing." In *Connected Viewing: Selling, Streaming, & Sharing Media in the Digital Age*, edited by Jennifer Holt and Kevin Sanson, 19–39. New York: Routledge, 2014.

Holt, Jennifer, and Steven Malčić. "The Privacy Ecosystem: Regulating Digital Identity in the United States and European Union." *Journal of Information Policy* 5 (2015): 155–178. https://doi.org/10.5325/jinfopoli.5.2015.0155.

Holt, Jennifer, and Lisa Parks. "The Labor of Digital Privacy Advocacy in an Era of Big Tech." *Media Industries* 8, no. 1 (2021): 1–25. https://doi.org/10.3998/mij.93.

Holt, Jennifer, and Patrick Vonderau. "Where the Internet Lives: Data Centers as Digital Media Infrastructure." In *Signal Traffic: Critical Studies of Media Infrastructures*, edited by Lisa Parks and Nicole Starosielski, 71–93. Urbana: University of Illinois Press, 2015.

Horwitz, Robert Britt. *The Irony of Regulatory Reform: The Deregulation of American Telecommunications*. New York: Oxford University Press, 1989.

Howard, Philip N. *Lie Machines: How to Save Democracy from Troll Armies, Deceitful Robots, Junk News Operations, and Political Operatives*. New Haven, CT: Yale University Press, 2020.

Hu, Tung-Hui. *A Prehistory of the Cloud*. Cambridge, MA: MIT Press, 2016.

Igo, Sarah E. *The Known Citizen: A History of Privacy in Modern America*. Cambridge, MA: Harvard University Press, 2018.

Irion, Kristina. "Government Cloud Computing and National Data Sovereignty: Government Cloud Computing and National Data Sovereignty." *Policy & Internet* 4, no. 3/4 (2012): 40–71. https://doi.org/10.1002/poi3.10.

Jackaway, Gwenyth. "America's Press-Radio War of the 1930s: A Case Study in Battles between Old and New Media." *Historical Journal of Film, Radio and Television* 14, no. 3 (1994): 299–314. https://doi.org/10.1080/01439689400260211.

Janson, Michael A., and Christopher S. Yoo. "The Wires Go to War: The U.S. Experiment with Government Ownership of the Telephone System During World War I." *Texas Law Review* 91 (2013): 983–1050. https://doi.org/10.2139/ssrn.2033124.

John, Richard R. *Network Nation: Inventing American Telecommunications*. Cambridge, MA: Harvard University Press, 2010.

John, Richard R. "Recasting the Information Infrastructure for the Industrial Age." In *A Nation Transformed by Information: How Information Has Shaped the United States*

Bibliography

from Colonial Times to the Present, edited by Alfred D. Chandler and James W. Cortada, 55–106. New York: Oxford University Press, 2000.

Johnson, Alix, and Mél Hogan. "Introducing Location and Dislocation: Global Geographies of Digital Data." *Imaginations Journal* 8, no. 2 (September 2017): 4–7. http://dx.doi.org/10.17742/IMAGE.LD.8.2.1.

Johnson, David R., and David Post. "Law and Borders: The Rise of Law in Cyberspace." *Stanford Law Review* 48 (1996): 1367–1402.

Johnson, Nicholas. "Carterfone: My Story." *Santa Clara High Technology Law Journal* 25, no. 3 (2008): 677–700. https://digitalcommons.law.scu.edu/chtlj/vol25/iss3/5.

Jones, Meg Leta. *Ctrl + Z: The Right to Be Forgotten.* New York: New York University Press, 2016.

Khan, Lina M. "Amazon's Antitrust Paradox." *Yale Law Journal* 126, no. 3 (January 2017): 710–805. https://www.yalelawjournal.org/note/amazons-antitrust-paradox.

Khan, Lina M. "The Separation of Platforms and Commerce." *Columbia Law Review* 119 (May 15, 2019): 973–1098. https://scholarship.law.columbia.edu/faculty_scholarship/2789/.

Kimball, Danny. *Net Neutrality and the Battle for the Open Internet.* Ann Arbor: University of Michigan Press, 2022.

Klatt, Tyler. "The Streaming Industry and the Great Disruption: How Winning a Golden Globe Helps Amazon Sell More Shoes." *Media, Culture & Society* 44, no. 8 (2022): 1541–1558.

Klobuchar, Amy. *Antitrust: Taking on Monopoly Power from the Gilded Age to the Digital Age.* New York: Alfred A. Knopf, 2021.

Klonick, Kate. "The New Governors: The People, Rules, and Processes Governing Online Speech." *Harvard Law Review* 131, no. 6 (April 2018): 1598–1670. https://harvardlawreview.org/print/vol-131/the-new-governors-the-people-rules-and-processes-governing-online-speech/.

Kobayashi, Kōji. *Computers and Communications: A Vision of C&C.* Cambridge, MA: MIT Press, 1986.

Kosseff, Jeff. *The Twenty-Six Words That Created the Internet.* Ithaca, NY: Cornell University Press, 2019.

Landau, Susan. *Listening In: Cybersecurity in an Insecure Age.* New Haven, CT: Yale University Press, 2017.

Larkin, Brian. *Signal and Noise: Media, Infrastructure, and Urban Culture in Nigeria.* Durham, NC: Duke University Press, 2008.

LeBlanc, Travis, and Lindsay DeFrancesco. "The Federal Communications Commission as Privacy Regulator." In *The Cambridge Handbook of Surveillance Law*, edited by David Gray and Stephen E. Henderson, 727–756. Cambridge, UK: Cambridge University Press, 2017.

Lefebvre, Henri. *The Production of Space*. Cambridge, MA: Blackwell, 1991.

Lentz, Becky, and Allison Perlman, eds. "Net Neutrality." Special issue, *International Journal of Communication* 10 (2016).

Lepore, Jill. *If Then: How the Simulmatics Corporation Invented the Future*. New York: Liveright Publishing, 2020.

Lessig, Lawrence. *Code and Other Laws of Cyberspace, Version 2.0*. New York: Basic Books, 2006.

Levin, Harvey J. "Television's Second Chance: A Retrospective Look at the Sloan Cable Commission." *Bell Journal of Economics and Management Science* 4, no. 1 (Spring 1973): 343–365. https://ideas.repec.org//a/rje/bellje/v4y1973ispringp343-365.html.

Li, Luzhou. "How to Think about Media Policy Silence." *Media, Culture & Society* 43, no. 2 (2021): 359–368. https://doi.org/10.1177/0163443720948004.

Lobato, Ramon. *Netflix Nations: The Geography of Digital Distribution*. New York: New York University Press, 2019.

MacDougall, Robert. "Long Lines: AT&T's Long-Distance Network as an Organizational and Political Strategy." *Business History Review* 80, no. 2 (July 2006): 297–327. https://doi.org/10.1017/S0007680500035509.

MacKinnon, Rebecca. *Consent of the Networked: The Worldwide Struggle for Internet Freedom*. New York: Basic Books, 2012.

MacKinnon, Rebecca, Elonnai Hickok, Allon Bar, and Hai-in Lim. *Fostering Freedom Online: The Role of Internet Intermediaries*. Reston, VA: UNESCO, 2014.

Marchand, Roland. *Creating the Corporate Soul: The Rise of Public Relations and Corporate Imagery in American Big Business*. Los Angeles: University of California Press, 1998.

Marvin, Carolyn. *When Old Technologies Were New: Thinking about Electric Communication in the Late Nineteenth Century*. New York: Oxford University Press, 1990.

Marwick, Alice. "Silicon Valley and the Social Media Industry." In *The Sage Handbook of Social Media*, edited by Jean Burgess, Alice E. Marwick, and Thomas Poell, 314–329. Thousand Oaks, CA: Sage Publications, 2018.

Mattern, Shannon. "Deep Time of Media Infrastructure." In *Signal Traffic: Critical Studies of Media Infrastructures*, edited by Lisa Parks and Nicole Starosielski, 94–112. Chicago: University of Illinois Press, 2015.

Mattern, Shannon Christine. *Code and Clay, Data and Dirt: Five Thousand Years of Urban Media*. Minneapolis: University of Minnesota Press, 2017.

Mayer, Jane. *Dark Money: The Hidden History of the Billionaires behind the Rise of the Radical Right*. New York: Random House, 2016.

Mazzucato, Mariana, Josh Entsminger, and Rainer Kattel. "Reshaping Platform-Driven Digital Markets." In *Regulating Big Tech: Policy Responses to Digital Dominance*, edited by Martin Moore and Damian Tambini, 17–34. New York: Oxford University Press, 2021.

McCabe, Katie. "Making History in a Segregated Washington." *Journal of the Bar Association of the District of Columbia* 42, no. 1 (May 2011): 67–97.

McChesney, Robert W. "Press-Radio Relations and the Emergence of Network, Commercial Broadcasting in the United States, 1930–1935." *Historical Journal of Film, Radio and Television* 11, no. 1 (1991): 41–57. https://doi.org/10.1080/01439689100260031.

McChesney, Robert W. *Rich Media, Poor Democracy: Communication Politics in Dubious Times*. Champaign: University of Illinois Press, 1999.

McChesney, Robert W., and Victor Pickard, eds. *Will the Last Reporter Please Turn Out the Lights: The Collapse of Journalism and What Can Be Done to Fix It*. New York: The New Press, 2011.

Medine, David, and Esteban Morin. "Privacy and Civil Liberties Oversight Board." In *The Cambridge Handbook of Surveillance Law*, edited by David Gray and Stephen E. Henderson, 687. New York: Cambridge University Press, 2017.

Miller, Arthur Raphael. *The Assault on Privacy: Computers, Data Banks, and Dossiers*. Ann Arbor: University of Michigan Press, 1971.

Minow, Martha. *Saving the News: Why the Constitution Calls for Government Action to Preserve Freedom of Speech*. New York: Oxford University Press, 2021.

Minow, Newton N. "Commemorative Messages." In *A Legislative History of the Communications Act of 1934*, edited by Max Paglin. New York: Oxford University Press, 1989.

Minow, Newton, and Craig LaMay. *Abandoned in the Wasteland: Children, Television, & the First Amendment*. New York: Hill and Wang, 1995.

Minow, Newton N., Nell Minow, and Martha Minow. "Social Media, Distrust, and Regulation." In *Social Media, Freedom of Speech, and the Future of Our Democracy*, edited by Lee C. Bollinger and Geoffrey R. Stone, 285–300. New York: Oxford University Press, 2022.

Moore, Martin, and Damian Tambini. "Conclusion: Without a Holistic Vision, Democratic Media Reforms May Fail." In *Regulating Big Tech: Policy Responses to Digital Dominance*, edited by Martin Moore and Damian Tambini, 338–348. New York: Oxford University Press, 2021.

Moore, Richard K. "Cyberspace Inc. and the Robber Baron Age: An Analysis of PFF's 'Magna Carta.'" *Information Society* 12, no. 3 (1996): 315–323. https://doi.org/10.1080/019722496129503.

Mosco, Vincent. *To the Cloud: Big Data in a Turbulent World.* Boulder, CO: Paradigm Publishers, 2014.

Mueller, Milton. *Universal Service: Competition, Interconnection and Monopoly in the Making of the American Telephone System.* Cambridge, MA: MIT Press, 1997.

Mumford, Lewis. *The City in History: Its Origins, Its Transformations, and Its Prospects.* Orlando, FL: Houghton Mifflin Harcourt, 1961.

Napoli, Philip, and Robyn Caplan. "Why Media Companies Insist They're Not Media Companies, Why They're Wrong, and Why It Matters." *First Monday* 22, no. 5 (2017). https://doi.org/10.5210/fm.v22i5.7051.

Napoli, Philip M. *Social Media and the Public Interest: Media Regulation in the Disinformation Age.* New York: Columbia University Press, 2019.

Napoli, Philip M. "Treating Dominant Digital Platforms as Public Trustees." In *Regulating Big Tech: Policy Responses to Digital Dominance*, edited by Martin Moore and Damian Tambini, 151–168. New York: Oxford University Press, 2022.

Napoli, Philip M. "What If More Speech Is No Longer the Solution? First Amendment Theory Meets Fake News and the Filter Bubble." *Federal Communications Law Journal* 70, no. 1 (2017): 55–104.

Naughton, John. "The Evolution of the Internet: From Military Experiment to General Purpose Technology." *Journal of Cyber Policy* 1, no. 1 (2016): 5–28. https://doi.org/10.1080/23738871.2016.1157619.

Newman, Russell A. *The Paradoxes of Network Neutralities.* Cambridge, MA: MIT Press, 2019.

Nichols, Philip. "Redefining 'Common Carrier': The FCC's Attempt at Deregulation by Redefinition." *Duke Law Journal* 1987:501–520. https://doi.org/10.2307/1372565.

Nieland, Andrew. "National Security Letters and the Amended Patriot Act." *Cornell Law Review* 92, no. 6 (September 2007): 1201–1238. https://scholarship.law.cornell.edu/clr/vol92/iss6/4.

Noble, Safiya Umoja. *Algorithms of Oppression: How Search Engines Reinforce Racism.* New York: New York University Press, 2018.

O'Mara, Margaret. *The Code: Silicon Valley and the Remaking of America.* New York: Penguin Books, 2019.

Packard, Vance. *The Naked Society.* New York: Pocket Books, 1964.

Parkhill, Douglas F. *The Challenge of the Computer Utility.* Palo Alto, CA: Addison-Wesley Publishing Company, 1966.

Bibliography

Parks, Lisa. "Around the Antenna Tree: The Politics of Infrastructural Visibility." *Flow*, March 6, 2009. https://www.flowjournal.org/2009/03/around-the-antenna-tree-the-politics-of-infrastructural-visibilitylisa-parks-uc-santa-barbara/.

Parks, Lisa. "Stuff You Can Kick: Toward a Theory of Media Infrastructures." In *Between Humanities and the Digital*, edited by Patrik Svensson and David Theo Goldberg, 355–373. Cambridge, MA: MIT Press, 2015.

Parks, Lisa, and Nicole Starosielski, eds. *Signal Traffic: Critical Studies of Media Infrastructures*. Chicago: University of Illinois Press, 2015.

Pasquale, Frank. *The Black Box Society: The Secret Algorithms That Control Money and Information*. Cambridge, MA: Harvard University Press, 2015.

Perzanowski, Aaron, and Jason Schultz. *The End of Ownership: Personal Property in the Digital Economy*. Cambridge, MA: MIT Press, 2016.

Peters, Benjamin. *How Not to Network a Nation*. Cambridge, MA: MIT Press, 2016.

Peters, John Durham. *The Marvelous Clouds: Toward a Philosophy of Elemental Media*. Chicago: University of Chicago Press, 2015.

Phillips, Charles Franklin. *The Regulation of Public Utilities: Theory and Practice*. Arlington, VA: Public Utilities Reports, 1988.

Phillips, Whitney. *This Is Why We Can't Have Nice Things: Mapping the Relationship between Online Trolling and Mainstream Culture*. Cambridge, MA: MIT Press, 2015.

Pickard, Victor. "A New Social Contract for Platforms." In *Regulating Big Tech: Policy Responses to Digital Dominance*, edited by Martin Moore and Damian Tambini, 323–337. New York: Oxford University Press, 2022.

Pickard, Victor. *Democracy without Journalism?: Confronting the Misinformation Society*. New York: Oxford University Press, 2019.

Pickard, Victor, and David Elliot Berman. *After Net Neutrality: A New Deal for the Digital Age*. New Haven, CT: Yale University Press, 2019.

Picker, Randal. "The Arc of Monopoly: A Case Study in Computing." *University of Chicago Law Review* 87, no. 2 (March 2020): 523–552. https://chicagounbound.uchicago.edu/uclrev/vol87/iss2/9.

Piraino, Thomas. "Reconciling the Harvard and Chicago Schools: A New Antitrust Approach for the 21st Century." *Indiana Law Journal* 82, no. 2 (Spring 2007): 345–409. https://www.repository.law.indiana.edu/cgi/viewcontent.cgi?article=1354&context=ilj.

Pistor, Katharina. "Rule by Data: The End of Markets?" *Law and Contemporary Problems* 83, no. 2 (2020): 101–124. https://scholarship.law.duke.edu/lcp/vol83/iss2/6.

Plantin, Jean-Christophe, Carl Lagoze, Paul N. Edwards, and Christian Sandvig. "Infrastructure Studies Meet Platform Studies in the Age of Google and Facebook." *New Media & Society* 20, no. 1 (2018): 293–310. https://doi.org/10.1177/1461444816661553.

Popiel, Pawel. "The Tech Lobby: Tracing the Contours of New Media Elite Lobbying Power." *Communication, Culture and Critique* 11, no. 4 (2018): 566–585. https://doi.org/10.1093/ccc/tcy027.

Posner, Richard. "The Decline and Fall of AT&T: A Personal Recollection." *Federal Communications Bar Journal* 61 (2008): 11–19. https://chicagounbound.uchicago.edu/journal_articles/6780.

Reilly, Robert. "Mapping Legal Metaphors in Cyberspace: Evolving the Underlying Paradigm." *Journal of Information Technology & Privacy Law* 16, no. 3 (Spring 1998): 579–596. https://repository.law.uic.edu/jitpl/vol16/iss3/3.

Renda, Andrea. "Cloud Privacy Law in the United States and the European Union." In *Regulating the Cloud: Policy for Computing Infrastructure*, edited by Christopher S. Yoo and Jean-François Blanchette, 135–164. Cambridge, MA: MIT Press, 2015.

Ritter, Jeffrey, and Anna Mayer. "Regulating Data as Property: A New Construct for Moving Forward." *Duke Law & Technology Review* 16, no. 1 (2018): 220–277. https://scholarship.law.duke.edu/dltr/vol16/iss1/7.

Roberts, Sarah T. *Behind the Screen: Content Moderation in the Shadows of Social Media.* New Haven, CT: Yale University Press, 2019.

Rohde, David. *In Deep: The FBI, the CIA, and the Truth about America's "Deep State."* New York: W. W. Norton & Company, 2020.

Rozenshtein, Alan Z. "Surveillance Intermediaries." *Stanford Law Review* 70 (January 2018): 99–189.

Sandvig, Christian. "Network Neutrality Is the New Common Carriage." *Info* 9, no. 2/3 (2007): 136–147. https://doi.org/10.1108/14636690710734751.

Sandvig, Christian. "The Internet as Infrastructure." In *The Oxford Handbook of Internet Studies*, edited by William H. Dutton, 86–108. Oxford: Oxford University Press, 2013.

Schiller, Dan. *Crossed Wires: The Conflicted History of US Telecommunications, from the Post Office to the Internet.* New York: Oxford University Press, 2023.

Schiller, Dan. "Reconstructing Public Utility Networks: A Program for Action." *International Journal of Communication* 14 (2020): 4989–5000. https://ijoc.org/index.php/ijoc/article/view/16242.

Schiller, Dan. "The Hidden History of US Public Service Telecommunications, 1919–1956." *Info* 9, no. 2/3 (2007): 17–28. https://doi.org/10.1108/14636690710734625.

Schivelbusch, Wolfgang. *The Railway Journey: The Industrialization of Time and Space in the 19th Century.* Berkeley: University of California Press, 1986.

Schneier, Bruce. *Data and Goliath: The Hidden Battles to Collect Your Data and Control Your World.* New York: W. W. Norton & Company, 2015.

Bibliography

Schulze, Matthias. "Clipper Meets Apple vs. FBI—A Comparison of the Cryptography Discourses from 1993 and 2016." *Media and Communication* 5, no. 1 (March 2017): 54–62. https://doi.org/10.17645/mac.v5i1.805.

Schwartz, Paul M. "Legal Access to the Global Cloud." *Columbia Law Review* 118, no. 6 (October 2018): 1681–1762. https://columbialawreview.org/content/legal-access-to-the-global-cloud/.

Schwartz, Paul M. "The EU-U.S. Privacy Collision: A Turn to Institutions and Procedures." *Harvard Law Review* 126, no. 7 (May 2013): 1966–2009. https://www.jstor.org/stable/23415063.

Segall, Sasha. "Jurisdictional Challenges in the United States Government's Move to Cloud Computing Technology." *Fordham Intellectual Property, Media and Entertainment Law Journal* 23, no. 3 (Spring 2013): 1105–1153. https://ir.lawnet.fordham.edu/iplj/vol23/iss3/7.

Selbst, Andrew D., and Solon Barocas. "Unfair Artificial Intelligence: How FTC Intervention Can Overcome the Limitations of Discrimination Law." *University of Pennsylvania Law Review* 171 (Forthcoming). https://papers.ssrn.com/abstract=4185227.

Selby, John. "Data Localization Laws: Trade Barriers or Legitimate Responses to Cybersecurity Risks, or Both?" *International Journal of Law and Information Technology* 25, no. 3 (September 2017): 213–232. http://resolver.scholarsportal.info/resolve/09670769/v25i0003/213_dlltbortcrob.xml.

Simon, Leslie David. *NetPolicy.Com: Public Agenda for a Digital World.* Washington, DC: Woodrow Wilson Center Press, 2000.

Snowden, Edward. "Introduction." In *Little Brother & Homeland*, Cory Doctorow. New York: Macmillan Publishing Group, 2020.

Snowden, Edward. *Permanent Record.* New York: Metropolitan Books, 2019.

Solomun, Sonja, Maryna Polataiko, and Helen A. Hayes. "Platform Responsibility and Regulation in Canada: Considerations on Transparency, Legislative Clarity, and Design." *Harvard Journal of Law & Technology Digest* 34 (Spring 2021): 1–18. https://jolt.law.harvard.edu/digest/platform-responsibility-and-regulation-in-canada-considerations-on-transparency-legislative-clarity-and-design.

Spencer, Shaun B. "The First Amendment and the Regulation of Speech Intermediaries." *Marquette Law Review* 106, no. 1 (Fall 2022): 1–71. https://scholarship.law.marquette.edu/mulr/vol106/iss1/2.

Srinivasan, Dina. "Why Google Dominates Advertising Markets: Competition Policy Should Lean on the Principles of Financial Market Regulation." *Stanford Law School* 24, no. 1 (December 2020): 55–175. https://law.stanford.edu/publications/why-google-dominates-advertising-markets/.

Starosielski, Nicole. *The Undersea Network*. Durham, NC: Duke University Press, 2015.

Starosielski, Nicole. "'Warning: Do Not Dig': Negotiating the Visibility of Critical Infrastructures." *Journal of Visual Culture* 11, no. 1 (April 2012): 38–57. https://doi.org/10.1177/1470412911430465.

Stoller, Matt. *Goliath: The 100-Year War between Monopoly Power and Democracy*. New York: Simon and Schuster, 2019.

Strassburg, Bernard. "Competition and Monopoly in the Computer and Data Transmission Industries." *Antitrust Bulletin* 13 (1968): 991–997.

Streeter, Thomas. *Selling the Air: A Critique of the Policy of Commercial Broadcasting in the United States*. Chicago: University of Chicago Press, 1996.

Streeter, Thomas. *The Net Effect: Romanticism, Capitalism, and the Internet*. New York: New York University Press, 2011.

Suzor, Nicholas P. *Lawless: The Secret Rules That Govern Our Digital Lives*. New York: Cambridge University Press, 2019.

Teachout, Zephyr. *Break 'Em Up: Recovering Our Freedom from Big Ag, Big Tech, and Big Money*. New York: All Points Books, 2020.

Temin, Peter. "Fateful Choices: AT&T in the 1970s." *Business and Economic History* 27, no. 1 (Fall 1998): 61–77. https://www.jstor.org/stable/23703063.

Temin, Peter. *The Fall of the Bell System: A Study in Prices and Politics*. New York: Cambridge University Press, 1987.

Thibault, Ghislain. "Bolts and Waves: Representing Radio Signals." *Early Popular Visual Culture* 16, no. 1 (2018): 39–56. https://doi.org/10.1080/17460654.2018.1472621.

Thierer, Adam D. "Unnatural Monopoly: Critical Moments in the Development of the Bell System Monopoly." *Cato Journal* 14, no. 2 (1994): 267–285. https://www.cato.org/sites/cato.org/files/serials/files/cato-journal/1994/11/cj14n2-6.pdf.

Troesken, Werner. "Regime Change and Corruption: A History of Public Utility Regulation." In *Corruption and Reform: Lessons from America's Economic History*, edited by Edward L. Glaeser and Claudia Goldin, 259–281. Chicago: University of Chicago Press, 2006.

Turner, Fred. *From Counterculture to Cyberculture: Stewart Brand, the Whole Earth Network, and the Rise of Digital Utopianism*. Chicago: University of Chicago Press, 2006.

Turner, Fred. "Machine Politics." *Harper's Magazine*, December 30, 2020. https://harpers.org/archive/2019/01/machine-politics-facebook-political-polarization/.

Tushnet, Rebecca. "Power without Responsibility: Intermediaries and the First Amendment." *George Washington Law Review* 76, no. 4 (June 2008): 986–1016. https://www.gwlr.org/wp-content/uploads/2012/08/76-4-Tushnet.pdf.

Bibliography

Vaidhyanathan, Siva. *Antisocial Media: How Facebook Disconnects Us and Undermines Democracy*. New York: Oxford University Press, 2018.

Van Schewick, Barbara. *Internet Architecture and Innovation*. Cambridge, MA: MIT Press, 2010.

Velkova, Julia. "Data Centers as Impermanent Infrastructures." *Culture Machine* 18 (2019). https://culturemachine.net/vol-18-the-nature-of-data-centers/data-centers-as-impermanent/.

Velkova, Julia. "Data That Warms: Waste Heat, Infrastructural Convergence and the Computation Traffic Commodity." *Big Data & Society* 3, no. 2 (December 2016): 1–10. https://doi.org/10.1177/2053951716684144.

Vonderau, Asta. "Technologies of Imagination: Locating the Cloud in Sweden's North." *Imaginations Journal of Cross-Cultural Image Studies* 8, no. 2 (2017): 8–21. https://doi.org/10.17742/IMAGE.LD.8.2.2.

Warren, Samuel D., and Louis D. Brandeis. "The Right to Privacy." *Harvard Law Review* 4, no. 5 (1890): 193–220.

Weber, Rolf H. "Transborder Data Transfers: Concepts, Regulatory Approaches and New Legislative Initiatives." *International Data Privacy Law* 3, no. 2 (May 2013): 117–130. https://doi.org/10.1093/idpl/ipt001.

Weinberg, Steve. *Taking on the Trust: How Ida Tarbell Brought Down John D. Rockefeller and Standard Oil*. New York: W. W. Norton & Company, 2008.

Werbach, Kevin. "The Federal Computer Commission." *North Carolina Law Review* 84, no. 1 (2005): 1–75. https://scholarship.law.unc.edu/nclr/vol84/iss1/3.

Werbach, Kevin. "The Network Utility." *Duke Law Journal* 60, no. 8 (May 2011): 1761–1840. https://scholarship.law.duke.edu/dlj/vol60/iss8/3.

Westin, Alan F. *Privacy and Freedom*. New York: Atheneum, 1967.

Whitehouse, Sheldon. "Section 230 Reforms." In *Social Media, Freedom of Speech, and the Future of Our Democracy*, edited by Lee C. Bollinger and Geoffrey R. Stone, 103–118. Oxford: Oxford University Press, 2022.

Winseck, Dwayne. "The Broken Internet and Platform Regulation: Promises and Perils." In *Digital Platform Regulation: Global Perspectives on Internet Governance*, edited by Terry Flew and Fiona R. Martin, 229–258. London: Palgrave Macmillan, 2022.

Woods, Andrew Keane. "Against Data Exceptionalism." *Stanford Law Review* 68, no. 4 (April 2016): 729–789. https://www.stanfordlawreview.org/print/article/against-data-exceptionalism/.

Woods, Andrew Keane. "Litigating Data Sovereignty." *Yale Law Journal* 128, no. 2 (November 2018): 328–406. https://www.yalelawjournal.org/article/litigating-data-sovereignty.

Wu, Tim. *The Attention Merchants: The Epic Scramble to Get inside Our Heads*. New York: Alfred A. Knopf, 2016.

Wu, Tim. "Is the First Amendment Obsolete?" *Michigan Law Review* 117, no. 3 (2018): 547–581. https://doi.org/https://doi.org/10.36644/mlr.117.3.first.

Wu, Tim. *The Curse of Bigness: Antitrust in the New Gilded Age*. New York: Columbia Global Reports, 2018.

Wylie, Christopher. *Mindf*ck: Inside Cambridge Analytica's Plot to Break the World*. London: Profile Books, 2019.

Yoo, Christopher S. "Common Carriage's Domain." *Yale Journal on Regulation* 35 (2018): 991–1026. https://scholarship.law.upenn.edu/faculty_scholarship/2016.

Zuboff, Shoshana. *The Age of Surveillance Capitalism: The Fight for a Human Future at the New Frontier of Power*. New York: PublicAffairs, 2018.

Select Cases

American Civil Liberties Union v. National Security Agency, 493 F.3d 644 (6th Cir. 2007).

Calhoun, et al. v. Google LLC, 20-CV-05146-LHK (N.D. Cal. 2021).

Clarksburg Publishing Company v. Google LLC, et al., 1:21-00051 (N.D.W. Va. 2021).

Cubby, Inc. v. CompuServe Inc., 776 F. Supp. 135 (S.D.N.Y. 1991).

Data Protection Commissioner v. Facebook Ireland Ltd., ECLI:EU:C:2020:559 (C.J.E.U. 2020).

Eisenstadt v. Baird, 405 U.S. 438 (1972).

Gonzalez v. Google, 598 U.S. 617 (2022).

Google Spain SL, Google Inc. v Agencia Española de Protección de Datos, Mario Costeja González C-131/12 (C.J.E.U. 2014).

Hepting v. AT&T Corp., 439 F. Supp. 2d 974 (N.D. Cal. 2006).

Hush-A-Phone Corp v. United States, 238 F. 2d 266 (D.C. Cir. 1956).

In the Matter of Use of the Carterfone Device in Message Toll Telephone Service, 13 FCC 2d 420 (1968).

Katz v. United States, 389 U.S. 347 (1967).

Lawrence v. Texas, 539 U.S. 558 (2003).

Bibliography

Lopez v. United States, 373 U.S. 427 (1963).

Munn v. Illinois, 94 U.S. 113 (1877).

National Association of Regulatory Utility Commissioners v. Federal Communications Commission, 525 F.2d 630 (D.C. Cir. 1976).

National Cable and Telecommunications Association v. Brand X Internet Services, 545 U.S. 967 (2005).

Olmstead v. United States, 277 U.S. 438 (1928).

Reno v. American Civil Liberties Union, 521 U.S. 844 (1997).

Roe v. Wade, 410 U.S. 113 (1973).

Schrems v. Data Protection Commissioner and Digital Rights Ireland, Ltd., C-362/14, ECLI:EU:C:2015:650 (C.J.E.U. 2015).

Smith v. Maryland, 442 U.S. 735 (1979).

Stratton Oakmont, Inc. v. Prodigy Services Company, 23 Media L. Rep. 1794 (N.Y. 1995).

Twitter v. Taamneh, 598 U.S. 471 (2023).

United States v. Microsoft Corp., 584 U.S. ___ (2018).

United States v. Microsoft Corporation, 65 F. Supp. 2d 1 (D.D.C. 1999).

United States v. Miller, 425 U.S. 435 (1976).

United States v. Reicherter, 647 F.2d 397 (3d Cir. 1981).

United States v. Western Elec. Co., 1956 Trade Cas. (CCH) P68, 246 (D.N.J. January 24, 1956).

Western Union Tel. Co. v. Call Publishing Co., 181 U.S. 92 (1901).

Laws and Statutes

Clarifying Lawful Overseas Use of Data Act, Pub. L. No. 115–141, Div. V, 132 Stat. 1213 (2018).

Communications Act of 1934, 47 U.S.C. § 151 (1934).

Communications Assistance for Law Enforcement Act, Pub. L. No. 103–414, 108 Stat. 4279 (1994).

Electronic Communications Privacy Act of 1986, Pub. L. No. 99–508, 100 Stat. 1848 (1986).

The Communications Decency Act of 1996, 47 U.S.C. § 230 (1996).

Foreign Intelligence Surveillance Act of 1978, Pub. L. No. 95–511, 92 Stat. 1783 (1978).

Foreign Intelligence Surveillance Act of 1978 Amendments Act of 2008, Pub. L. No. 110–261, 122 Stat. 2436 (2008).

FISA Amendments Reauthorization Act of 2017, Pub. L. No. 115–118, 122 Stat. 2474 (2017).

General Data Protection Regulation, 2016/679 of the European Parliament and of the Council (2016).

Protect America Act of 2007, Pub. L. No. 110–55, 121 Stat. 552 (2007).

Stored Communications Act, Pub. L. No. 99–508, 18 §§ 2701–2712 (1986).

Telecommunications Act of 1996, Pub L. No. 104–104, 110 Stat. 56 (1996).

Uniting and Strengthening America by Providing Appropriate Tools Required to Intercept and Obstruct Terrorism (USA PATRIOT) Act of 2001, Pub. L. No. 107–56, 115 Stat. 272 (2001).

Hearings

US House of Representatives Committee on Energy and Commerce. *Facebook: Transparency and Use of Consumer Data.* 115th Congress, 2d sess., April 11, 2018.

US House of Representatives Committee on Energy and Commerce. Subcommittee on Consumer Protection and Commerce. *Mainstreaming Extremism: Social Media's Role in Radicalizing America.* 116th Cong., 2d sess., September 24, 2020.

US House of Representatives Committee on Government Operations. Special Subcommittee on Invasion of Privacy. *The Computer and Invasion of Privacy.* 89th Cong., 2d sess., July 26, July 27, and July 28, 1966.

US House of Representatives Committee on International Relations. Subcommittee on Africa, Global Human Rights and International Operations and the Subcommittee on Asia and the Pacific. *The Internet in China: A Tool for Freedom or Suppression?* 109th Cong., 2d sess., February 15, 2006.

US House of Representatives Committee on the Judiciary. *Data Stored Abroad: Ensuring Lawful Access and Privacy Protection in the Digital Era.* 115th Cong., 1st sess., June 15, 2017.

US House of Representatives Committee on the Judiciary. *International Conflicts of Law and Their Implications for Cross Border Data Requests by Law Enforcement.* 114th Cong., 2d sess., February 25, 2016.

Bibliography

US House of Representatives Committee on the Judiciary. Subcommittee on Antitrust, Commercial, and Administrative Law. *Online Platforms and Market Power, Part 1: The Free and Diverse Press*, 116th Cong., 1st sess., June 11, 2019.

U.S Senate Committee on Commerce, Science and Transportation. Subcommittee on Consumer Protection, Product Safety, and Data Security. *Protecting Kids Online: Testimony from a Facebook Whistleblower.* 117th Cong., 1st sess., October 4, 2021.

US Senate Committee on the Judiciary. *Breaking the News: Censorship, Suppression, and the 2020 Election.* 116th Cong., 2d sess., November 17, 2020.

US Senate Committee on the Judiciary. *The Electronic Communications Privacy Act: Promoting Security and Protecting Privacy in the Digital Age.* 111th Cong., 2d sess., September 22, 2010.

US Senate Committee on the Judiciary Subcommittee on Administrative Practice and Procedure. *Computer Privacy.* 90th Cong., 1st sess., March 14 and March 15, 1967.

US Senate Committee on the Judiciary Subcommittee on Antitrust and Monopoly. *The Industrial Reorganization Act. Part 6: The Communications Industry.* 93d Cong., 2d sess., July 9, July 30, and July 31, 1974.

US Senate Committee on the Judiciary Subcommittee on Crime and Terrorism. *Law Enforcement Access to Data Stored across Borders: Facilitating Cooperation and Protecting Rights.* 115th Cong., 1st sess., May 24, 2017.

US Senate Committee on the Judiciary and Committee on Commerce. Subcommittee on Constitutional Rights and the Special Subcommittee on Science, Technology, and Commerce. *Surveillance Technology.* 94th Cong., 1st sess., June 23, September 9, and September 10, 1975.

US Senate Committee on the Judiciary and Committee on Commerce, Science and Transportation. *Facebook, Social Media Privacy, and the Use and Abuse of Data.* 115th Cong., 2d sess., April 10, 2018.

US Senate Committee on Small Business. Subcommittee on Retailing, Distribution, and Fair Trade Practices. *The Role of Private Antitrust Enforcement in Protecting Small Business.* 85th Cong., 2d sess., March 3, 1958.

Government Reports

Brannon, Valerie C., and Eric N. Holmes. *Section 230: An Overview.* CRS Rep. No. R46751. Washington, DC: Congressional Research Service, 2021.

Cabinet Committee on Cable Communications, Office of Telecommunications Policy. *Cable: Report to the President.* Washington, DC: US Government Printing Office, 1974.

Commission on CIA Activities within the United States. *Report to the President by the Commission on CIA Activities within the United States*. June 6, 1975.

European Parliament. Policy Department C: Citizens' Rights and Constitutional Affairs. Directorate-General for Internal Policies. *Fighting Cyber Crime and Protecting Privacy in the Cloud*. Rep. No. PE 462.509. October 2012.

Federal Communications Commission. *Amendment of Section 64.702 of the Commission's Rules and Regulations (Third Computer Inquiry)*. Report and Order, 104 FCC 2d 958. 1986.

Federal Communications Commission. *First Report and Order and Further Notice of Proposed Rulemaking. Communications Assistance for Law Enforcement Act (CALEA) and Broadband Access and Services*. 20 FCC Rcd 14989. September 23, 2005.

Federal Communications Commission. *Computer II Final Decision*. Final Decision and Order, 77 FCC 2d 384. 1980.

Federal Communications Commission. *Investigation of the Telephone Industry in the United States. Letter from the Chairman of the Federal Communications Commission*. 76th Cong., 1st sess., June 14, 1939. H. Doc. 340.

Federal Communications Commission. *Regulatory and Policy Problems Presented by the Interdependence of Computer and Communication Services and Facilities*. Final Decision and Order, 28 FCC 2d 267. 1971.

Federal Communications Commission. *Regulatory and Policy Problems Presented by the Interdependence of Computer and Communication Services & Facilities*. Notice of Inquiry, 7 FCC 2d 11. 1966.

Federal Communications Commission. *Report and Order and Notice of Proposed Rulemaking*. Rep. No. FCC 05-150. August 5, 2005.

Federal Trade Commission. *Privacy Online: Fair Information Practices in the Electronic Marketplace. A Report to Congress*. May 2000.

Federal Trade Commission. *Protecting Consumer Privacy in an Era of Rapid Change. Recommendations for Businesses and Policymakers. FTC Report*. March 2012.

Furman, Necah S. *Contracting in the National Interest: Establishing the Legal Framework for the Interaction of Science, Government, and Industry at a Nuclear Weapons Laboratory*. Rep. No. SAND87-1651 UC-13. Albuquerque, NM: Sandia National Laboratories for the United States Department of Energy, April 1988.

Kundra, Vivek. *Federal Cloud Computing Strategy*. White House Report by US Chief Information Officer. February 8, 2011.

Offices of Inspectors General. *Unclassified Report on the President's Surveillance Program*. Rep. No. 2009-0013-AS. July 10, 2009.

Bibliography

Rostow, Eugene V. *President's Task Force on Communications Policy. Final Report.* Washington, DC: President's Task Force on Communications Policy, December 7, 1968.

Thompson, Richard M., II, and Jared P. Cole. *Stored Communications Act: Reform of the Electronic Communications Privacy Act (ECPA).* CRS Rep No. R44036. Washington, DC: Congressional Research Service, May 19, 2015.

United Kingdom National Infrastructure Commission. *Data for the Public Good.* December 2017.

US Department of Health, Education, and Welfare. Secretary's Advisory Committee on Automated Personal Data Systems. *Records, Computers, and the Rights of Citizens.* Rep. No. (OS) 73–94. Cambridge, MA: DHEW Publication, July 1973.

US Department of Justice. Office of the Inspector General. *A Review of the Federal Bureau of Investigation's Use of National Security Letters (U).* March 2007.

US House of Representatives Committee on Government Operations. *Privacy and the National Data Bank Concept.* 9th Cong., 2d sess., July 1968.

US House of Representatives Committee on the Judiciary Antitrust Subcommittee. *Consent Decree Program of the Department of Justice.* H.R. Rep. 33261, 86th Cong., 1st sess., January 30, 1959.

US House of Representatives Committee on the Judiciary Subcommittee on Antitrust, Commercial and Administrative Law. *Investigation of Competition in Digital Markets: Majority Staff Report and Recommendations: Part I.* H.R. Rep. CP 117–8, 117th Cong., 2d sess., July 2020.

US Office of the Director of National Intelligence. *Section 702 Overview.* n.d.

US Senate *Senate Report No. 92.* 102d Cong., 1st sess., 1991. S. Rep. 92.

US Senate Committee on Commerce. *Appointments to the Regulatory Agencies: The Federal Communications Commission and the Federal Trade Commission (1949–1974).* S. Rep. 62–119, 94th Cong., 2d sess., April 1976.

US Senate Committee on Commerce, Science, and Transportation. *Telecommunications Competition and Deregulation Act of 1995.* S. Rep. 104–123, 104th Cong., 1st sess., March 30, 1995.

US Senate Committee on the Judiciary. *Electronic Communications Privacy Act of 1986 Report.* S. Rep. 99–541, 99th Cong., 2d sess., October 17, 1986.

US Senate Select Committee to Study Governmental Operations. *Intelligence Activities and the Rights of Americans: Book II.* S. Rep. 94–755, 94th Cong., 2d sess., April 26, 1976.

US Senate Select Committee to Study Governmental Operations. *Supplementary Detailed Staff Reports on Intelligence Activities and the Rights of Americans: Book III.* S. Rep. 69–684, 94th Cong., 2d sess., April 23, 1976.

Other Government Documents

Aid, Matthew M., and William Burr, eds. *"Disreputable If Not Outright Illegal": The National Security Agency versus Martin Luther King, Muhammad Ali, Art Buchwald, Frank Church, et al.* The National Security Archive Electronic Briefing Book no. 441, November 14, 2008.

Cicilline, David. "Online Platforms and Market Power, Part 6: Examining the Dominance of Amazon, Apple, Facebook, and Google." Press release, July 29, 2020.

Clinton, Bill. "Technology: The Engine of Economic Growth." Clinton-Gore National Campaign Headquarters, September 21, 1992.

Copps, Michael J. "Statement of Commissioner Michael J. Copps on Chairman Genachowski's Announcement to Reclassify Broadband." Washington, DC: Federal Communications Commission, May 6, 2010.

European Commission. "Digital Services Act: Commission Welcomes Political Agreement on Rules Ensuring a Safe and Accountable Online Environment." Press release, April 23, 2022.

Federal Communications Commission. *Policy Statement*. FCC 05-151. Washington, DC: Federal Communications Commission, August 5, 2005.

Hundt, Reed. "Speech by Reed Hundt." Speech, Center for National Policy, Washington, DC, May 6, 1996.

"The Many Lives of Herbert O. Yardley." *NSA Cryptologic Spectrum* 11, no. 4 (Fall 1981).

Roosevelt, Franklin D. "Acceptance Speech for the Renomination for the Presidency." Speech, Philadelphia, PA, June 27, 1936.

Roosevelt, Franklin D. *Message from the President of the United States Recommending That Congress Create a New Agency to Be Known as the Federal Communications Commission*. S. Doc. No. 144, 73d Cong., 2d sess., February 26, 1934.

Roosevelt, Franklin D. "Power: Protection of the Public Interest." Speech, Portland, OR, September 21, 1932.

Sherman, John. "Trusts." Speech, Senate of the United States, Washington, DC, March 31, 1890.

Truman, Harry S. Harry S. Truman Letter to Leroy A. Wilson, President of AT&T, May 13, 1949.

US Post Office Department. Postmaster General. *Government Ownership of Electrical Means of Communication: Letter from the Postmaster General, Transmitting, in Response to a Senate Resolution of January 12, 1914, a Report Entitled "Government Ownership of Electrical Means of Communication."* S. Doc. 399, 63d Cong., 2d sess., January 31, 1914.

US Senate. *Telecommunications Act of 1996.* S. 652. 104th Cong., 2d sess., *Congressional Record* 142, No. 14., February 1, 1996.

Wheeler, Tom. *Statement of Chairman Tom Wheeler: FCC Releases Open Internet Order. GN Docket No. 14–28.* Washington, DC: Federal Communications Commission, March 12, 2015.

Index

Aaron, Craig, 140
Abbate, Janet, 67–68
Access Now, 199
ACLU (American Civil Liberties Union), 39, 171, 176–177, 199
Activision Blizzard, 195
Activism, 81, 87, 199–202. *See also* Advocacy organizations
Adaptability, 203
Advanced Research Projects Agency (ARPA), 8–9
Advertising revenue, 131–132
Advocacy organizations, 122–123, 197, 199–202
Agur, Colin, 23
Albright, Madeleine, 6
Alden Global Capital, 134
All Writs Act of 1789, 184–185
Alphabet. *See* Google
Amazon
    advertising revenue, 236n5
    antitrust policy and, 30–31
    cloud infrastructure and, 30 (*see also* Amazon Web Services (AWS))
    as cloud service provider, 11–12
    Defense Department contract and lawsuit, 265–266n148
    Digital Markets Act and, 121
    fines, 209
    foreign governments and, 108
    founding of, 69
    labor activism and, 200
    lawsuits against, 209
    market consolidation, 128
    as monopoly, 123–124, 127
    national clouds and, 189
    ownership of content and platforms, 87–88
    power of as a platform, 92–93
    Prime Video, 97
    privacy and, 107
    railroads, compared to, 127
    Ring doorbells, 111
    structural separation, 138
    as technology vs. media company, 97, 98
    tracking online activity, 107
"Amazon's Anti-trust Paradox" (Khan), 30
Amazon Studios, 97
Amazon Web Services (AWS), 30
    in China, 191
    jurisdiction requirements in TOS, 182–183
    market share, 181
    Pentagon contract, 265n148
American Civil Liberties Union (ACLU), 39, 171, 176–177, 199, 244n94
American Edge, 130
American Telephone and Telegraph. *See* AT&T
American Tobacco, 49, 61
America Online (AOL), 67

Index

Antitrust laws and policies
   AT&T and, 48–50, 51–53, 55, 57–58, 60–62
   Amazon and, 30–31
   of Biden, 130–131
   breaking up companies, calls for, 138
   Carterfone case, 58–60, 87
   changing interpretation of, 60–62
   Chicago School, 62, 67, 126
   cloud policy and, 30–32
   congressional hearings, 31, 92, 97–98, 127–130, 134, 139–140
   corporate power and, 31–32
   Department of Justice and, 49–50, 52–53, 208
   enforcement of, 125, 126–127
   Federal Communications Commission (FCC) and, 50–51
   Harvard School, 126
   IBM and, 60, 66
   industry competition, thresholds for, 201
   platforms and, 123–131
   during Reagan administration, 66
   Roosevelt on, 125
   Sherman Antitrust Act, 60, 61, 91, 126, 247n126, 249n153
   telegraph companies and, 49–50
   telephone service and, 49–50
   US-led cases vs. Big Tech companies, 31, 129–130, 207–211
Antitrust Paradox, The (Bork), 66
Apple
   in China, 187, 191
   Digital Markets Act and, 121
   foreign governments and, 108
   founding of, 63
   House Judiciary Antitrust Subcommittee (2020–2021), 127–128
   labor activism and, 200
   lawsuits against, 210
   market consolidation, 128
   as monopoly, 124
   NSA's mass surveillance program, 173
   power of as a platform, 92
   PRISM program and, 108
   privacy and, 108
   privacy and branding, 186–187
   standoff with FBI, 183–184
   structural separation, 138–139
Applebaum, Anne, 138
ARPANET, 8–9, 10, 11, 60, 63, 68
Assault on Privacy, The (Miller), 157
AT&T, 83
   antitrust lawsuit (1913), 49–50
   antitrust lawsuit (1949), 51–53
   antitrust lawsuit (1974), 60–62
   antitrust policy and, 48–50, 52–53, 55, 57–58, 60–62
   break up of, 65–66
   Carterfone case, 58–60, 87
   cloud policy blueprint and, 48–54
   common carrier, restricted to, 53
   consent decree (1956), 52–53, 54, 55, 56, 57–58, 66, 71, 139
   consent decree (1982), 65–66, 72
   corporate strategy of, 43–47
   in early Internet development, 10–11
   enhanced services and, 64
   history of, 44–48
   Hush-A-Phone case, 227n107
   Internet, miscalculations regarding, 53–54
   Johnson administration and, 70
   later government investigations, 54
   legislation proposed by, 71
   lobbying of, 39
   marketing of, 45–46
   motto of, 46
   national defense work, 51–53
   1910 Annual Report, 47, 68
   NSA's mass surveillance program, 175–176
   nuclear weapons and, 51–52
   origin of, 19

Index

ownership of content and platforms, 88
political action committee of, 72
PRISM program, 108
structural separation, 64, 229n145
Time Warner takeover, 12
treated as natural monopoly, 43–48, 49–50, 58–59
Walker Report and, 50–51
Western Electric and, 228n137
Western Union, divestment of, 49
Atlantic magazine, 160
Atomic Energy Commission, 51, 52
Attention economy, 94–95
Attwell Baker, Meredith, 218n71
Aufderheide, Patricia, 72
Australian News Media and Digital Platforms Mandatory Bargaining Code, 135

Baby Bells, 65, 66, 71–72, 228n135, 229n145, 231n169
Baldrige, Malcolm, 61
Balkin, Jack, 39, 112, 137–138
Banks, David A., 19
Baran, Paul, 11, 53–54
Barbrook, Richard, 101–102
Barlow, John Perry, 75, 150
Basic services, 63–64, 77
Baxter, William, 61–62, 66, 126
Belfort, Jordan, 113
Bell Labs, 51, 54, 66
Bell Operating Companies (BOCs), 229n145
BellSouth, 176
Bell System, 50, 52. *See also* AT&T
Bennet, Michael, 136
Berman, Jerry, 73
Bezos, Jeff, 127, 181, 265–266n148. *See also* Amazon
Biden, Joe, 130–131
Big Tech, 29, 87–88, 96, 98, 140, 167, 173, 198, 200, 220n4. *See also* Amazon; Apple; Digital Services Act (DSA); Facebook; Google; Microsoft; Platforms; Twitter; X
antitrust, and, 30, 32, 125, 129, 130, business models, 107–108, 136
Congressional hearings, 31, 108–109
data, and, 108, 149, 154, 189–192
informal policy, and, 3, 143, 181
investigations of, 127–129, 207–211
lobbying, 139
news industry, and, 131–133
privacy, and, 28, 84, 91, 183, 184, 186–187
Section 230, and, 113, 117, 119–120
Bill of Rights, 123
Black-boxing of infrastructures, 28, 95, 143
Black Chamber, 23
Blanchette, Jean-François, 22
Blumenthal, Richard, 120
Bork, Robert, 62, 66, 105, 126, 127
"The Bosses of the Senate" (Keppler), 124
Braman, Sandra, 6, 58
Brandeis, Louis, 24, 104
Branding, privacy and, 186–187
Brand X case, 78
Broadband Internet
common carriage and, 38, 81–82
digital divide, 28–29, 39–40, 42–43, 84–85, 88–89, 195, 235n227
public utility and, 39, 86
regulation as information service, 77
on tribal lands, 85, 195
Brooks, John, 49
Brownell, Herbert, Jr., 53
Browsers, 69, 97, 123
Bush, George H. W., 68
Bush, George W., 175, 259n78

Cabinet Committee on Cable Communications, 5
Cable Act of 1992, 134

Cable Communications Policy Act of 1984, 72
Cable industry, 5–6, 72, 76, 231n169
California, net neutrality (2018), 83–84
California Consumer Privacy Act (CCPA), 180–181
Californian ideology, 101–102
Cambridge Analytica data breach, 99–101, 117, 129, 207–208, 245n106
Cameron, Andy, 101–102
Campaign finance reform, 72, 201
Cannon, Robert, 57
Carrier hotels, 20, 217n56
Carter, Thomas, 58–59, 60, 206, 227n107
Carterfone case, 58–60, 87
Censorship, corporate, 73
Center for Democracy & Technology, 171, 199
Challenge of the Computer Utility, The (Parkhill), 8
Charter cable, 76, 83
Chicago School of antitrust, 62, 67, 126
Chicago Tribune, 251n176
Children's Online Privacy Protection Act of 1998 (COPPA), 105
China
  Apple and, 187, 191
  congressional hearing on Internet freedom in, 108–109
  Cybersecurity Law of 2017, 191
  mass surveillance and censorship in, 108–109
  Microsoft Azure, 190–191
  national clouds and, 191
Chrome browser, 97, 123
Church, Frank, 141, 142, 163, 206, 255n11, 271n16
Church Committee, 141–143, 165, 254n4
Cicilline, David, 96
Cipher Bureau, 23

Civil liberties, digital, 1, 4, 93–94, 95, 131, 143, 156
Civil society, role of in policymaking, 122–123, 199
Clarifying Lawful Overseas Use of Data (CLOUD) Act. *See* CLOUD Act
Clinton, Bill, 68–69, 74–76, 106
Clipper Chip, 185
Clopton, Zachary, 150
CLOUD Act, 172–173, 262n111
Cloud computing, origin of, 8
Cloud infrastructure. *See also* Cloud pipelines; Cloud policy; Platforms; *specific companies*
  Amazon Web Services (AWS), 30, 181, 182–183, 191, 265n148
  analog frameworks and predecessors, 18–22, 23–24
  black boxing, 28, 95, 143
  coexistence of old and new technologies, 18–20
  decentralization, 3, 57, 59–60, 78, 85, 89, 98, 101, 180, 203
  definition of, 2
  design of, 8–12
  foreign governments and, 190–191
  fragility of, 13
  marketing of, 13, 48, 116, 143–146, 181
  privatization, 3, 36, 54, 57, 76, 99, 102, 106, 146, 180, 196
  providers of (*see* Amazon Web Services (AWS); Platforms; *specific companies*)
  public-private surveillance partnerships, 23, 23–27, 104–112, 173–180, 264n132
  public understanding of, 13, 15, 215n37
  scarcity and, 21–22
  visibility/invisibility of, 143–149, 202, 204
  visualization of, 12–18

Index

Cloud pipelines
  common carriage, 37–38
  definition of, 35
  history of, 35–36
  natural monopoly, 43–48
  net neutrality, 77–84
  policy history, 67–68
  principles of, 36–48, 84–89
  private control, 67–70
  public utility, 39–43
  regulatory frameworks, 48–54
  Section 230, 113, 115–116
  Telecommunications Act of 1996, 70–77
  universal service, 43–48
Cloud policy. *See also* Cloud infrastructure; Cloud pipelines
  AT&T and, 48–54
  access and speech rights, 28–29
  accountability, growing demands for, 197
  antitrust policy and (*see* Antitrust laws and policies)
  competition and antitrust, 30–32
  components of, 4–5
  convergent era, 54–67
  creative solutions, 202–205
  disciplinary perspectives in, 4, 7, 32–33, 199, 204
  environmental protection model, 138
  foundational issues to confront, 201–202
  future of, 195–206
  global nature of, 3–4
  history of, 1–8
  jurisdictions of, 149–156
  lack of affordable public access to Internet, 28–29, 39–40, 43, 88–89
  legacy constructs, 30, 170–171, 202 (*see also* Natural monopoly)
  localized policies, 187–192, 198–199
  missed opportunities, 205–206
  negative policy, 196–197
  path dependencies of, 18–22, 63–64
  platforms' power over, 92
  power shifts in, 3
  privacy (*see* Privacy)
  privacy and surveillance, 22–28
  private surveillance infrastructure, reliance on, 112, 177–178, 202
  privatization, 3, 36, 54, 57, 76, 99, 102, 106, 146, 180, 196
  reach of, 6
  regulatory capture, 22, 40, 48, 79, 94, 199, 204
  regulatory hangover and, 5–7
  siloed nature of, 7, 95, 103–104, 199
  speech rights and, 28–29, 93–95
  stakes of, 22–32
  Telecommunications Act of 1996 and, 77
  trajectory of, 33
Cloud services, 11–12
Code and Other Laws of Cyberspace (Lessig), 98
Cohen, Julie, 6, 86, 116
Cold War, 23, 24–27, 51, 59, 69, 178
Colorado, 181
Comcast
  lawsuit against FCC, 79
  lobbying of, 39, 195
  market share, 76, 83
  ownership of content and platforms, 87
  role in blocking nomination of Gigi Sohn to FCC, 195
  subsidiaries of, 87
Comey, James, 184
Common carriage, 37–38
  AT&T and, 53
  basic vs. enhanced services, 63–64, 77
  Brand X case, 78
  broadband pipelines and, 38, 81–82
  cloud pipelines, 37–38
  Computer I and, 55–57
  Computer II and, 63–64

Common carriage (cont.)
  data services and, 56–57
  FCC and, 38
  Johnson administration and, 70
  legal definition of, 38
  net neutrality and, 77, 78, 81, 82, 205
  vs. public utility, 39
  Roosevelt and, 41–42
  structural separation, 138
  Telecommunications Act of 1996, 70
  telegraph and, 38, 220n9
  telephone companies and, 38
  US Supreme Court on, 37, 38
Communications Act of 1934
  broadband Internet and, 77
  Interstate Commerce Act and, 37
  net neutrality and, 78
  as outdated, 70
  passage of, 38
  telecommunication services vs. information services, 78, 80
  universal service and, 43
  update of, 205
  wiretapping and, 23, 24
Communications Act of 1978, 70–71
Communications Assistance for Law Enforcement Act (CALEA), 69, 185
Communications Decency Act (CDA), 244n94. *See also* Section 230
Competition
  cloud policy and, 30–32
  platforms and, 123–131
  regulation and, 76
  thresholds for, 201
Competition and Markets Authority (CMA), 209, 210
Competition in Digital Markets hearings, 92, 97–98, 127–128, 134, 139–140
CompuServe, 67, 113–114
"The Computer and the Invasion of Privacy" Hearings (1966), 58, 158–161, 206

Computer I Inquiry (1966–1971), 55–58
Computer II Final Decision, 64–65
Computer II Inquiry (1976–1980), 63–67
Computer II Final Decision, 227n114
Computer III Inquiry (1985–1986), 67, 229n145
Computer Inquiries, 54–57, 63–67
Computers and Communications (Kobayashi), 11, 67
Computer utility model, 8–9, 11, 197
Congressional hearings and investigations, 207–208
  on anticompetitive abuses, 127–130, 139
  Cambridge Analytica data breach hearing, 99–101
  Church Committee, 141–143, 165, 254n4
  Competition in Digital Markets hearings, 92, 97–98, 127–128, 134, 139–140
  "The Computer and Invasion of Privacy," 58, 158–161, 206
  Facebook and, 92, 97, 120, 127–128
  House Judiciary Antitrust Subcommittee investigation, 127–129
  on Internet Freedom in China, 108–109
  Online Platforms and Market Power, 140, 207
  Pike Committee, 141–142, 254n4
  on privacy, 158–159
  Rockefeller Commission, 141–142
  Section 230 reform, 120
  social media, understanding of, 31
  on surveillance technology, 163–164
  Watergate hearings, 142
  Zuckerberg and, 99–101, 120, 127
Connecticut, 181
Consent decrees. *See* AT&T
Consumer Privacy Bill of Rights, 109

Index

Consumer welfare, 62
Consumer welfare standard, 30, 126, 127, 132, 201, 202
Content
   lawful, blocking, 81
   liability for, 96
   platforms, ownership of, 87–88
Content delivery networks (CDNs), 86
Content moderation, 114–117, 122
Convergence, 54–55
Convergent era, 54–67
   antitrust policy, changing interpretation of, 60–63
   Carterfone case, 58–60
   Computer I (1966–1971), 55–58
   Computer II Inquiry (1976–1980), 63–67
Conyers, John, 74
Cook, Tim, 127, 183, 184, 186–187
Cookies, 69, 198, 209, 210
Copps, Michael J., 79–80, 206
Copyright Directive (2021), 135
Cord cutting, 12, 215n35
Corporate gatekeepers, 75, 82, 92–93
Corporate power, antitrust and, 31–32
Counterspeech, 112
Court of Justice of the European Union (CJEU), 110
COVID-19 pandemic, 28–29, 117
Cox, 76
Cox, Chris, 114
Crain, Matthew, 69, 106
Crandall, Robert, 47
Crawford, Susan, 37, 39, 88
Crypto-wars, 185
Cubby, Inc. v. CompuServe, 113
The Curse of Bigness, The (Wu), 31–32
Customer proprietary network information (CPNI), 84
Cybersecurity Law of 2017 (China), 191
"Cyberspace and the American Dream," 73

Daskal, Jennifer, 150, 189
Data
   "at rest", 170–171
   black boxing of infrastructures, 28, 95, 143
   breaches, 242–243n72, 245n106 (*see also* Cambridge Analytica data breach)
   CLOUD Act, 173
   cultural understanding of, 204
   Electronic Communications Privacy Act (ECPA) of 1986, 168–172
   email and, 152, 170–171
   end-user license agreements and, 143, 181–183
   FISA, 165–168, 186, 259n79
   government control of, 157–164
   immaterial and invisible nature of, 143–149
   jurisdiction of, 149–156, 183, 188–192
   national clouds and, 188, 189–192
   National Data Center, 28, 157–164
   ownership of in the cloud, 149
   PATRIOT Act, 164–165, 168, 260n87
   policy evolution, 198
   private control of, 181–187
   regulation, history of, 156–181
   sovereignty and localization, 187–193, 198, 269n190
   state legislation and, 180–181
   Stored Communications Act (SCA), 170–171, 173
   terms of service agreements and, 107–108, 143, 181–183

Data caps, 83
Data centers, 2, 143
   electricity costs, 154–155
   Facebook and, 154
   images of, 14, 144–145, 147–148
   locations of, 152–154
   Project Natick, 152, 153, 154
   structures of, 146–148

Data colonialism, 95, 188
Data controllers, 110
Data encryption. *See* Encryption
Data localization, 188–190, 198, 269n190
Data processing services, 56–58, 63, 64, 65, 139
Data sovereignty, 151, 187–189
Dataveillance, 84, 107, 131. *See also* Surveillance
DC Comics, 88
Debs, Eugene V., 41, 206
Decentralization, 3, 57, 59–60, 78, 85, 89, 98, 101, 180, 203
"A Declaration of the Independence of Cyberspace" (Barlow), 75
Defense, Department of, 8–9, 12, 52, 61. *See also* ARPANET
Democracy, 44–46, 103
DeNardis, Laura, 78, 94
Denmark, 94
Deregulation
   electricity market, 40, 221n22
   ideology of, 62–63, 65, 73
   legislation in 1978 and 1980, 71
   public utility and, 40
   Telecommunications Act of 1996 and, 72, 75–76
de Sola Pool, Ithiel, 6, 57
Devices, government forcing unlocking of, 183–185
Digital civil liberties, 1, 4, 93–94, 95, 131, 143, 156
Digital constitutionalism, 136
Digital divide, 28–29, 39–40, 42–43, 84–85, 88–89, 185, 235n227
Digital Due Process, 171
Digital Markets Act (DMA), 121
Digital media literacy, 140, 202
Digital packet switching network, 11, 53
Digital Platform Commission Act of 2022, 136–137
Digital Services Act (DSA), 121, 197
Dijck, José van, 107

Discovery, 12, 88
Discrimination, 37. *See also* Net neutrality
Disinformation/misinformation
   in 2016 election, 117
   content moderation and, 116
   digital media literacy and, 202
   DMA/DSA and, 121
   platforms and, 91, 117, 130, 133–134, 136
   prevalence of, 29
   Section 230 and, 114, 115, 120
Disney, 12, 93
Dita Beard affair, 228n120
Dobbs v. Jackson, 240n56
Doctorow, Cory, 94
Do Not Track browser setting, 109
Do Not Track legislation, 109
Dorsey, Jack, 120
DoubleClick, 131–132
Douglas, Susan, 15, 18
Dreamworks Animation, 87
Dublin, Ireland, 154

Easterbrook, Frank, 62
e-Commerce Directive of 2000, 96
ECPA. *See* Electronic Communications Privacy Act (ECPA) of 1986
Edwards, Paul, 8
Einstein, Mara, 21
Eisenhower, Dwight, 53
Election of 2016, disinformation and, 117
Electricity market, deregulation of, 40, 221n22
Electronic Communications Privacy Act (ECPA) of 1986
   creation of, 168–169
   extraterritorial rights and, 172–173
   gag orders under, 186
   provisions of, 261n97
   reform of, 171–172
   third-party doctrine, 170–171

Index

Electronic Frontier Foundation (EFF), 73, 170, 171, 176–177, 179, 186, 199
Electronic Privacy Information Center (EPIC), 171, 199
Elizabeth, Queen, 63
Encryption, 69, 183–185
End-to-end design, 9, 75, 78, 85, 89
End-user license agreements (EULAs), 143, 181–183
English common law, common carriage and, 37
Enhanced services, 63–65, 67, 77, 229n145
Equifax, 242n72
European Court of Justice (CJEU), 179, 180
European Union
    anticompetitive behavior of platforms, 124
    Copyright Directive (2021), 135
    data localization and, 190
    data protection in, 265n139
    Digital Markets Act and Digital Services Act, 121, 197
    e-Commerce Directive of 2000, 96
    EU-US Privacy Shield, 179–180
    General Data Protection Regulation (GDPR), 110, 190
    investigations and lawsuits, 207–210
    privacy protections in, 110
    Safe Harbour agreement, 178–180
EU-US Privacy Shield, 179–180
Executive Order 12333, 166, 259n78
Expression, freedom of, 112–120
Extraterritorial rights, 172–173

Facebook
    advertising revenue, 131
    American Edge and, 130
    Cambridge Analytica data breach, 99–101, 117, 129, 207–208, 245n106
    Civic Integrity Team, 103
    congressional hearings and investigations, 92, 97, 120, 127–128
    cultural power, 93
    data breaches, 243n72, 245n106
    data centers and, 154
    Digital Markets Act and, 121
    fact-checking, 117
    fines, 209, 210, 211
    Free Basics program, 93
    House Judiciary Antitrust Subcommittee (2020–2021), 92, 97, 127–128
    initial public offering, 240n47, 240n48
    Instagram purchase, 128
    lawsuit over data sharing, 186
    lawsuits against, 129, 208, 209, 210
    market consolidation, 128
    misinformation/disinformation and, 133
    news producers, compensation to, 135
    NSA's mass surveillance program, 173
    Oversight Board, 117
    ownership of content and platforms, 87–88
    power of as a platform, 92, 93
    PRISM program, 108
    privacy and, 106–107, 108
    privacy policy of, 107, 179, 182
    privacy violations, financial consequences for, 195
    Safe Harbour agreement and, 178–179
    as technology vs. media company, 96–97
Fact-checking, 117–119
Fake News. *See* Disinformation/misinformation
Federal Bureau of Investigation (FBI), 183–184
Federal Communications Commission (FCC), 22
    antitrust policy and, 50–51
    Carterfone case, 58, 87

Federal Communications Commission (FCC) (cont.)
  common carriage and, 38
  Computer II Final Decision, 64–65
  Computer II inquiry (1976–1980), 63–67
  Computer III Inquiry (1985–1986), 67
  Computer Inquiries, 54–57, 63–67
  data services, regulation of, 57–58
  Final Decision and Order (1971), 56–57
  Internet Policy Statement (2005), 78–79
  information privacy and, 58
  maximum separation rule, 57, 64
  net neutrality and, 78–83
  Open Internet Order of 2010, 79, 80
  Open Internet Order of 2015, 81–82, 89
  proposals to eliminate, 71
  Restoring Internet Freedom Order, 82
  Roosevelt and, 42
Federal Power Act, 42
Federal Trade Commission (FTC), 109, 130–131, 208, 210
Feld, Harold, 39, 40, 203
Ferris, Charles, 64–65
Ferry boats, 42
Fiber, 88–89
Fifth Amendment, 104, 165
Final Decision and Order (1971), 56–57
Financial Interest and Syndication Rules, 129, 139
First Amendment, 29, 112, 164–165, 186
First-party tracking, 198
FISA. *See* Foreign Intelligence Surveillance Act (FISA)
FISA Amendments Act of 2008, 167, 168, 176, 179, 259n79
FISA Court, 167–168
Flew, Terry, 122, 134
Florida, 198
Fly, James, 23

Focus Features, 87
FollowTheMoney.org, 271n16
Foreign Intelligence Surveillance Act (FISA), 165–168, 186, 259n76, 259n79
Foreign Intelligence Surveillance Court (FISC). *See* FISA Court
Fourth Amendment, 104, 164–165, 167, 170, 230n157
"Framework for Global Electronic Commerce" (1997), 76
France, 209
Francis, Pope, 85
Free Basics program, 93
Freedman, Des, 196, 204
Freedom of expression, 112–120
Free Press, 39, 79
Freevee, 97
Fuchs, Christian, 199

Gag orders, 186
Gallagher, Cornelius, 141, 158–159, 160–161, 206
Garland, David, 2–3
Gazette-Mail, 135
Genachowski, Julius, 79
General Data Protection Regulation (GDPR), 110, 190
Germany, Network Enforcement Act (NetzDG), 121–122
Gertner, John, 51–52
Ghost work, 200
Gilder, George, 76, 232n185
Gillespie, Tarleton, 115
Gingrich, Newt, 73, 76
Giphy, 210
Gitelman, Lisa, 19, 20
Gonzalez v. Google, 120
Google
  advertising revenue, 131–132
  Chrome, 97, 123
  as cloud service provider, 11–12
  cultural power, 93

Index 309

data centers and, 143, 144–148
Defense Department and, 266n148
Digital Markets Act and, 121
fines, 209, 210
foreign governments and, 108
labor activism and, 200
lawsuit over data sharing, 186
lawsuits against, 129, 208–209, 210–211
Maps, 97
market consolidation, 128
marketing, 143–146
as monopoly, 123
news producers, compensation to, 135
NSA's mass surveillance program, 173
ownership of content and platforms, 87–88
power of as a platform, 92, 93
PRISM program, 108
privacy and, 106–107, 108, 109
privacy policy of, 182
privacy violations, financial consequences for, 109, 195
proposed legislative framework, 79–80
Right to Erasure, 110
U.S. v. Google, 129, 208
Google Cloud Platform, 181
Gore, Al, 68, 69, 73–74
Governance, vs. regulation, 98–99
Government, private sector partnerships, 69, 108, 173–178
Graham, Lindsey, 99, 120
Gray, Horace, 40
Griswold v. Connecticut, 104
GTE (General Telephone and Electronics Corp.), 58, 64, 227n108
Guizhou, China, 155–156
Guizhou-Cloud Big Data (GCBD), 187

Hafner, Katie, 67, 85
Hale, Judge Lord, 42
Harmful speech, 117. *See also* Disinformation/misinformation
Harvard School of antitrust, 126
Hatch, Orrin, 31, 99
Haugen, Frances, 133, 206
Hayden, Michael, 185
Hazlett, Thomas Winslow, 21
HBO, 88
Helsinki, Finland, 154, 155
Henck, Fred, 47
High Performance Computing and Communication Act of 1991, 68, 230n154
History of the Standard Oil Company (Tarbell), 201
Hofstadter, Richard, 62
Holmes, Oliver Wendell, 24
Hoover, J. Edgar, 23, 159
Horwitz, Robert, 22, 38
House Antitrust Subcommittee, 53, 93
House Judiciary Antitrust Subcommittee "Investigation of Competition in Digital Markets" (2020–2021), 92, 97–98, 127–129, 134, 139
Howard, Philip, 117
Hu, Tung-Hui, 7, 12–13, 15
Hundt, Reed, 22
Hush-A-Phone case, 227n107

IBM, 8, 60, 62, 66
IBM Cloud, 181
Igo, Sarah, 104, 163
Information services, 77, 78, 80, 233n190
Information superhighway, 68. *See also* National Information Infrastructure
"Innovation Delayed Is Innovation Denied" (Gore), 73
Instagram, purchase of, 128
Insurrection. *See* January 6, 2021, insurrection and riot
Intelligence gathering abuses, 141–143, 167. *See also* Surveillance; Wiretapping

Internet. *See also* Broadband Internet
  access inequality, 28–29, 39–40,
    42–43, 84–85, 88–89, 195
  development of, 9–11, 68–69
Internet browsers, 69, 97, 123
Internet Research Agency (Russia), 117
Internet service providers (ISP), 67,
    76–77, 78, 82, 83, 84, 89, 105, 121,
    170, 185, 205
Internet Slowdown Day, 81
Interoperability, 59–60, 85, 89, 189, 203
Interstate Commerce Act of 1887, 37,
    247n126
Interstate Commerce Commission
    (ICC), 22, 37–38, 49, 218n70
Irion, Kristina, 188
iTunes, privacy policy of, 182

January 6, 2021, insurrection and riot,
    103, 117, 118–119
Jedi Blue, 135, 252n183
Jefferson, Thomas, 123
Jewel v. NSA, 176–177
John, Richard, 33, 41
Johnson, David, 150
Johnson, Lyndon, 28, 70, 158
Johnson, Nicholas, 87, 202–203, 206
Joint Declaration on Freedom of
    Expression and the Internet, 96
Joint Warfighting Cloud Capability
    (JWCC), 266n148
Journalism industry
  activism and, 200–201
  advertising revenue, loss of, 132
  compensation from platforms,
    134–135
  congressional recommendations, 134
  financial firms, control by, 134,
    251n176
  radio vs. newspapers, 132
  revitalizing, ideas for, 140
Jurisdiction
  of cloud policy, 149–156

  data and, 149–156, 183, 188–192
  end-user license agreements and,
    182–183
  terms of service agreements and,
    182–183
Jurisdiction shopping, 155–156
Justice, US Department of
  AT&T and, 49, 52–53, 60–62, 65
  antitrust policy and, 49–50, 52–53, 208
  Google and, 129, 208–209, 210–211
  IBM and, 60, 66
  lawsuits over gag orders, 186
  Microsoft case, 172–173, 249n153

Kanter, Jonathan, 130
Kapor, Mitch, 9, 73
Karr, Timothy, 140
Katzenbach, Nicholas, 60
Katzenberg, Jeffrey, 91
Katz v. United States, 24, 230n157
Kelly, Mervin, 51, 52
Kennedy, Robert F., 58
Keppler, Joseph, 124
Kerr, Orin, 172
Khan, Lina, 30–31, 127, 130, 138, 182
Kimmelman, Gene, 80, 84, 137
Kingsbury, Nathan, 49
Kingsbury Commitment, 49–50, 51
Klein, Mark, 175–176
Kobayashi, Kōji, 11, 67
Kristof, Nicholas, 108

Labor activists, 200
Landau, Susan, 184
Larkin, Brian, 19
Lasar, Matthew, 58
Last-mile pipelines, 36, 220n4
Leahy, Patrick, 169
Legal frameworks, 202
Lepore, Jill, 75–76, 178, 206
Lessig, Lawrence, 79, 98–99, 106–107
Leta Jones, Meg, 110, 111
Levin, Harvey J., 5

Li, Luzhou, 204
Libertarian Internet, 240n45
Lichtenstein, Nelson, 131
Licklider, J. C. R., 10
LinkedIn, 186, 243n72
Lloyd, H. D., 123, 200, 205
Lobbying
  Dita Beard affair, 228n120
  by Internet service providers, 83
  platforms and, 94, 130
  reform, need for, 201
  revolving door, 218n71
  Telecommunications Act of 1996 and, 72, 74
  telecommunication services and, 22, 39–40
Loebsack, Dave, 100
Loevinger, Lee, 125
Longevity principle, 85, 89, 203
Lucent Technologies, 228n137

Maher, Bill, 201
Manila Principles on Intermediary Liability, 122
Mann-Elkins Act, 38
Marchand, Roland, 44
Market competition. *See* Competition
Market consolidation, 102, 128–129
Marketing
  AT&T and, 45–46
  cloud infrastructure and, 13–15, 48, 116, 143–146, 181
  Google and, 143–146
Marriott International, 242n72
Martin, Kevin, 218n71
Marvin, Carolyn, 19
Mattern, Shannon, 19, 20
Maximum separation rule, 57, 64, 139
Mazzucato, Mariana, 140
McCarthy, John, 8
McChesney, Robert, 132
McClure's magazine, 201
McDougall, Robert, 89

MCI, 175, 176
Meta. *See* Facebook
MGM+ streaming services, 97
MGM Studios, 97, 111
Microsoft
  and Activision Blizzard, 128, 195
  as cloud service provider, 11–12
  data sharing, lawsuit over, 186
  Defense Department and, 266n148
  extraterritorial rights and, 172–173
  gag orders, lawsuit over, 186
  lawsuits against, 249n153
  market consolidation, 128
  national clouds and, 189
  NSA's mass surveillance program, 173
  Pentagon contract, 265–266n148
  power of as a platform, 92
  PRISM program, 108
  privacy and, 108
  Project Natick, 152, 153, 154
Microsoft Azure, 181, 190–191
Military-industrial complex, 51
Miller, Arthur, 157, 159
Minow, Newton, 21, 71, 195, 196, 206
Misinformation. *See* Disinformation/misinformation
Modems, connecting to telephone network, 60
Modern infrastructural ideal, 35–36, 38, 76, 196
Monopolies
  Amazon as, 123–124, 127
  Apple as, 124
  Google as, 123
  history of, 123, 125
  infrastructure management and, 35–36
  Internet service providers and, 83
  "natural," 40, 43–48, 49–50, 58–59, 85
  public utilities and, 39, 40
  Roosevelt on, 125
Monopoly capital, 3

Monopoly capitalism, 40, 102
Morgan, J. P., 44
Morse, Samuel, 41
MOSAIC browser, 69, 230n154
Mozilla, 81
Mueller, Milton, 46, 49
Mumford, Lewis, 146
Musk, Elon, 195

Naked Society, The (Packard), 157
Napoli, Philip, 98, 137
National Broadband Plan, 21
National clouds, 188, 189–192
National Data Bank (1965), 28, 158–161
National Information Infrastructure, 68
National Science Foundation, 9
National Security Administration (NSA)
   crypto-wars, 185
   Executive Order 12333, 259n78
   FISA and, 166
   in-house posters, 24, 25f, 26f, 27f
   PATRIOT Act and, 165
   precursor to, 23
   PRISM program, 108, 168, 173, 174, 175, 178, 179, 185, 241n65, 263n117
   Project SHAMROCK, 23, 253n1
   propaganda from, 24–27
   warrantless mass surveillance program, 173–180, 264n132
National Security Letters (NSLs), 168, 260n87, 260n88
National Telecommunications and Information Administration, 15, 230n162
Native American tribal land, Internet access inequality, 85, 195
Natural monopoly, 40
   AT&T and, 43–48, 49–50, 58–59
   cloud pipelines, 43–48
   history of, 43–44
   rejection of, 85
Navalny, Aleksei, 191

NBC, 87
NBC Universal, 87
Negative policy, 196–197
Netflix, 81, 105
Netflix effect, 12
Net neutrality
   in California, 83–84
   common carriage and, 77, 78, 81, 82, 205
   Communications Act of 1934 and, 78
   critiques of, 86
   expansion of, 86–87
   FCC and, 78–83
   history of, 77–78
   information services and, 78, 80
   Internet Slowdown Day, 81
   Open Internet Order of 2015, 81–82
   public service rationale, 81–82
   public understanding of, 82
   repeal of by Trump administration, 82–83
   state legislation and, 83–84, 198, 234n208
   Telecommunications Act of 1996, 205
   telecommunication services and, 78, 80
   US Supreme Court and, 78–79
Netscape, 69
Network Enforcement Act (NetzDG), 121–122
Network resiliency, 82
Neutrality principle, 85, 89
Newman, Russell, 81, 87
News Corp, 195
News deserts, 132
New York Times, 82, 175
Nextel, 176
Nichols, Philip, 37
Nixon, Richard, 5, 60, 70
Nuclear weapons, AT&T and, 51–52

Obama, Barack, 80, 151
Office of Indigenous Communications and Technology, 195

Index

Office of Telecommunication Policy, 1, 230n162
Oil industry, 49, 61, 123, 200, 201
Olmstead v. United States, 24
Olsen, James, 71
O'Mara, Margaret, 106, 161, 163
Online Platforms and Market Power Hearings (2019–2020), 96, 140, 207
Open Internet Order of 2010 (FCC), 79, 80
Open Internet Order of 2015 (FCC), 81–82, 89
Opensecrets.org, 271n16
Oracle, 266n148

Packard, Vance, 157, 158
Pai, Ajit, 82, 83, 218n71
Paid prioritization, 77, 81
Parkhill, Douglas, 1, 8, 205–206
Parks, Lisa, 13, 21
Pasquale, Frank, 107
PATRIOT Act, 164–166, 168, 190, 259n71, 259n79, 260n87
Paypal, privacy policy of, 182
Peacock, 87
Peters, Benjamin, 9
Peters, John Durham, 13
Pichai, Sundar, 127
Pickard, Victor, 78, 86, 137, 200
Pike, Otis, 141–142
Pike Committee, 141–142, 254n4
Pistor, Katharina, 149, 187
"Platformized Internet," 240n45
Platforms
  advocacy organizations and, 122–123, 197, 199–202
  alternative visions, 136–140
  antitrust policy and market competition, 123–131
  calls for breaking up, 138
  competition and, 123–131
  European Union governance of, 121–122
  freedom of expression and, 112–120
  investigations and lawsuits, 127–130, 207–211
  journalism industry and, 131–140
  legislative proposals for, 136–137
  liability for content, 96
  lobbying of, 94, 130
  market consolidation, 102, 128–129
  monopoly capitalism and, 40, 102
  news producers, compensation to, 134–135
  policy, ideologies foundational to, 101
  policy lag, 98
  political influence, 94
  power of, 91–96
  PRISM program and, 108, 168, 173, 174, 178, 241n65
  privacy and surveillance, 93–95, 102–103, 104–112
  regulation, alternative proposals, 136–140
  regulation challenges, 92–93
  regulation vs. governance, 98–99
  role of in their own governance, 99–101
  Section 230 (*see* Section 230)
  settings, privacy and, 106–107
  speech rights and, 93–95
  structural separation, 138–139
  as technology vs. media company, 96–97
Political action committees (PACs), 72
Pomerantsev, Peter, 138
Popiel, Pawel, 94
Posner, Richard, 62
Post, David, 150
Postal service, 33, 41, 43, 44, 86, 197
Postal Telegraph Company, 23
Post Roads Act of 1866, 220n9
Powell, Michael, 218n71
Prehistory of the Cloud (Hu), 7
Prime Video, 97

PRISM program, 108, 168, 173, 174, 175, 178, 179, 185, 241n65, 263n117
Privacy
  Amazon and, 107
  Apple and, 108, 186–187
  Biden executive order (2021), 131
  branding and, 186–187
  Cambridge Analytica data breach, 99–101, 117, 129, 207–208, 245n106
  congressional hearings and investigations (1966–1967), 58, 158–159
  early concerns, 157–161
  Electronic Communications Privacy Act (ECPA) of 1986, 168–172
  end-user license agreements and, 143, 181–183
  EU legal protections, 110, 178–180
  FCC and, 58
  financial consequences for violations, 109, 207–211
  FISA and, 165–168
  intelligence gathering abuses, 23, 141–143, 253n1. *See also* PATRIOT Act, Foreign Intelligence Surveillance Act (FISA), and PRISM program
  legislation affecting, 105–106, 109, 240n57
  National Data Bank, 158–161
  PATRIOT Act and, 164–165, 168, 260n87
  personal information, selling of, 84
  platforms and, 93–95, 102–103, 104–112
  in popular press, 159–160
  reclaiming, ideas for, 139–140
  right to in U.S., 104–105
  state legislation and, 180–181
  surveillance and (*see* Surveillance)
  terms of service agreements and, 107–108, 143, 181–183
  trust, international community and, 178–180
Privacy Act of 1974, 105–106, 163, 241n58
Privacy and Freedom (Westin), 111, 157
Privacy Shield of 2016, 179–180
Privatization, 3, 36, 54, 57, 76, 99, 102, 106, 146, 180, 196
Prodigy, 113
Production Code, 246n109
Profit over Privacy (Crain), 106
Progress and Freedom Foundation (PFF), 73, 76
Progressive Era, 49, 62, 125, 196, 200
Project Natick, 152, 153, 154
Project SHAMROCK, 23, 253n1
Public good, infrastructure as, 36, 196
Public interest values, 5, 40, 73–74, 76, 122, 137, 140, 196, 197, 199, 205
Public Knowledge, 39, 171, 199
Public utility
  broadband Internet and, 32, 38–39, 43, 86
  vs. common carriers, 39, 82
  definition of, 39
  deregulation and, 40
  Johnson administration and, 70
  monopolies and, 39, 40
  platforms as, 197
  profit motivation vs. public service requirements, 40–42
  public ownership of, 41–42
  regulation and, 36, 40, 42
  reimagining, 203–204
  scandals among, 40
  Telecommunications Act of 1996 and, 70
  telegraph companies and, 41
Public visibility, 202
Putin, Vladimir, 191
Pyle, Christopher, 141, 206

Quello, James, 64

Index   315

Radio, newspapers and, 132
Radio spectrum, 15, 16–17, 21
Railroad industry, 35, 37, 49, 123, 127, 130, 139, 160, 205
Ranking Digital Rights, 119, 122, 198, 199
Reagan, Ronald, 61, 71, 102, 126, 166
Records, Computers, and the Rights of Citizens (1973), 161–163
Reddit, 81
Redemption, digital, 111
Reeve Givens, Alexandra, 123, 199–200
Regional Bell Operating Companies (RBOC), 65. *See also* Baby Bells
Regulatory capture, 22, 40, 48, 79, 94, 199, 204, 281n71
Regulatory hangover, 5–7, 156
Regulatory lag, 6
Reinvention, digital, 111
Renda, Andrea, 165
Reno v ACLU, 150
Reston, James, 141
Restoring Internet Freedom Order (FCC), 82
Right to Be Forgotten, 110–111
Right to Erasure, 110–111
"The Right to be let alone" (Warren and Brandeis), 104, 110
Ring doorbells, 111
Ring Nation, 111
Roberts, John, 167
Roberts, Sarah, 116
Rockefeller, John D., 61, 201
Rockefeller Commission, 141–142
Roe v. Wade, 240n56
Roosevelt, Franklin D., 23, 41–42, 50, 125
Rostow, Eugene, 231n163
Rozenshtein, Alan, 94, 187
R.R. Donnelley building, 19–20
Ruiz, Raul, 100
Rural areas, Internet access inequality, 85. *See also* Digital divide

Rural electrification, 36
Rural Electrification Administration, 42
Russia, 117, 190, 191–192

Sadowski, Jathan, 204
Safe Harbour agreement, 178–180
Sandvig, Christian, 20, 86
Santa Clara County, 83
Santa Clara Principles on Transparency and Accountability in Content Moderation, 122
Sarbanes, John, 100
Saturday Night Live, 48
Schaake, Marietje, 202
Schiller, Dan, 41, 50, 51, 58, 203–204
Schivelbusch, Wolfgang, 19
Schmidt, Eric, 8, 94
Schneier, Bruce, 177, 189
Schrems, Max, 178–180, 206
Schrems I decision, 179
Schrems II decision, 180
Schwartz, Paul, 150, 189
Scotland, 152
Section 230, 195, 122
  content moderation and, 115–117
  debate over, 115
  free speech and, 116–117
  future of, 197, 198
  Good Samaritan provision, 115
  origin of, 112–114
  reform, 117, 119–120
  Supreme Court and, 120
Securities and Exchange Commission (SEC), 207
September 11, 2001, terrorist attacks, 165
Sherman, John, 91
Sherman Antitrust Act, 60, 61, 91, 126, 247n126, 249n153
Shi Tao, 108
Silicon Valley, tech culture, 101–102
Sinclair, Upton, 200
Sloan Commission, 5

Smart Voting app, Russia, 191–192
Smith, Brad, 171–172, 189
Smith, Chris, 108–109
Snowden, Edward, 108, 141, 168, 173, 175, 177, 178, 179, 192, 206
Soghoian, Christopher, 176
Sohn, Gigi, 127, 270n1
Sony Pictures, 243n72
Sovereign clouds, 188, 190, 198. *See also* Data sovereignty, National clouds
Sovereign Internet Law (Russia), 191, 192
Spectrum scarcity, 21–22
Speech rights, 1, 28–29, 78, 91, 93–95, 112–120, 122, 191, 196, 199, 202, 244n94, 245–246n109
Spotify, 139
Sprint, 19, 175, 176
Standard Oil, 49, 61, 123, 200, 201
Stanford Research Institute, 9
Starosielski, Nicole, 146
Starr, Paul, 32
State legislation, 83–84, 180–181, 198, 234n208
Steffens, Lincoln, 200
Stephenson, Randall, 88
Stimson, Henry, 23
Stoller, Matt, 62, 93, 126, 127
Stored Communications Act (SCA), 170–173
Strassburg, Bernard, 47, 56
Stratton Oakmont, 113–114
Stratton Oakmont, Inc. v. Prodigy Services Co., 114–115
Streaming services, 12, 97, 239n29. *See also* Netflix, Peacock, Prime Video
Streeter, Thomas, 15, 102
Structural separation, 64, 138–139, 229n145.
Supreme Court, U.S., 24, 240n56
  FISA Court and, 167
  Jewel v. NSA and, 176–177
  on common carriage, 37, 38
  on cyberspace, 150
  on free speech, 244n94
  net neutrality and, 78–79
  on privacy, 24, 104, 157
  Section 230 and, 120
  "third-party doctrine" and, 170
Surveillance. *See also* Privacy
  in China, 108–109
  cloud policy stakes and, 22–28
  congressional hearings and investigations, 158–159, 163–164
  early tensions over, 157–164
  FISA and, 165–168
  loss of trust, international community and, 178–180, 187–188,
  NSA's mass surveillance program, 108, 168, 173–180, 185, 264n132
  origins of, 23–28
  PATRIOT Act and, 164–165, 168, 260n87
  personal information, selling, 84
  public-private partnership, 23, 23–27, 104–112, 173–180, 202, 264n132
  telephone service and, 23–27, 264n132
  wiretapping and, 23–24, 69, 169, 176, 185, 230n157, 261n97
Surveillance capitalism (Zuboff), 28, 91
Suzor, Nicholas, 136
Sweden, 154

Takedown provision, 115
Tarbell, Ida, 200–201, 206
Teachout, Zephyr, 32
Technologies of Freedom (de Sola Pool), 6
"Technology and Freedom" (Gallagher), 159
"Technology: The Engine of Economic Growth," 68–69
Telecommunications Act of 1980, 70–71

Index

Telecommunications Act of 1996, 38, 69, 205
   amendment to, 112–114. *See also* Section 230
   cloud pipelines and, 70–77
   deregulation and, 75–76
   net neutrality and, 78–79
   passage of, 74–75
   private sector partnerships, 76
   reaction to, 75–76
Telecommunication services, 39–40, 77, 78, 80
Telegraph companies, 23, 38, 41, 44, 49–50, 220n9
Telemundo, 87
Telephone surveillance, 23–27. *See also* Wiretapping
Temin, Peter, 61
Tennessee Valley Authority, 42
Terms of service agreements (TOS), 3, 84, 107–108, 116, 143, 149, 181–183
Texas, 198, 221n22
Thibault, Ghislain, 15
Thiel, Peter, 128
Thierer, Adam, 46, 47
Third-party cookies, 198
Third-party doctrine, 170–171, 202
Throttling, 77, 81, 83, 86
Time Warner, 12, 88
Tobacco industry, 49, 61, 127
Toll broadcasting, 88
Tomlin, Lily, 48, 75
Trans-Atlantic Data Privacy Framework, 180
Truman, Harry, 51, 53
Trump, Donald, 82–83, 117–119, 265n148
Tunney, John, 163
Turkey, 188–189, 190
Turner, Fred, 73, 103, 203
Turner Broadcasting, 88
21st Century Fox, 12

Twitter. *See also* X (formerly Twitter)
   AWS and, 181
   Congressional hearings and investigations, 120
   fact-checking, 117–119
   lawsuit over data sharing, 186
   misinformation/disinformation and, 133, 134
   Musk purchase of, 195
Twitter, Inc. v. Taamneh, 120

UCLA, 9
UC Santa Barbara, 9–10
Undersea cables, 146, 220n4
Universal service, 28, 43–49, 73–74, 85
Universal Studios, 87
University of Illinois, 69
University of Utah, 10
Unterberger, Klaus, 199
Upstream program, 177
Upton, Fred, 100
Urban media (Mattern), 19–20
USA Freedom Act, 264n132
USA PATRIOT Act. *See* PATRIOT Act
USA Today, 176
Uspenski Cathedral, 154, 155
U.S. v. Google, 129, 208
University of Utah, 10
Utah, 181

Vail, Theodore, 43–44, 46–47, 197
Van Deerlin, Lionel, 71
Velkova, Julia, 152, 154
Verizon, 227n108
   lawsuit against FCC, 80
   net neutrality and, 83
   NSA's mass surveillance program, 175, 176
   PRISM program, 108
   proposed legislative framework (2010), 79–80
Verveer, Phil, 137
Video Privacy Protection Act (VPPA), 105

Virginia, 181
Virtualization, 15, 18
Vonderau, Asta, 15
von der Leyen, Ursula, 121
Vonnegut, Kurt, 195

Walberg, Tim, 100
Walker, Paul, 50
Walker Report, 50–51
Warner Bros. Studios, 88
WarnerMedia, 88
Warren, Earl, 157
Warren, Samuel D., 104
Washington, George, 185
Watergate hearings, 142
Weinberger, Caspar, 61
Welch, Peter, 100
Wells, Ida B., 200
Werbach, Kevin, 56, 64, 227n107
Western Electric, 51–53, 61, 65–66, 215n30, 228n137
Western Union Telegraph Company, 23, 44, 49, 156
Westin, Alan F., 111, 157, 202
Wheeler, Tom, 6, 35, 78, 80, 82, 89, 130, 137, 197
Whistleblower protections, 178
Whitehead, Clay, 5–6, 70
Whitehouse, Sheldon, 119
White House Office of Telecommunications Policy, 70
Wicker, Roger, 99
Wiesner, Jerome, 164
Wilson, Charles, 52, 61
Wilson, Woodrow, 41
Winseck, Dwayne, 97
Wiretapping, 23–24, 69, 169, 176, 185, 230n157, 261n97. *See also* Surveillance
Woods, Andrew Keane, 151
World War II, 51
World Wide Web, 68

Wu, Tim, 30, 31, 50, 62, 77, 126, 130
Wyden, Ron, 114

X (formerly Twitter), 195

Yahoo!
  Chinese government and, 108–109
  data breach, 242n72
  lawsuit over data sharing, 186
  PRISM program, 108, 173–175, 263n117
  privacy and, 108
Yang, Jerry, 108
YouTube, 117, 208

Zero rating, 83
Zuboff, Shoshana, 3, 28, 93, 112, 116–117, 133, 203
Zuckerberg, Mark
  antitrust hearings and, 31
  congressional hearings and investigations, 99–101, 120, 127
  on Facebook as technology vs. media company, 96–97
  on privacy, 28
  self-promotion, 91, 102–103
Zuckerman, Ethan, 204